ABOUT ISLAND PRESS

Island Press is the only nonprofit organization in the United States whose principal purpose is the publication of books on environmental issues and natural resource management. We provide solutions-oriented information to professionals, public officials, business and community leaders, and concerned citizens who are shaping responses to environmental problems.

In 1998, Island Press celebrates its fourteenth anniversary as the leading provider of timely and practical books that take a multidisciplinary approach to critical environmental concerns. Our growing list of titles reflects our commitment to bringing the best of an expanding body of literature to the environmental community throughout North America and the world.

Support for Island Press is provided by The Jenifer Altman Foundation, The Bullitt Foundation, The Mary Flagler Cary Charitable Trust, The Nathan Cummings Foundation, The Geraldine R. Dodge Foundation, The Ford Foundation, The Vira I. Heinz Endowment, The W. Alton Jones Foundation, The John D. and Catherine T. MacArthur Foundation, The Andrew W. Mellon Foundation, The Curtis and Edith Munson Foundation, The National Fish and Wildlife Foundation, The National Science Foundation, The New-Land Foundation, The David and Lucile Packard Foundation, The Surdna Foundation, The Winslow Foundation, The Pew Charitable Trusts, and individual donors.

ABOUT ANDROPOGON ASSOCIATES

Since its beginning in 1975, Andropogon Associates, Ltd., has been effecting fundamental change in the management of the landscapes that support us by demonstrating ecologically sound alternatives to conventional practices. As a landscape architecture, planning, and design firm, Andropogon makes a habit of restoration in its work, which ranges from national parks to former landfills and slag piles, from college campuses to corporate headquarters. Andropogon has pioneered a participatory process that engages the community directly in project planning, site monitoring, training workshops, and hands-on restoration because the community is ultimately responsible for sustaining the landscapes it occupies. By involving the many individuals and groups who impact the landscape, Andropogon builds consensus among users and caretakers alike and promotes a creative partnership between people and their environment.

THE ONCE AND FUTURE FOREST

The Once and Future Forest

A Guide to Forest Restoration Strategies

Leslie Jones Sauer
and Andropogon Associates

ISLAND PRESS
Washington, D.C. • Covelo, California

Library of Congress Cataloging in Publication Data

Sauer, Leslie Jones.
    The once and future forest : a guide to forest restoration strategies / Leslie Jones Sauer.
        p.   cm.
    Includes bibliographical references and index.
    ISBN 1-55963-552-5 (cloth). — ISBN 1-55963-553-3 (pbk.)
    1. Restoration ecology—East (U.S.)   2. Forest ecology—East (U.S.)   3. Forest conservation—East (U.S.)   I. Title.
    QH541.15.R45S358   1998
    333.75′153′0974—dc21   97-14843
                                                                                          CIP

Printed on recycled, acid-free paper ∞

Manufactured in the United States of America

10   9   8   7   6   5   4   3   2   1

CONTENTS

PART III   MANAGEMENT MANUAL

In the interstices of cities, at the margins of suburban development, are common areas that we call parks, forest preserves, undeveloped land, open space. In communities across the East these special places are in jeopardy of disappearing forever, the orphans of deferred maintenance, budget cuts, and misuse. Many parks were developed a hundred or more years ago, so that most of the trees are the same age, and they have reached advanced age. Replanting, when it occurs, often is unsuccessful. Too often these depleted landscapes bear only the most rigorous, often invasive vegetation. In the interests of security, park managers in many communities have removed the midstory of shrubs, so often the park consists of diseased and dying trees and weedy lawn. Many other areas—roadsides, vacant lots—are in even more obvious distress.

Yet these spaces remain important havens for their denizens, wildlife and human alike, and within them still remains the possibility to regenerate their original native flora—the habitat that existed before development and modern use. And they offer opportunities for communities to come together to achieve important goals while preserving a bit of the natural world in their midst.

I know of no other document as crucial to fulfilling these roles as *The Once and Future Forest* by Leslie Jones Sauer. This extraordinary book envisions the restoration of the natural processes of succession—the patterns by which land undergoes its own innate stages of development—to form habitat for diverse species. It goes beyond rhetoric, providing knowledge of the operation of natural systems and practical information to undertake beneficial interventions and eventual restorations. Its principles can be applied to any place, whether it is a forested glade in a city park or a weedy roadside.

The level of scientific discourse here is at once impeccable but thoroughly comprehensible. Leslie Sauer, an experienced field ecologist, and her colleagues at Andropogon—Carol and Colin Franklin and Rolf Sauer—and their staff have no equal in the performance of ecological planning and design. They, more than any other group, have formulated the theory and practice of ecological landscape architecture and developed and applied them. This book, a vast compendium of essential knowledge of forest restoration, represents both their long experience and current thoughts on the subject.

Without doubt *The Once and Future Forest* should have a powerful and salutary influence on many professions and agencies. Sauer observes that few professional projects occur in pristine environments; most, indeed, are in thoroughly degraded areas. It follows, then, that ecological restoration and its attendant skills are central to the practices of landscape architecture, land management, city planning, and civil engineering. Sauer's emphasis upon the reestablishment of native vegetation and natural drainage patterns compels a reevaluation of conventional practices by anyone concerned with or affecting environmental health through their decisions and actions. More, reestablishing ecological health will necessitate a reorientation of education and practice. Restoration must become a habit in our practices, a consistent and continual process occurring over the long term.

One of Sauer's central themes is that extensive investigation and close observation—of climate, geology, hydology, soil, plants, and animals—are required for an understanding of the environment, and therefore, for a successful restoration of habitat. The complementary pursuit, knowledge of the systematic relationships between people and places, leads to the conclusion that restoration must occur as empowered public action, and that means it must be based upon consensus and cooperation. The book's splendid case study of the Central Park renovation demonstrates these theses most dramatically.

Global environmental destruction was accomplished in less than three centuries. Can we reverse the path we have taken; can we restore our habitats? I once thought that natural regeneration—allowing the environment to heal by means of its own natural processes—would be sufficient, but apparently that is not so. Climatic change; soil depletion; loss of microorganisms, plants, and animals; and all the other degradations to the environment have exceeded Nature's ability to regenerate. So we must participate, with action and all the knowledge and experience we can bring to the task. You'll find *The Once and Future Forest* an essential guide to this all-important work.

Ian McHarg

Professor Emeritus

Department of Landscape Architecture and Regional Planning

Graduate School of Fine Arts

University of Pennsylvania

ACKNOWLEDGMENTS

A complete list of the people who contributed to this book would comprise another chapter. Rather than try to list them all, we would like to single out a few people and organizations that are emblematic of the kind of support we have received. We owe a very special debt of gratitude to James Amon, the director of New Jersey's Delaware and Raritan Canal Commission. He collaborated on Chapter 1 and his words can be found throughout this document. Without him, this book might never have happened.

Every page of this book stems in part from our clients, both public and private, who haved worked with us to set new standards for how we use the land. They have accepted the challenge to change conventional approaches with eagerness and often courage in the face of uncertainty about any innovation. Behind every restoration effort there are a few extraordinarily dedicated and creative people who make it all happen. Central Park, for example, has many heroes, but none has been so important to forest restoration as Maryanne Cramer, the Central Park Planner, who has championed the North Woods Project since its inception. We thank her and the many others like her with whom we have worked.

We have also relied upon many organizations and volunteer groups. One in particular, the Society for Ecological Restoration, has served as the most diverse and creative forum for exchanging ideas and information about restoration. The Wissahickon Rebuilding Committee has shown us the potential of grassroots initiatives. We also wish to express our gratitude to Island Press, a very exceptional publisher whose books fill our library. It has been a pleasure to work with Barbara Dean, Barbara Youngblood, and Fran Haselsteiner, who together brought this project to fruition.

We can never adequately thank the many individuals at Andropogon who have had a hand in this book and the projects that inspired it. In addition to providing invaluable help with graphics and manuscript preparation, they have been instrumental in developing our approach to landscape planning and design. Lastly, Leslie Sauer wishes to express her deepest gratitude to her partners, Carol and Colin Franklin, Rolf Sauer, Noëline Mills, José Almiñana, and Yaki Miodovnik, for their willingness to undertake this effort and for making Andropogon a firm with a mission.

Making a Habit of Restoration

For many of us, urban and suburban forests are the closest we can come to nature. Sadly, these beloved places are deteriorating throughout the country. Some forests are destroyed in a moment—cut over and built upon. Others, especially urban parks and remnant woodlands, die more slowly, their destruction is caused not by a single act but by an accumulation of daily assaults—by public use of the landscape as well as by the public agencies responsible for their care.

Protection of the land has not necessarily protected the landscape. We all contribute to this deterioration—from the mountain biker gouging a rutted trail up a steep slope to the birder who steps off the path for a better view. Damage occurs when a police car, for example, compacts the soil on either side of a woodland trail meant only for pedestrians or when uncontrolled stormwater careens downslope, eroding the forest floor. Less visible but no less serious is the damage done daily by atmospheric pollutants from vehicles, industry, and other energy consumption.

For many species of wildlife, these forest fragments are habitat vital to their survival. In our sprawling, developed landscapes, every patch of green has become an increasingly important remnant in an ever more tattered fabric. Today, those responsible for the care of protected landscapes are expressing growing concern about the accelerating deterioration of this resource. The negative impacts of use and abuse, already apparent in urban parks, are becoming more visible in suburban and rural areas as well. For millions of people, contact with the natural world is a progressively diminished experience. Our own observations confirm the gravity of our environmental condition: We are losing the rich variety of native plants and animals that once typified our regional landscapes. The biodiversity crisis is here in our backyards and parks.

Biodiversity is the variety of forms of life. In addition to the 30 million or more species of plants and animals on Earth, the term "biodiversity" embraces highly specialized subspecies, which may be far more numerous as well as more vulnerable to extinction. The diversity of our living world also includes the information of evolution, the bonds of interdependency that have evolved over millions of years, such as predator–prey and plant–pollinator relationships. Pattern is an aspect of diversity as well, including the landscape mosaics we see around us.

Endemic species—that is, those found nowhere else but in a given area, have been the hardest hit, in part because they are the most specialized and poorly suited to a world that is becoming increasingly homogenized. Indigenous species, those that were native to an area before European settlement in this hemisphere, are dying out at about the same rate that exotic or alien species, those introduced by people to a region, are establishing themselves.

Human-induced disturbances to the landscape are now of such great scope and scale that they overshadow the patterns of natural disturbances. Natural disturbance is, of course, part of the natural cycle, the result of climatic extremes, fire, the death of a tree, a flood, or countless other common phenomena. Indeed, these cyclical events are a vital stimulus to change and integral to sustaining regional diversity within great forest expanses. What most distinguishes natural disturbance from human-induced disturbance is the extent to which it falls within the historic range of its occurrence. Events at the limit of the natural range shape the landscape profoundly, such as the blowdown of 1938 in New England or the fire of 1963 in the Pine Barrens of New Jersey. Events that extend well beyond their naturally occurring variability, however, exceed the recoverability of many plant and animal communities. Complex, long-established ecosystems are collapsing after repeated disturbance. For example, repeated clearcutting inflicts far more serious and long-term impact on natural forest regeneration than was previously recognized. At the same time, a few supercompetitive and generalist species are thriving at the expense of almost all others in the landscapes created by human settlement. Now unchecked suburbanization and resource extraction are consuming ever more of the remaining wild and rural lands. The living systems around us are losing their richness and resilience, and we sense the implications for our own lives.

Few of us can fully imagine or appreciate the grandeur and intricacy of the original forest encountered by the first settlers. But most of us remember a forest we knew once that we have watched decline or disappear altogether. The lands we saved for their rich landscapes are changing before our eyes as the impacts of the last few centuries become more visible.

While park users and land managers are becoming more aware of the urgency of the problem, they are hampered by lack of information and experience in dealing with the management and restoration of disturbed landscapes. Natural resource managers typically study intact ecosystems and may have little experience with disturbed landscapes, and horticulturists are usually inadequately trained in large-scale natural systems. Ecologists and biologists in the past often devoted relatively little energy to solving on-the-ground management problems and sought instead to find and document the most pristine sites. Today the scientific community is shifting its focus toward restoration, but there are no consistent policies or proven methodologies that reliably result in restoration. Perhaps the most difficult aspect of all is that restoration is a long-term effort requiring a high degree of expertise and commitment rather than a quick fix.

Despite the challenges facing them, many landscape managers are attempting forest restoration and getting some good results for their efforts. These concerned managers are developing the art and science of caring for fragmented forests by monitoring, studying, maintaining, replanting, and experimenting in woodlands and forests. They are aware that restoration is an ongoing job and that natural systems are often so compromised we cannot expect them to recover if they are simply left on their own. Progress is not necessarily smooth and transformations are not instantaneous, but these landscape managers are monitoring the landscape, limiting further impacts, and initiating improvement in the management of the natural systems under their care.

We undertake to restore indigenous communities and ecosystem function in the face of great uncertainty. We do not know very much about how natural systems work, and we do not even have all the component pieces. The concept of restoration, taken literally, might presume that we can replace missing parts or remove added ones. But while we can eliminate invasive alien species on a specific site, we cannot necessarily take away all new elements. How does the restorationist, for example, remove the large amounts of nitrogen raining down on the landscape from air pollution, seriously modifying one of the most basic processes, the nitrogen cycle? Nor is it any easier to add the lost pieces. Where do we find the huge flocks of migratory passenger pigeons whose numbers collapsed from billions to extinction with the first great wave of deforestation or the once-numerous but now extinct Carolina parakeets of the eastern forest? We simply do not know enough about our ecosystems; nor are we yet able to modify our lifestyles and land use to re-create those conditions necessary to truly "restore" a prior state, extinctions aside.

The management of complex living systems necessarily involves many interrelated natural processes and functions. Some of these natural processes and functions we may seek to replace or emulate; others we may try to rehabilitate or reestablish. The cumulative result is intended to move toward restoration.

This is a heuristic process in which we will learn as we go along. If we are committed to sustaining indigenous plants and animals, we will, over time, discover new approaches and techniques that cannot be implemented, or even imagined, today.

We have written this guidebook primarily as a stimulus to those individuals and groups who want to do something constructive about our deteriorating forest patches. We hope it will assist in this restoration process, providing an approach for assessing each site, guidelines for determining management goals, and an overview of appropriate management and restoration techniques. The objective is to provide a framework for action rooted in the idea that those who use and care for a landscape should be responsible for sustaining its value over time. The goal is to develop programs to ensure that most of our actions will be restorative and not destructive. We have set ourselves on a critical path, and the direction will be shaped by our goals.

How This Book Is Organized

The format of this guide has three major sections. The first part, "The Forest Today," is a discussion of disturbance and the larger issues affecting the health of our remaining forest fragments. It is about the broader context in which all our efforts occur. It also discusses grasslands and meadow environments, transitional areas that may precede forest establishment or border woodlands. The next section, "The Restoration Process," describes broad strategies for restoration that give communities and agencies a context for decision making. The last section, "Management Guide," describes basic approaches for implementing a restoration program. The discussion and examples center primarily on the eastern United States, although the lessons and perspective are applicable to all regions.

We have used common names of plants and animals throughout except in a few special instances. A complete list of the scientific names of species mentioned in the text appears at the end of this book along with a list of nonnative (exotic) plants that are at least locally invasive in the Northeast.

The Forest Today

While there are books about the great forests in national parks and national forests, there is little information on the fragments of forest that surround our homes and businesses. Similarly, there are many books and courses on identifying native plants and animals as well regional natural area guides, but the general public has very little understanding about the plight of regional forest systems and the adverse impacts, often unnecessary and inadvertent, that we have upon native plants and wildlife daily.

The goal of Part I is to help the reader see the forest as a system that is not static but that changes both locally and pervasively, to help the reader read the landscape and recognize the patterns of both health and degradation. Despite our living in a culture that distances us from the natural world, we have actually seen more of the landscapes around us and across the face of the planet than we might be aware of, through the media and travel. With a slight change of focus, we can learn a lot about an environment simply by the way it looks.

The chapters in Part I provide essential background for understanding the condition of woodland habitats today—the fragmentation of the landscape caused by expanding growth and suburbanization, our habit of continuously altering natural patterns, and the failure of many native plants and animals to regenerate themselves sufficiently to survive, at the expense of vital natural functions and services that will be very costly to replace or restore. The chapters in Part I also examine how our management of the landscape, or lack thereof, affects the entire ecosystem as well as the particular site being used or maintained. Lastly, they discuss the impacts of our changing atmosphere and climate, which may have the most far reaching consequences of all for the forest.

CHAPTER 1

A Landscape Overview

Earth is the "water planet," and on the one-fifth of its surface area that is land, water is the most crucial factor in determining the character of the landscape. Where water is abundantly available for much of the year forests can grow, while grasslands occur where conditions become too droughty to support tree growth. Deserts are even drier landscapes. Sometimes cold restricts the availability of water by holding it as ice, unavailable to plants. Climate, and hence the availability of water, is determined in part by latitude, the position of a place between the equator and the poles, as well as by ocean currents and wind patterns and the ways in which they interact with the different sizes, forms, and positions of the landmasses.

These factors combined produce somewhat loosely banded patterns of similar landscapes around the globe (Figure 1.1). If we could look down from a satellite circling the Earth, we would be able to distinguish the forest landscape from the expanses of grassland. We could see where the northernmost forest ends and the low tundra begins or where grasslands give way to deserts. Scientists call these different bioclimatic zones "biomes," a word that literally means "life-groups."

These landscapes look dramatically different, and the way they look reveals their environment. As we observe plants more closely, their appearance tells us a lot about how water occurs in their habitat. and the patterns in which they grow. In the case of forests, differing kinds grow along a gradient of changing temperatures and moisture availability.

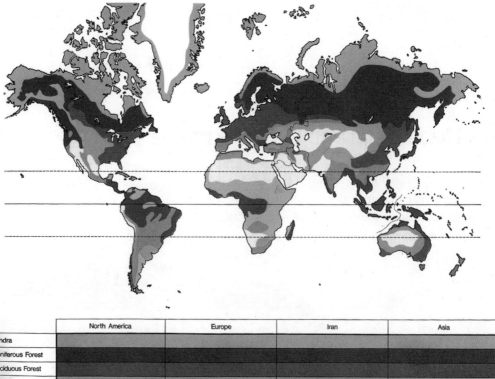

Figure 1.1. *All landscapes occur in bands that correspond to climatic zones around the Earth reflecting the amount of energy and water available for development of vegetation.*

Tropical Forest

The tropical forest is generally considered to be the richest terrestrial landscape. Like the oceans, tropical forests are cauldrons of life, out from which come most of the plants that have colonized, at least to some extent, almost every place on the planet. All landscapes are layered, but in the tropical forest the pattern of stratification reaches its greatest expression. Abundant rainfall and intense light year-round nurture forests that bind the environment's nutrients into a lush tangle of plants from the forest floor to the tops of its 200-foot-tall trees. There are up to fifteen discrete layers, each supporting its specialized community of plants and animals, many of which live their entire lives in a single layer of the canopy, never reaching into layers above or below. Other aspects of life in the tropics are revealed by the look of these plants. We find thick and waxy leaves and thick, hard seed husks that provide some protection from rotting in the high humidity of the tropics.

Temperate Forest

In the middle of the global forest continuum are the temperate forests, the primary subject of this book. The temperate landscape is characterized by seasonal change in a year that includes a prolonged growing period that is warm and well watered, followed by a dormant period of cold and drought when water is frozen and hence unavailable to plants. Rather than having evergreen leaves that are adapted to extended drought, the trees of the temperate forests are predominantly deciduous, with leaves that are shed each fall and replaced each spring.

The complexity of the aboveground structure of the temperate landscape is limited by its shorter growing season, even where there is high rainfall, and as a result its forests have fewer layers than a tropical forest. A temperate forest generally has four layers. The canopy layer is composed of trees that are typically taller than 35 feet and often more than 100 feet. The understory is also largely woody and may include saplings of canopy trees and smaller trees, such as dogwood, that do not reach canopy height. The shrub layer also is woody and may include juvenile canopy and understory trees as well as shrubs, which are usually multistemmed and smaller than understory trees (less than 25 feet in height). The ground layer consists of herbaceous (nonwoody) plants, such as grasses, sedges, rushes, ferns, mosses, and seedlings of trees and shrubs.

Woody vines appear in every layer of the forest. Occasionally there is also a super-canopy, of higher, and usually older, trees. Another layer in the landscape is the litter layer, which is composed of the debris of leaves and tree limbs and other parts of vegetation as well as small animals, both living and dead, throughout this organic debris on the surface of the ground.

The predominance and distribution of species within the temperate forest has been influenced by land use as well as by the natural environment. Deciduous forest trees such as maple, basswood, and beech tend to predominate under the more mesic (moist) conditions in the temperate forest biome, while oaks and hickories are more typical at the drier end of the spectrum. The relative predominance of oaks today over presettlement conditions is due to past fire and logging history as well as the loss of the American chestnut in the early part of the twentieth century. Droughtier soils in temperate landscapes often support a mix of deciduous and evergreen trees, such as the mixed oak-and-pine forests of the sandy soils along the Atlantic coastal plain landscapes. The abundance of pine in these areas is maintained by fire, which in turn makes the landscape more fireprone. Fire suppression since settlement in areas with sandy soils is resulting in greater predominance of oaks and hickories. Beech and maple may become more numerous as fire is further suppressed by suburbanization.

The eastern deciduous forest, which is the temperate-forest biome type that covers most of the United States east of the Mississippi, reaches its most complex and diverse type in the southern Appalachians. Until European settlement, vegetation there persisted without interruption, over 300 million years, since the origin of flowering plants, when the land was still part of a supercontinent. While adjacent lands were either under water or scraped by glaciers for extended periods of time, the sheltered coves of what is called the "mixed mesophytic forest" served as both refuges for and sources of biodiversity in the region. The Ozark Mountains are nearly as rich biologically although somewhat drier.

Boreal Forest

At the other end of the global forest gradient is the boreal forest. Moisture is scarce in the frozen winter but is sufficient in the rest of the year to support a dense cover of pine, fir, and spruce trees. The long winters make their mark on the appearance of the landscape. The coniferous trees, whose name, incidentally, means "cone-bearing," reflect the relatively dry conditions. Their compact, spiky needles serve to limit the surface area from which moisture can be lost. The persistence of these needles for several seasons is also an adaptation to the harsh environment. It requires a great deal of energy to produce a new leafy canopy every spring, and when the fuel for that energy is in short supply it is better directed to growth and reproduction. Even the shape of the coniferous trees reflects the conditions of their environment. We can hardly look at the low downward-sloping boughs without seeing them bending even farther under the weight of winter snowfall.

The trees in the boreal forest are also shorter, and the ground layer is far less dense than is usually seen in milder climates. The relatively bare forest floor is partly the result of the low levels of light and limited physical space that remain year-round beneath the dark evergreen boughs, but the paucity of herbaceous (nonwoody) plants, shrubs, and saplings also reflects the effect of the long droughty winter and the difficulty the smaller plants have in competing with the large established trees for the limited resources.

Grassland

Trees will grow where there is adequate moisture and a long enough growing season. The character of the grassland biome is shaped by a critical period of drought during the growing season (in addition to the winter drought) that precludes tree growth except along water courses.

Like trees, grasses seem to be little affected by extremes of temperature, growing abundantly from northern Canada to the tropics. It is interesting to note that grasses do not *require* droughty conditions; it is just that they survive them better than trees can generally. We have more than enough evidence that grasses will grow perfectly well in regions where resources will support a forest, but, as most homeowners know, grass does not grow well in competition with trees. The extensive root and shoot systems of trees are more effective than grass at marshaling available light, water, and nutrients, so grasses persist in the more humid areas only where forests are held back by human interventions. Often grasslands border woodland or precede forest establishment.

In the great prairies of western North America, flat, open grassland mixed with a rich variety of forbs (broadleaved herbaceous plants) once extended for miles on end. The land here is sometimes so flat and so devoid of trees that it often seems we can see the Earth curving away. The open land also allows the wind to sweep across the vast spaces, blowing and howling relentlessly, further desiccating the landscape. Only along the corridors of streams or in the occasional low, wet areas do trees break the view of the oceanlike grass.

Grasses can prosper in such habitats. Their buds are safely underground, rather than on aboveground branches, during the winter freeze as well as during the frequent summer fires that have had a strong hand in shaping this landscape. Their canoe-shaped blades capture the rain when it comes, sending it directly to the roots.

The root systems of the great prairies formed a mat that became legendary among the early pioneers known as the "sodbusters," who left the eastern deciduous forest to farm the midwestern lands. This mat held the soil in place despite droughts and high winds and was so thick it inhibited the germination of trees. It even resisted breaking up when a great herd of buffalo passed, a trampling that was reported to lower the level of the ground in places by as much as 4 feet.

Just as there are major differences between boreal, temperate, and tropical forests, similar distinctions can be seen in the grassland biome. The American grassland is usually divided into three types: a tallgrass prairie (with grasses between 6 and 10 feet tall) in its eastern range, a shortgrass prairie (in which the grasses are about 1 to 2 feet tall) to the west, and a mixed section in the middle.

Landscapes in Transition

Some areas in the landscape are in fact called transitional between two different biomes. The boundary between two biomes is generally quite broad, with elements of both biomes blended together. For example, a savanna, which is a mix between a grassland and woodland, often occurs between the forest and grassland biomes, where there are woody thickets and small woodlands woven into a prairie fabric. Trees thin out and vanish in the drier areas of a savanna and become more dense at a water source and along streams. The boundary of the transition area shifts over time and is also affected by human activities such as burning and animal activities such as grazing. We are presently seeing a critical process of desertification taking place in sub-Saharan Africa, where grasslands and forests at the edge of their climatic range are being exploited, destroying a fragile balance and causing the desert to expand.

The transitional area will also shift locations with long-term climatic changes. During periods of abundant rainfall, for instance, forest may become established in areas that had been grasslands. Then, when droughts return, some of the trees may survive because they draw moisture from deep in the ground, but new trees cannot grow and the region will gradually return to grassland. Evidence indicates that several thousand years ago (but since the last glaciers) the United States was drier than it is today. At that time the great prairie extended as far east as mid-Ohio. Since then, the forest has been expanding and now has reached as far west as Manhattan, Kansas. With global warming we are seeing the ranges of plants generally migrating northward.

The landscape changes not only with climate but also with time. Landscapes, like people and all other living comunities, mature and age. Change may occur suddenly and in ways that are very visible, such as after a great fire, flood, volcanic eruption, clearcutting, or rapid suburbanization. Secondary growth after events such as fire or clearcutting as well as young primary landscapes developing on new land such as volcanic ash are called "successional landscapes," and the process of changing community types over time is called "succession."

Change also occurs slowly. Where the general patterns of the landscape have persisted for an extended period of time with limited change, very unique and complex interrelationships between species and place develop. Plants and animals that can be found only in highly specialized environments, called "conservative species," are becoming increasingly rare in today's rapidly changing environments that favor generalists. "Keystone species," those whose presence holds the whole system together, such as bison and wolves or prairie dogs, have been especially diminished over the centuries of European-style land use. When keystone species are lost, the whole ecosystem can collapse. Weeds, or "ruderals," as they are sometimes called, as well as invasive species, are at the other end of the gradient of conservatism and are distinguished by their ability to thrive in many habitat types within their range. Where natural patterns are disrupted and damage goes unheeded until well under way, such as with increasing air pollution and the introduction of nonnative plants and animals, indigenous species are threatened or lost. Even when species per se are not threatened, the rich diversity of subspecies varieties adapted to particular environmental factors such as frequent fire, called "ecotypes," may be disappearing as the landscape is rendered ever more uniform. In the habitat we call the eastern deciduous forest, that uniformity is increasingly the case today.

CHAPTER 2

The Once and Future Forest

Before settlement by Europeans, most of the eastern United States was covered by what is known as the eastern deciduous forest, a dense and multilayered forest interrupted only by rocky outcrops, large rivers, and coastal wetlands. Native peoples cleared some land in their use of forest resources and for agriculture and burned even larger areas to manage for game and other resources, but the forest remained largely intact. The early European colonists, however, carved towns, pastures, and croplands out of the forest and maintained them by constant control of natural growth. The early forest industries, which harvested timber for firewood, charcoal, and building, further fragmented the landscape (Whitney 1994).

The forest of five centuries past is largely gone, and the recoverability of its remnants is, in fact, very much in question. Areas once thought to have regenerated naturally after logging operations, scientists now recognize, are more damaged than previously believed. Just to cite one example, recent studies show that even a century after clearcutting, salamander populations and woodland wildflowers have not returned to previous levels. The impact of this great wave of deforestation no doubt goes well beyond anything we yet understand. Great numbers of plant and animal species were lost, many not yet documented.

During that era extraordinary amounts of soil were lost to erosion and sedimentation. Poor land-use practices that increase the amount and velocity of runoff have continued to this day, hindering the recovery of the landscape. Deforestation exposed huge expanses of soil to erosion, leaving behind mineral subsoils that favored the reproduction of plants that were not characteristic of the historic forests. These changes in many areas were gradual and barely perceptible, but in others they were rapid, even spectacular.

On the high eastern edge of the Appalachian plateau, in what is now West Virginia, for instance, there once grew an extensive forest dominated by great red spruce trees beneath which an organic peat soil had accumulated over millennia to depths of many feet. The timber industry began to harvest the trees in the last century but left large amounts of waste slash that fueled huge fires, burning much of the rest of the forest and even much of the soil over large areas. Today, more than a century later, parts of this landscape—one that supported some of the most productive forests—are still completely deforested and support only shrublands, called "Dolly Sods," with only pockets of the once deep, peaty soil remaining.

Before they were cleared, fire was a recurrent form of natural disturbance in these former sprucelands as it was in most forests. Landscape communities are, in part, determined by their ability to adapt to local fire frequency. Human presence increased the frequency of fire in precolonial times as well as later. It is only in the last century that we have reduced the frequency of fire in many areas through suppression strategies and techniques. Forest managers now recognize that prolonged absence of fire can often result in a more severe fire cycle. Fire control itself is a kind of disturbance that changes the nature of forest vegetation.

In the mid-nineteenth century, industrialization and the associated worker migration to the cities as well as emigration to newly opened rich farmland of the Midwest temporarily diverted the intense development in the easternmost part of the forest. Farms were abandoned and the total amount of forested land in the East slowly increased until just recently. Today these remnant landscapes are experiencing a rate of destruction that parallels the first era of deforestation as development sprawls outward from cities along highways that now crisscross the continent. Morris County, New Jersey, is a typical example of these repeating scenarios. From 1970 to 1985, as much forest was cleared in the area as had regrown in the previous seven decades (DiGiovanni and Scott 1987).

Today, there is virtually no regulatory protection of privately owned forests and little incentive for private landowners to sustain woodlands. The Northern Forest Lands Council, after years of public input and extensive study of the forests of New York and New England by the U.S. Forest Service, reports that development pressures are intense, especially in scenic areas where forestland is most valuable, and that taxes on forestlands are excessive. There is little coherent policy on conservation easements and acquisitions, and government funding to sustain forest programs is inadequate. They conclude that public–private partnerships will be necessary to sustain these resources. We can no longer separate natural areas from their regional context. We now need to see the landscape as a whole and integrated system.

Islands of Forests

Today's remaining forest fragments are very different from the landscape that greeted eighteenth-century settlers. Unbroken forest expanse has been replaced by small islands, each with little or no forest interior, and most of the forest has been cleared more than once. Virtually all remaining native and volunteer landscapes occur within the fabric of developed land and have experienced rapid changes in environmental conditions, including major alterations in the hydrologic cycle; soil disturbance from vegetation clearance, increased erosion, and trampling; and air and water pollution.

Forests today generally are restricted to less easily used land, the rocky outcrops and steep slopes or drainage ways not suitable for farming and most other human uses. These landscapes are also younger, recovering from past clearings, and are bordered by many miles of "edge," the seam between two different types of landscapes, such as forest and field. Meanwhile the amount of interior forest habitat, and the species it supports, is diminishing. Hundreds of new plant and animal species have been introduced to eastern forests (and in fact most landscapes) on a large scale, both deliberately and accidentally. Some, such as Norway maple, kudzu, Japanese honeysuckle, and Japanese knotweed, have experienced population explosions in the absence of their natural controls, and are spreading so aggressively that they are overwhelming many stressed native plant communities. In Pennsylvania, for example, almost one-third of the native plants are listed "of special concern," and 15 percent are "endangered" or "threatened."

Similarly, diseases and pests introduced from abroad have found unresisting hosts here, nearly eliminating some important native species. The chestnut and American elm now make token appearances in landscapes they once dominated, and their disappearance has left gaps that have shifted the composition of the forest community. Before the blight hit, one in every four trees in the Appalachian forest was a chestnut and the regional economy depended on chestnut. Many other species are in serious jeopardy from disease and stress. Dogwood blight, beech bark disease, ash yellows, and hemlock woolly adelgid, for example, are other introduced major pests and diseases of trees. The realities of the blight sweeping this continent chronicled by Charles Little (1995) in *The Dying of the Trees* will be felt in every park and natural area for decades and centuries to come:

> I see a world of dying trees; dying because the trunks have been bored into and the leaves have been stripped by adventitious pests; dying because fungi are girdling their bases and branches and turning their leaves to black corpses; dying because their shrunken roots can no longer absorb enough nutrients and water to keep them alive; dying from the direct effects of too much ozone in the troposphere and not enough in the stratosphere; dying because neighbor trees have been clear-cut,

allowing unwonted cold and heat and drying winds into their precincts; dying because of being bathed too often in the sour gases of industry; dying because the weather patterns have changed and they cannot adapt quickly enough to them; dying, in fact, because they are dying (225).

Disturbance occurs on a continuum ranging from minimal levels to severe degradation. Once initiated, deterioration often accelerates. One of the most disturbed of the forest fragments is the urban woodland, or "metroforest." It's supposed to be a woodland, but the ground beneath it is glaringly different in appearance from what we think of as forest floor. Trash is often ubiquitous and may range from dumped cars and construction rubble to larger particles of urban "soil," including broken glass, paper bits, and cigarette butts. Bare soil may be exposed over a large area from trampling, stormwater runoff, filling, or excavation. Various hydrocarbons, including motor oil, fuels, and their derivatives, may be present, at times in large amounts, from leakage, improper disposal, and deposition from engine, furnace, and incinerator exhaust. In urban areas, woodlands may also provide temporary shelter to the homeless and others who use them as bathrooms, especially in the many areas where few or no public alternatives exist to meet these quite human needs.

In the metroforest, where disturbance is chronic, nonnative or exotic vegetation usually prevails in the absence of the native vegetation that it has replaced. Large, remnant native canopy trees dating back to a more rural time in the landscape may still be alive but have long ceased reproducing successfully because of the proliferation of alien plants, such as the Norway maple, which frequently appears in every layer of the landscape. Plants that establish themselves without being planted by people are called "volunteers." Native saplings of cherry and locust, usually typical of open field landscapes rather than older forests, also volunteer to fill the large gaps in the canopy. Vines, both native and exotic, occur in heaping mounds on the ground, draped over shrubs and saplings, and even cloaking trees in the canopy. Ferns and woodland wildflowers may be conspicuously absent or reduced to one or two species. Species diversity has usually been declining for years, and human use, especially for recreation, may be steadily increasing.

If we could step back in time, we would not need to look at tree size to see how different the forest was before Europeans arrived. The ground we walked upon would immediately tell us how the landscape has changed. For millennia in the "primary," or original, forest, fallen ancient trees left behind an uneven terrain called "pit and mound" as well as many great trunks. Traveling through the landscape before the plow homogenized the ground surface was often slow because one needed to climb over masses of roots and soil traversing an undulating forest floor. In the spring there would be a thick carpet of woodland wildflowers, also called "spring ephemerals" or "vernal herbs"—as many as fourteen different species in a single square yard (Duffy and Meier 1992).

Pits and mounds create varied microhabitats, each favoring slightly different species. The terrain, combined with thick layers of leaves in the canopy and the litter of leaves and dead wood in various stages of decay on the ground, held water like a sponge, even through periods of drought. Typically, in that same square yard of ground one or two salamanders would be living in the permanent dank of the deep litter and humus layers (Petranka, Eldrige, and Haley 1993). And the forest floor would be visibly alive with insects of all kinds.

While the metroforest may be an extreme example to compare with the primary forest, it demonstrates that the natural processes that once sustained local biodiversity have been significantly altered. That is also true in what we think of as our wildest lands. Conditions now are very different from those under which indigenous species evolved and thrived. In the eastern forests, no extensive old growth remains, nor even a roadless area large enough to truly be called wilderness or sustain wilderness into the future. In a recent study for the National Biological Service, the Department of the Interior concluded that natural habitats over nearly half the United States, including Alaska and Hawaii, have declined to the point of endangerment. The most imperiled ecosystems are found clustered in the eastern half of the country (Noss, LaRoe, and Scott 1995). The eastern temperate forest, like the tallgrass prairie, the oak savanna, and the longleaf pine forests, is critically endangered and has been reduced by more than 98 percent of its original area.

A recently completed study by Brian Drayton and Richard Primack (1996) shows that a conservation area in metropolitan Boston has been losing native species in the last century at the rate of about one a year. Exotics, on the other hand, are colonizing the site at the rate of one new species every five years. The conservation area, Middlesex Fells, a 1,080-acre park isolated by surrounding urbanization and heavily influenced by human use, is probably typical of many other natural areas. In 1894 a team of botanists surveyed the area and documented 422 species, of which 83 percent were native. The recensus taken in 1993 showed only 267 species present and the proportion of natives dropping to 74 percent. None of the species lost was listed as endangered or threatened, but all have nonetheless suffered local extinction in what many had perceived to be a good conservation area. Many factors are responsible, from trampling and increased incidence of fire at the local level, to acid rain at the regional scale.

Once we start restoring a landscape we quickly find ourselves increasingly involved in the lands beyond the site's boundaries and, in particular, the watershed. The conditions of the remaining fragments of natural areas reflect our management of the whole landscape. What we do in our higher-use areas affects protected places, and their fates are inextricably linked. A site will be affected by local disturbances as well as regional and global phenomena. The mismanagement of stormwater, for example, reveals itself locally in eroded and gullied slopes and regionally in long-term changes in wetland plant communities.

In other words, we cannot talk about restoring particular habitats without addressing the larger issues driving change within the regional ecosystem, including human behavior and values. Our attitudes toward the forest must also change, and we must come to know the landscapes around us. Indeed, the act of restoration itself is one of the most powerful vehicles for fostering awareness of place and environment.

This book deals with the landscape of today, the inheritor of a long history of use and abuse. It is largely about the remaining fragments of our forests and how to restore, expand, and link them to re-create a whole landscape fabric. The next sections examine human-induced disturbances that typify what is occurring in the modern landscape and the consequences they have in natural areas at a variety of scales. Some factors of disturbance can be addressed immediately, such as ceasing to plant exotics on publicly owned land. Some issues, such as changing patterns of access and use, will require a high degree of community involvement and environmental education before any real progress can be made.

CHAPTER 3

Fragmentation

What is a forest fragment? By definition, the landscape of a forest fragment is not whole; it is diminished in scale and discontinuous with the larger regional landscape system. All fragmentation is relative to what once was wilderness. In the East, fragmentation is relative in scale and pattern to the intact forest ecosystems that existed at the arrival of European settlers. A fragment cannot support wilderness even if it is well managed because large scale is an integral quality of wilderness.

In a recent study of the world's remaining wildernesses (McCloskey and Spalding 1990), the authors identified fragmentation as the defining trait for the loss of wilderness. Their definition of "wilderness" is a land area of almost one million acres (400,000 hectares) without prominent or marked evidence of significant human impact. McCloskey and Spalding used aviation navigation charts that depict roads, paved or otherwise, clearings, and other such features to indicate significant fragmentation of the landscape. According to their definition, the eastern United States has no wilderness at all and the Great Basin in the West has only a few areas of minimum size.

Many human activities result in the division of natural habitats. But what are the consequences of fragmentation? With each parcelization, not only is the overall amount of habitat reduced in each local area, but there also is a greater loss in the number and total area of remaining large-scale tracts of similar habitat. Once fragmented, the entire habitat may change in character, such as the eastern forest did after the initial wave of deforestation that doomed species like the passenger pigeon. Fragments are not comparable to formerly continuous large areas, making them inadequate habitat for many species that depend upon larger systems that meet their needs for survival. We usually think in terms of extinction with regard to rare species. However, sweeping changes in the scale of the forest after European settlement eradicated the most abundant species because they had depended upon forest expanses to survive. People once counted days when multitudinous passenger pigeons blackened the skies overhead, but that species depended upon large-scale presettlement forests, much like the once-numerous bison depended upon our formerly continuous prairie lands.

Edges Instead of Patches

All landscapes are now and have been for some time affected by human modifications, but human impacts do not always fragment the landscape in permanent ways. The difference between management by indigenous peoples and much human-caused disturbance today hinges to a large degree on the persistence of fragmentation in landscape patterns. Small-scale slash-and-burn agriculture, for example, creates a shifting pattern of openings in the forest, while the construction of a road and permanent settlements along it bifurcates a landscape more permanently. Fragmentation, at its simplest level, eliminates wilderness because it changes the most basic patterns of the landscape and the way it functions.

Patterns of indigenous land use and settlement were strikingly different from those of more modern settlements. Indigenous peoples used intentionally set fires and limited horticultural activities such as coppicing (cutting back sprouting woody plants to the roots), clipping, planting, selecting and dispersing seed, and weeding to shape the landscapes we now think of as wilderness. The management practices of indigenous peoples fostered medicinal and edible plants, many of which are herbaceous plants or low shrubs. Their impact was largely patchy—that is, they created relatively small holes in the landscape fabric, many of which reclosed quickly, and the fabric remained intact. Within the forest expanse patches of both natural and human disturbance shifted over time, thereby providing micro-opportunities for many plants and animals that would otherwise be limited in a more continuous canopy cover. There are often very few gaps in today's younger, more even-aged woodlands, such as in a second-growth or commercial forest, and therefore fewer opportunities for regeneration. The return of a rich and patchy pattern to the landscape may take centuries.

Patterns created by fragmentation are very different from the rich and shifting patterns created by gaps resulting from management by indigenous people or from windfalls and other natural disturbances. In the fragmented forest, there is no continuous forest matrix to block competing species and predators; rather, there is a continuous and connected edge that gives access to all places. In the old-growth forest gaps are not connected; it is the forest that is intact. Exotic weed seeds, for example, are poorly adapted to the closed canopy of an intact forest and do not travel easily from patch to patch, although they travel easily along the edges of the fragmented forest. A predator in one gap in an old-growth forest is not necessarily led to all the other gaps as it is when it travels along an edge. Edge effects are not confined to forests but occur in all landscape transitions. Fragmented prairie patches also suffer reduced nesting success because predators are sheltered by adjacent woodlands up to 150 feet distant.

The patches created by natural gaps are usually rich in species and serve as sources of diversity. Fragments, or islands, as they are also called, usually experience local extinctions. Birds are often attracted to the edge of the fragmented forest because of its similarities to a natural gap but then suffer reduced reproductive success and greater mortality from predators because they are not protected by deep forest (Gates and Gysel 1978). In a fragmented forest the amount of interior forest that is far from an edge is reduced. Many species, virtually confined to the once-extensive habitat of forest interior, are now rapidly disappearing, even where the total amount of forest has increased. So too are many other species that require interior habitat at least seasonally at some stage of their life cycles, such as during breeding or nesting. A few species, like white-tailed deer, adapted to younger landscapes, have benefited from fragmentation at the expense of other species. Deer populations have exploded with the spread of suburbia, a habitat in which they have found abundant food and shelter.

Estimates of the distance into the forest before edge effects diminish are variable, from 100 yards to over a mile. Therefore, even a squarish block of forest 200 yards on each side, roughly 8 acres in size, is, even by the most conservative estimates, all edge. Where edge impacts extend for a mile, forest plots that are more or less equidimensional in shape—that is, square or circular, for instance—must be considerably larger than 2,500 acres to support any amount of interior habitat. In irregular tracts, the proportion of interior is far less. By these standards, corridors 1,000 feet wide would have only limited interior, or none. Studies by the University of Illinois, University of Missouri, Indiana University, the U.S. Forest Service, and the Illinois Natural History Survey have demonstrated that no forest remaining in Illinois was large enough to support successfully reproducing populations of many locally declining birds. In response, The Nature Conservancy, the state of Illinois, and the U.S. Fish and Wildlife Service have begun a program to restore 60,000 acres of cypress swamps and rich bottomland forests in a continuous corridor.

Fragmented Migrations

As fragmentation proceeds, as has been occurring in most parts of the world, and the remaining large habitat areas are divided again and again, the average distance between remaining fragments increases. Not surprisingly, migratory herds cannot adapt to fragments. With the loss of their powerful influence on the landscape, everything else changes, too.

For migratory species of animals the longer distances between fragments of habitat pose enormous, perhaps insurmountable, difficulties. Successful migrations require a certain amount of food and other habitat resources. These include shelter, cover, and freedom from heavy predation in some minimal proportion, not just at one or a few places but all along routes of movement. At some point the distances between places that meet their needs becomes too great, and they cannot complete their circuits. For migratory forest songbirds, such as many warblers, the consequences of fragmentation have been devastating. Populations have plummeted, down 60 percent or more for many species since the last century. At the same time, populations of the cowbird skyrocketed with the increased access provided by forest fragmentation to once-isolated nest sites. These nomads lay their eggs in other birds' nests to be raised by them. Cowbird young mature very rapidly, which usually results in the starvation or elimination of the foster mother's true nestlings.

Once cowbirds fed upon the insects stirred up by bison. With the demise of the great herds, the cowbirds found a new niche that was facilitated by forest fragmentation, a landscape rapidly being transformed into islands of forest remnants by the expansion of settlers and their agriculture (Brittingham and Temple 1983). From its former range in the more open lands of the southern Midwest and south-central part of the country the cowbird has spread throughout the eastern forests.

Unlike robins and catbirds, which recognize cowbird eggs often enough to remove most from their nests, warblers, vireos, and other migratory, cavity-nesting birds of the forest interior that now are ever closer to the forest edge appear helpless to resist this takeover of their changing territory. As cowbirds expand their range, the numbers of tanagers and thrushes, vireos, and warblers fall. Each of these tiny birds consumes up to a thousand caterpillars a day, and without them the forest loses important protection from insect predation.

Without effective habitat protection and restoration the prospects are bleak for many interior bird species. A poor-quality site, one where not enough fledglings survive to replace their parents, might actually serve as a "sink" for regional species, diminishing overall numbers rather than replenishing them (Fahrig and Merriam 1990). Tracts of greater than 25,000 acres in size, as well as corridors connecting them, are necessary to sustain interior-dependent bird species. The largest remaining forested tract in the state of Illinois, for example, is only half that size.

Barriers Created by Fragmentation

Predation and parasitism are not the only problems associated with fragments. Seed dispersal among separated areas also becomes more difficult or even impossible for many plants. Extensive edge may create desiccating winds that dry out seeds as well as new seedlings, reducing the likelihood of successful germination and many species' survival. Plants better adapted to longer-range dispersal are favored, changing community composition. Many species, both native and exotic, that populate disturbed sites share common traits such as vigorous vegetative reproduction and abundant seed that is distributed by wind or birds.

For many animals as well, each separation lessens their ability to move between fragmented parts. This effect may be amplified by the nature of the intervening land use. Road kills are obvious evidence of the barrier that highways present. Many forest creatures simply cannot cross an open area for innate behavioral reasons or because they are much more vulnerable to predation and exposure to the elements where there is no vegetative cover. As one example of this at the small scale, studies of the rare Cheat Mountain salamander in the Monongahela National Forest of West Virginia show that it avoids crossing even hiking trails.

Studies of landscape revegetation following the retreat of the glaciers suggest that barriers created by fragmentation will strongly impede the adaptation of natural vegetation to global warming. As climate changes, so do the ranges of plants. They must move to new, more suitable, sites. At the University of Minnesota Margaret Davis and Catherine Zabinski (1992) have pointed out that past rates of plant colonization would not be likely today owing to the high degree of separation of vegetative stands. For this reason, native plants could become rarer or disappear from part or all of their present ranges at the same time that their spread into new areas is blocked by our highly disconnected landscapes.

The logical consequence is the loss of the rich variety of life forms. The full meaning of the crisis of biodiversity is not only seen in the reduced varieties of native forest plants but also in the animal populations that have coevolved with them. If we cannot sustain native flora, we cannot sustain native fauna either. They are dependent upon each other for survival.

Disrupted Water Systems

The same land-use changes that produce landscape fragmentation, such as forest clearance, agriculture, and settlement, are also linked to large-scale changes in the patterns of water in the landscape. Fragmentation has affected the aquatic landscape as well as the uplands. In the last century we have disrupted, more than probably any other natural system, the patterns of water to which plant and animal communities have adapted over millennia. A discontinuous patchwork of deteriorating pipes, ditches, channels, impoundments, and wells has replaced and significantly dismantled the natural infrastructure of streams, wetlands, and aquifers, which have been filled, drained, diverted, channeled, pumped, and dammed.

Small headwater streams are segmented by impoundments or eliminated altogether and buried in drainage pipes. Large rivers are channelized. At the coast

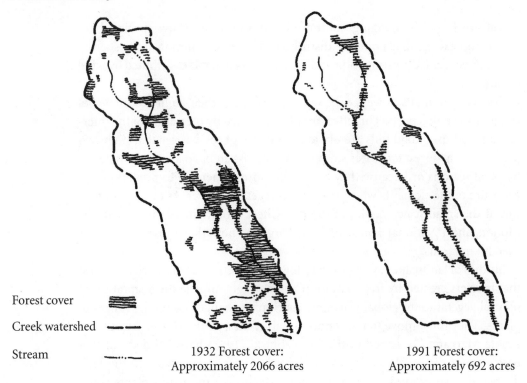

Forest cover

Creek watershed

Stream

1932 Forest cover:
Approximately 2066 acres

1991 Forest cover:
Approximately 692 acres

Figure 3.1. Forest loss since 1932 in the Sligo Creek Watershed, Maryland. Currently, Maryland is losing an estimated 10,000 acres of forest a year, more than half of which is due to construction activities. (Source: Maryland Forest Conservation Manual, 5. Reprinted by permission.)

bulkheading and weirs replace the continuity of tidal action. Compounding the impacts of fragmentation are the very severe problems of water pollution. Aquatic habitats now experience the highest rates of extinction as invaders— zebra mussels, lamprey eels, hydrilla, and many others—rapidly supplant native species.

Isolated by Exotics

At the same time that forests of all types have been reduced in total area and are ever more fragmented, the extent of forest that is still largely native is even more diminished. The predominance of landscapes dominated by invasive species also represents a kind of habitat fragmentation. Communities that were once connected by landscapes of similar species composition are now set in a fabric of alien species that are invading and expanding at the expense of native species. An ever-smaller area of native forest is highly fragmented within larger but still unconnected fragments of a former forest that is highly altered in character and composition by human influences. These landscapes are now the most common. In many places, even when we see trees all around us, we see none of the forest that once characterized the area.

Isolated by Development

In large areas of the world great division of the wilderness has occurred with extensive urban sprawl, such as the megalopolitan corridor that stretches from Boston to Washington, D.C. The Sligo River Watershed in Maryland illustrates this recent wave of forest clearance (Figure 3.1), where forest cover was reduced by two-thirds in just over half a century (Greenfeld et al. 1991). Many urban parks and woodland landscape remnants are set in this context. They are not just separated by one or a few major elements; they are fragments. If we wish to sustain the environments that persist today as remnants in this region, we must ultimately reweave this disconnected fabric, reconnecting them and their ecosystems into a coherent whole.

CHAPTER 4

Succession and Recruitment

Some might measure the success of a species in terms of its tenure on Earth prior to extinction. Although most are now gone, many of the great dinosaurs persisted for many millions of years. *Homo sapiens* has been around less than 1 million years. In order for a plant or animal to remain in the landscape it must reproduce itself in sufficient numbers to sustain a viable population over time. If it is very successful its population will increase and/or extend in range.

Plants and animals occur in related communities, like tribes and related groups of human families. A given group may die out or successfully continue itself for millennia. The core of the idea of sustainability is to reverse the trend we have set for progressive extinctions in the hopes of extending our tenure as well as that of the species that have historically been associated with us.

The Process of Succession

Many of us have watched, and many scientists have studied, the process of succession in the gradual emergence of a young forest from the early grasses and wildflowers of an abandoned agricultural field (Figure 4.1). "Succession" is a term used to describe how plant and animal communities develop over time. Landscapes, like human communities, change as they age or are affected by altered environmental conditions. They do not remain static. Succession is the process of changing plant and animal populations and developing communities over time.

Figure 4.1. An abandoned agricultural field is like a seed bed for growing a forest.

Not all species or communities are equally well adapted to different stages of landscape development. Very short lived species—herbaceous annuals and biennials (Figure 4.2), for example—are typical of earlier phases of succession. With time the longer-lived species, such as herbaceous perennials and later woody species, become increasingly predominant (Figure 4.3). As woody plants mature, they create islands within the remnant meadow grasses and wildflowers (Figure 4.4). The young woodland, at first, tends to be relatively even-aged (Figure 4.5). With time a more varied forest develops with more species, more complex structure, and varied ages of plants (Figure 4.6).

In juvenile landscapes such as regenerating forests, change may occur very quickly, whereas in old-growth landscapes it may proceed too slowly to be easily observable in human timeframes, except for occasional large-scale natural events such as a fire or tornado. Because plants are the basis of every food chain, they lead the way in succession. Many of the patterns of succession are frequently repeated from setting to setting although a wide array of factors may shift the direction taken.

While succession is not predictable, in temperate climates a forest will typically result after enough time has elapsed, except where water levels are too deep or where bare or moving rock is the only surface available. In the eastern United States, almost all landscapes are on a journey to temperate forest, the most complex system of vegetation that the limitations of climate permit. Natural landscapes typically become more diverse and complexly organized as they mature. A meadow with a few layers of herbaceous species may appear early in

Figure 4.2. The early meadow is dominated by annuals and biennials, growing like a crop.

Figure 4.3. The shifting mosaic patterns of long-lived herbaceous perennials characterize the established meadow, as well as developing trees.

Figure 4.4. The "savanna," or "woody oldfield," is the last stage of succession before a true woodland appears.

Figure 4.5. The grasses and showy wildflowers of the meadow find no place beneath the canopy of the young woodland.

Figure 4.6. As the woodland matures, trees, shrubs, and groundlayer plants gradually colonize the site to create a multiaged and multilayered landscape.

succession. A multilayered forest with a rich mosaic of gaps at different stages of succession throughout its fabric can develop over time. Diversity reaches its highest expression in the hotter, wetter tropical landscape, where there are up to fifteen or more distinct layers of the forest and hundreds of species of trees are found in a few acres.

Just as any plant may be confined to a narrow range of environmental conditions or be more of a generalist, some plant species occupy a site for a brief period and others for longer periods. Some trees like red cedar, for example, are typically restricted to the earlier phases of woodland development. Beech is associated with the later phases of forest development. The red maple is more versatile than either and can be found in many stages of succession as well as in a wide array of different environmental conditions.

Adaptations in plant structures and predominance of species in the composition of the developing forest also shift in succeeding natural areas, from smaller herbaceous species to large trees and shrubs. The animals change with the landscape. The succession of soil microbiota, for instance, follows the succession of vegetation types, and hence the food available to soil organisms.

The primary forest consisted of a shifting mosaic of widespread old growth and numerous gaps that ensured some habitat at every stage of succession dispersed throughout the landscape. After the period of deforestation the age of the landscape, and hence the successional communities, was suddenly and dramatically altered. The woodland wildflowers, which are also called vernal herbs, were especially diminished. The regrowing forest was everywhere a patchwork of relatively even aged stands that began a few decades after cutting as a dense cover of polelike saplings, with few if any gaps suitable for vernal herbs. In fact, the gap patterns of the original forest only begin to return more than a century and a half after clearcutting. Even then, the richness of wildflowers as well as many of their associates, such as ants, is still diminished.

Today, the successional changes around us entail more than shifts in the age of the landscape. All too often land use has arrested natural succession and shifted the competitive balance in favor of a few exotic invasives and generalist natives. In smaller forest fragments the patterns of succession sometimes seem to be going backward. All around us, forest systems that were once complexly layered are reverting to thickets and vinescapes. When once complex bonds are disrupted, such as when pollinators are eradicated, whole systems can collapse. The remaining components reorganize and start the cycle again, now under different conditions.

Each fragment retells the story of "progressive simplification" (Figure 4.7). Progressive simplification is one of the symptoms of what Sergei Yazvenko and David Rapport at the University of Guelph (1996) call "Ecosystem Distress Syndrome," a suite of traits that characterize the retrogression of heavily distressed ecosystems to opportunistic species such as weedy invasives, which they characterize as late-stage symptoms of distress.

Figure 4.7. Succession in some places seems to be going backward as woodlands revert to vines and brushlands.

Management is as important in buffer areas and corridors as in the more pristine landscapes they surround and connect. The ways we manage all of our landscapes affect our ability to sustain forests and other natural areas, but the ways we manage the areas bordering them will have the most direct effects. Much open space is composed of young, successional landscapes, recently re-leased from agriculture, turf, or paving. These areas are especially vulnerable to exotics invasion and other perturbations that disrupt the historic patterns of natural succession. They often, unfortunately, serve as pathways for distur-bance to less-degraded remnant landscapes. With appropriate management they can provide expanding habitat and vital connective greenways.

The Recruitment Process

In evaluating the stage and condition the land has reached, one of the most basic and interesting questions to ask about any landscape is "What is repro-ducing here?" In a very real way, this inquiry acts as a window into the future of the place and a key to sustaining or enhancing its biodiversity. Because the processes of successional change and reproduction in older landscapes, such as forests, are slow, they often go unobserved by a mobile human population. Trees also can live significantly longer than people, making it hard to observe trends directly. But in order to survive into the future, each species, no matter how long-lived, must reproduce at some point. It is not enough to simply produce viable seeds that germinate; plants must also persist through the various stages of their life histories, including successful reproduction. Not every seedling will live to be a great tree. Most, in fact, will not, but young must occur, survive, and mature somewhere in the habitat to ensure the species' survival.

Successful establishment of a new plant is referred to as "recruitment." Obviously, in order to be recruited, a tree must be present in the forest first as viable seed, then as a seedling, sapling, and small tree. It must survive competition with other plants for the same space. Ordinarily in a stable forest community there are enough gaps and a sufficient degree of recruitment that forest composition and biodiversity are sustained over extended periods of time. But conditions are never static. Quite the opposite is true. The processes of change and natural disturbance provide repeated opportunities for recruitment. Natural gaps occur at a variety of different scales that serve to provide a wide array of opportunities that favor different kinds of plants; so the processes occurring in gaps are important in determining the kinds of trees that will eventually grow to replace lost trees to form a canopy and close the gap. Under conditions that exceed the natural range of disturbance, however, the recruitment process may be strongly altered, leading to failure of replacement among some species and ultimately to a change of species composition.

When a canopy tree in a forest falls over, dramatic changes ensue. The impact of a great tree in death may be as strong and long-lasting as its influence while alive. Most obviously, there is a great deal more light and space where there once was shade and crowding. Not only do existing plants benefit, but for many new species this gap is also the opportunity to become established. The stump and the rotting trunk may serve as an important seedbed. The now-dead tree may previously have strongly inhibited the growth of adjacent plants, which are released and grow rapidly to fill the gap.

Impacts on Forest Recruitment Processes

Protection of the land and the landscape is not enough to sustain the conditions necessary for recruitment of all components of an ecosystem. Processes of rapid change and degradation in the landscape are generally well under way. Even when protected, the landscape may change over time into a forest that has lost all or most of its former plants and animals. To the casual observer it may not be obvious that the forest is being replaced by another, one quite different from any that grew there previously. Some knowledge is required to notice these differences. A person in a forest where most of the trees are fine specimens of native oak, hickory, and walnut may not be aware that the only plants growing on the floor of the forest are exotic groundcovers such as stilt grass and garlic mustard. He or she also might not notice that on the edges of that forest are a few young tree invaders such as Norway maple or Amur cork tree, still too young to produce the abundant seeds that will before long begin to spread in earnest. The forest often dies from the ground up; the ground layer reveals to the knowledgeable observer the failed recruitment.

Many disturbances affect recruitment and thereby alter the composition and structure of forest areas. Vine species may proliferate in gaps and affect recruitment over many years by covering and even strangling young trees (Figure 4.8). This is evident even with proliferation of native vines such as grape or greenbrier, which may become much more than ordinarily abundant as a result of chronic disturbance.

The worst problems are usually caused by exotic vine growth. Tangles of Japanese honeysuckle, multiflora rose, porcelain berry, kudzu, oriental bittersweet, and certain other prolific climbers overwhelm the early growth stages of forest trees and sometimes even the canopy trees as well. The description of these problems in Overton Park in Memphis by Guldin, Smith, and Thompson (1990) is an excellent example of an old-growth woodland in an urban park where honeysuckle, multiflora rose, and wisteria were blanketing gaps in the landscape and restricting the recruitment of almost all other species.

The effects of invasive vines can be rather spectacular, but herbaceous plant competition by exotics such as knotweed, garlic mustard, and lesser celandine buttercup can also negatively affect recruitment. An overpopulation of deer, which may consume virtually all seed, seedlings, and saplings of certain species,

Figure 4.8. Rampant vine growth in woodlands overwhelms native wildflowers and inhibits regeneration of trees.

making their recruitment impossible, is another example of a disturbance that impacts many once-predominant species as well as rare ones.

Opportunities for recruitment vary considerably in different landscapes. In older metropolitan areas, for instance, where much of the open space may consist of Victorian-era parks, the landscape is often dominated by century-old native and nonnative canopy trees, many of which are dead or dying. If turf is maintained beneath them, no replacement saplings are likely to have survived; elsewhere, recruitment may be limited by trampling and invasives. In Prospect Park in Brooklyn, New York, for example, the landscape management department mapped all the gaps in the canopy cover, which accounted for 25 percent of the total area. Many of these openings have few native species persisting and, without management, all are likely to become dominated by exotics. When a local tornado struck New York City in December 1992, some 105 major canopy trees were blown down in Central Park in Manhattan, 52 of which were in the woodlands at the northern end of the park. In the gap created by a downed red oak with a 40-inch trunk diameter, 11 Norway and sycamore maple saplings were counted by the first season, but not a single oak seedling was found in the immediate area.

Suburban woodlands often face recruitment problems different from those in urban forests. Many are set in landscapes that were, until recently, agricultural. The woodlands often are young, the result of recently abandoned farms. What is left of open space may be only a narrow linear network, typically including drainage channels and unused corners and strips of land. Such an open-space system is all fragments, unconnected to a larger natural system. All interior species may be in jeopardy because no extensive forest remains, and reproduction may be confined to a few, often invasive, abundant species that flourish in fragments. Even larger fragments of woodlands may never again achieve the rich mosaic of gaps found in old-growth forests.

Poor Herbaceous and Shrub Recruitment

Because of the ease with which we cultivate plants in our gardens, we may be unaware of how different the forest is. For example, rampant reproduction by seed is a very important phenomenon in the garden and yard, for both desirable plants and weeds. Not so in the forest, a landscape of very long-lived plants where reproduction is limited and where there are simply fewer chances for recruitment even under the best of conditions.

While popular interest often centers on old trees, stands of wildflowers, where they still persist, may be far older. What appears to be hundreds of mayapples, for example, may really be a single plant with myriad stems. The same is true for trout lily and many other forest ferns and wildflowers. Many forest wildflowers are in fact ancient clones that have been colonizing a site for centuries. Steve Handel, of Rutgers University's Department of Ecology, Evolution and Natural Resources, calls them the "immortals" (personal communication). But when they are eradicated by disturbance such as trampling or overgrazing or exotic invasion, the immortals do not return to the landscape and efforts to restore them are rarely successful. A huge clone takes as long to grow as an ancient tree. This kind of complexity in the landscape is almost impossible for us to comprehend. Relationships among plants may have taken centuries to establish and, like other complex systems, may take more than centuries to restore, if they ever do, unless we learn to assist their restoration.

The plight of native herbaceous plants is especially serious because reestablishing them is so difficult and takes so long. Once its ground is lost, a new plant has little chance in the highly structured environment of a forest. All the vacancies are filled and, when they do occur, such as when a great tree dies, many well-established plants are better poised to take advantage of the gap than a woodland wildflower.

Once established, an herbaceous plant must hold its spot; hence the strategy of living a long time. Living in the shadow of the trees makes flowering and setting seed a biologically expensive effort, so forest herbs tend to produce fewer and more infrequent seeds than their field-dwelling counterparts. Many often rely on "vegetative reproduction"; that is, a new plant is established from a sucker or other vegetative part of an existing plant rather than from seeds only. This is a form of asexual, or clonal, reproduction. Indeed, vegetative reproduction, not seeds, is often the rule in the eastern forest, and for woody species as well. The trees of the temperate landscape are often called the "sprout" hardwoods. For many species reproduction by seed may occur only infrequently.

Many forest shrubs also rely on vegetative reproduction rather than seeds, some to the extent that the majority of the seeds they produce are not even viable. The lowbush blueberry of New Jersey's Pine Barrens has recovered so slowly in the few areas plowed in a brief effort to make crops grow in these unsuitable soils that the outlines of old agricultural fields dating from colonial America are still discernible in the forest, somewhat softened but nonetheless rectangular. Similarly, in second-growth forest on former agricultural lands where the plow has disrupted all preexisting root systems, the return of shrubs to the landscape may start many decades after tree cover has been fully reestablished. The more rapidly spreading shrubs of suburbia's open landscapes, both native and exotic alike, such as the native arrowwood and the exotic shrub honeysuckles, often travel widely on bird-disseminated seeds. Herbaceous plants of the mature forest usually take far longer, if they reappear at all. A student of Henry Art's at Williams College, Katharine Nash, has documented that Dutchman's breeches, for example, appears to advance by the slow spread of its seeds at only about a hundred yards a *millennium,* which suggests what kinds of time periods may be necessary for the restoration of native communities naturally.

Disruptions to Natural Recruitment

Another obstacle to the reproduction of plants of the woodland ground layer is the loss of the means by which they are dispersed naturally. Many of the small herptiles (reptiles and amphibians) and mammals that are important agents for carrying seeds to new areas, where there may be greater opportunity than in the immediate vicinity of the mother plant, have been locally decimated or eliminated by the consequences of fragmentation. The eastern box turtle, for example, soon vanishes when a former woodland is crisscrossed by suburban roads. Another "immortal," often living 100 years or more like the plants that depend on it, this land turtle will not be easy to bring back into areas where roads are busier than ever. Many woodland ephemerals such as jack-in-the-pulpit and mayapple depend on it to assist in seed dispersal. Seeds of the ubiquitous exotic garlic mustard, on the other hand, travel easily on shoes and tires.

Ants are even more important to forest recruitment. Like people, they garden. Many woodland wildflowers are dependent upon ants to disperse their seed, including spring beauties (Figure 4.9), wild ginger, purple trillium, squirrel corn, bloodroot, liverleaf, bellwort, woodrush, many sedges, most violets, and trout lily. These plants lure the ant with an elaiosome, a nutritious morsel attached to the seed, which the ants take back to the anthill. In addition to facilitating the movement of seeds, ants boost the survival rate of seedlings. The anthill provides some protection from predators, and the colony's waste is

Figure 4.9. Woodland wildflowers that are distributed by ants, such as these spring beauties, are disappearing from the landscape.

a nutrient source for seedlings. It is no wonder that many plant species time their seasonal and daily seed release to periods when ant activity is greatest (Handel and Beattie 1990). Disturbance of the forest floor erodes this process. As hard as they are to eject from your kitchen, ant populations of the forest are exterminated by clearcutting and return very slowly, if at all.

Research undertaken at Williams College by Henry Art and his biology students illustrates just how difficult the restoration of woodland ground layers is likely to be, and how much restoration depends upon protecting existing diverse landscapes and primary forest patches. Art, like many forest watchers, had noted with interest that some forests teemed with a rich array of wildflowers while others were strikingly absent. In the course of reviewing maps of the extent of the original forest that remained in 1830, after the height of the first great wave of deforestation, Art and his students observed a remarkable correlation. The forest areas that had survived up until 1830 and had been continuously forested since that time had the richest herb layers. The landscapes that were the least disturbed over time support the most diverse and characteristic vegetation today. They then began to study these differences more intently, comparing and contrasting those areas forested in precolonial times with those reforested since 1830. Art recently said:

> What we found last spring is that all of the main reservoirs of rich spring flora . . . are in woodlots that have been in continuous forest cover. . . . Ecologists in the northeast too often have thought that it might take from 200 to 500 years to replicate the old-growth forest. Our studies indicate that if you want to replicate the entire system, including herbaceous layer, it would take considerably longer unless a local seed source of the ant-dispersed plants is in close proximity (personal communication).

Other younger forests, even those with the same tree and shrub species as the older landscape, simply are not rich in herbaceous species. The plants whose seeds travel by air—that is, by birds or wind—are generally present, and the ant-disseminated species are the most conspicuously absent. Today, suburban and rural development rivals the first great wave of deforestation in potential destructive impact. Obviously, it is vital that we save these richer landscapes and spare them the impacts from which other forests have not recovered, even after more than a century and a half of regrowth.

Even when dispersal is accomplished, seed may not find suitable conditions for germination (Figure 4.10). Many processes that once fostered seed reproduction, such as fire, are now disrupted by land-use and management practices. Fire often creates special opportunities for seed because it leaves behind a soil surface that is a perfect seedbed for many plants, but it is generally suppressed because it endangers homes and businesses or else occurs so frequently, as in derelict areas, that any natural cycle is completely disrupted. Because fire is comparatively infrequent in the eastern forest, we have underestimated its importance to the natural reproductive processes of this landscape.

Figure 4.10. The constant disturbance of stormwater runoff and trampling prevents reproduction of the next generation of forest in many parks.

Exploding Populations of White-Tailed Deer

The recovery of landscapes in the past occurred under much more favorable conditions than today. With each year, restoring the historical landscape becomes increasingly difficult. Those forests where a rich diversity of wildflowers still remains represent a vital genetic heritage, one that is not recognized in our ordinances and regulations unless an endangered species is present. Yet one consequence of regulatory protection has further jeopardized the forest: populations of white-tailed deer have exploded in the landscape of fragmented woodlands.

Pennsylvania's only stand of old-growth hemlock forest, Heart's Content, represents a biological lineage of millennia, but it has little hope of being passed on to future generations. From the perspective of evolution, the great hemlocks of Heart's Content are a landscape of the living dead because no progeny have survived browsing by deer. Yet this was historically one of the richest landscapes in the state.

In Heart's Content, a remnant of virgin forest preserved in the Allegheny National Forest, there was a rich understory of hobblebush and 87 other plant species in 1926, before the deer population irrupted [exploded]. Today the same area is nearly devoid of all understory growth. Woody stems more than one foot tall, which averaged nearly 1,600 per acre in 1929, now number only 13 per acre, nearly all of which are beech root suckers. Of the 87 different plant species found originally, only 6 or 8 remain there today. Such profound changes in understory vegetation have altered the entire ecological structure of many forest communities. If allowed to persist, excessive deer populations could result in many thousands of acres of productive forest being converted to treeless savanna and nonproductive brushland (Marquis 1993).

Throughout the eastern region, uncontrolled populations of white-tailed deer have literally browsed away the next generation of the forest by consuming all the seedlings of many trees, shrubs, and herbaceous species. Species that deer avoid increasingly dominate the landscape, including many ferns, mosses, and vines. In suburban areas deer numbers often reach ten times, or more, those in natural forest. Exclosure studies using fencing to eliminate deer browsing indicate that many once-common species such as white cedar and Canada hemlock are seriously inhibited by deer densities as low as 6 per square mile. Populations today exceed 60 per square mile in many communities. For example, in 1993–94, the Pennsylvania Game Commission estimated deer densities at 52 per square mile in Chester County, outside Philadelphia.

John Palmer, superintendent of the Allegheny National Forest, has reported that in studying more than "6,000 monitoring plots over 300,000 acres [they] found that only five percent of the area has adequate regeneration" due to deer browse primarily (Reidel 1995). Despite the remarkable resilience of nature and the repeated seeming recovery of the landscape, the forest is losing its ability to replenish itself.

CHAPTER 5

Water Systems

One way of looking at vegetation is as an interaction between soil and water: whatever affects one affects the others. The landscapes we see are the result of this interaction, from the old-growth forest to the most barren wasteland. The word "landscape" implies a visual aspect, so let us take a look at what we see around us and how our management of soil and water affects the landscape. Effective conservation and restoration of remaining soil and water resources will require that we explicitly recognize their value in the landscape.

The Changing Water Cycle

Water in the landscape is not a closed system; rather, it flows through a larger system. No matter what the scale of your site, you will always need to look at the watersheds of which it is part.

The water cycle begins with the snow and rain that falls each year, variable within a historical and climatic range. Some enters the soil and reaches groundwater, where it may be stored for millennia or released into streams in periods of low flow. Some precipitation runs off, reaching a stream almost immediately, and from there, the sea or another large body of water. We have interrupted this cycle profoundly by funneling water away from natural drainage systems into built waterways and continue to accelerate the rate of change by increasing our rate of water consumption and waste.

In the temperate landscape, where the extensive forest once held water, almost every aspect of civilization has effected a shift to increased runoff and diminished infiltration, beginning with deforestation. Deforestation is always accompanied by accelerated soil loss and runoff and reduced recharge of groundwater. Other human uses, such as agriculture or settlement, intensify the shift to runoff by increasing the exploitation of ground and surface waters. Meier (1990) has calculated, for instance, that the sea level rise we may expect from the melting of snow and ice from global warming may be less than what we can expect from the water we are dumping into the oceans as sewage and wastes, increased runoff, reduced storage capacity from wetland losses, and conversion of groundwater to runoff in the process of drawing it down and using it. He states that the total rise by the year 2050 might be as much as about 1 foot, which would result in a worldwide shoreline retreat averaging about 90 feet.

Changes of that magnitude have far-reaching impacts. Natural vegetation corresponds to differences in the balance and timing of seasonal precipitation and evapotranspiration, which is the water used by plants combined with that lost to evaporation (Stephenson 1990). The eastern deciduous forest, for example, is a mesic forest where the water lost by evapotranspiration seldom exceeds rainfall, so that water is rarely a limit to growth. Also, fire frequency is reduced because the forest is rarely parched. That is the way that it was, anyway, when the great hardwood forest emerged after the glacial retreat. Now, however, deforestation, agriculture, and settlement have changed the landscape over much of the land. As W. Clark Ashby (1987) describes it,

> The old-growth forests of today developed in environments that differed in many ways from those of the present. Upland habitats are commonly more xeric now than in presettlement times because of drainage entrenchment and accelerated runoff. Lowlands have experienced sedimentation, and many areas have a greater incidence of flooding because of levees and dams built to protect agricultural lands. Soil types have changed through erosion, deposition and altered hydrologic conditions (102).

Streams depend upon both runoff and baseflow for water. The runoff comes from precipitation that was not absorbed into the ground; the baseflow comes from the water that was. Streams occur at the low places in the landscape where the ground level sinks below the level of groundwater. When water that previously infiltrated the soil runs off, failing to replenish groundwater, the streamflow that comes from groundwater, the baseflow, is reduced. Streams that once ran year-round become more "flashy"—that is, more subject to periods of sudden flooding as well as drying out.

Typically a forest will absorb and recharge half or more of the precipitation falling upon it; on an urban site usually less than 1 to 5 percent is retained in the soil and groundwater. But groundwater levels do not necessarily drop locally because there is extensive impervious cover and recharge is reduced. The same land uses associated with impervious cover often entail importing large amounts of water that is harvested elsewhere—from surface impoundments or wells. This water may be discharged into the ground via septic drainfields or as wastewater into a stream, which, like baseflow, is available year-round. Therefore, actual groundwater impacts may be quite variable. Small tributaries, local wetlands, and seasonal pools may show the effects of diminished infiltration well before more pervasive impacts to water tables are apparent.

The effects of reduced infiltration and groundwater contamination may also be felt at some distance, threatening, for example, local water supply sources since aquifer recharge zones receive minimal regulatory protection and are presently subject to rapid suburbanization throughout the country.

Locally dropping groundwater levels, which reduce the baseflow of streams, may also severely affect vegetation. Many mesic species, such as beech and white oak, are dependent upon proximity to the water table to weather times of drought and cannot survive conditions of falling groundwater or shifts in the seasonality of groundwater. If adequate levels of recharge are not sustained over time, even larger changes in vegetation are likely. These imbalances will be further aggravated by the more prolonged periods of drought alternating with more intense rainfall that are associated with global warming.

Stormwater Mismanagement

Erosion from excessive runoff represents one of the most ubiquitous and costly sources of damage to wildlands. A single outfall from a storm sewer discharged onto a steep slope can rapidly cut a deep gully that, once incised, will drain away groundwater as well. Many water management problems in natural areas originate off-site with stormwater coming from developed landscapes and may require complicated negotiations with adjacent landholders and public regulatory agencies to resolve. Because many stormwater management regulations are nonexistent or not enforced in urban areas, runoff is often simply shunted to the nearest stream or storm sewer unchecked (Figure 5.1).

For new development, where regulations apply, stormwater design often focuses on the potential for major flooding, anticipating only the larger but less frequent storms. The "twenty-five-year storm and greater" is a common standard, occasionally the five-year storm, but almost never the one-year or more frequent rainfall volumes. Stormwater basins built for these levels may detain floodwaters ("detention" basins) or hold water permanently ("retention" basins), but in either case the standard designs tend to minimize habitat opportunities for plants and animals (Figures 5.2 and 5.3).

In the face of these losses to our soil and water resources, we need a much more comprehensive approach to water management. Factors that are not regulated or are underregulated, such as the maintenance of baseflow and groundwater recharge and protection of stream channel stability and natural hydrologic regimes, simply become nonproblems on paper although they are very real problems on the ground. A frequent storm with the statistical likelihood of occurring about every year is the channel-shaper, but it is almost never considered in regulations by state and municipal agencies. The one-year storm, as it is

Figure 5.1. Stormwater runs off countless paved surfaces directly into woodlands and other natural areas despite regulation.

Figure 5.2. Muddy-bottomed and mown detention basins are mandated at the expense of a rich diversity of intermittently wet places.

Figure 5.3. Designed only with volume in mind, retention basins fringed with rip-rap are a poor habitat substitute for natural wetlands.

called, is not large and in the past might have brought a stream to a bank-full condition. Today, however, even a small amount of rain becomes a torrent when it courses across pavement and cleared land and no longer fits within the natural stream course that was shaped over centuries under a different regimen. The greater velocity gives the stream greater erosive force, cutting into the streambed in the upper watershed and bringing sediment downstream. As the stream incises, it traps high-velocity runoff in the stream channel, thereby causing further streambank erosion.

The channel of the modern stream is typically migrating rapidly, undercutting and eroding steeply cut banks (Figure 5.4). In places, the channel may be deeply incised (Figure 5.5), further draining groundwater as it cuts deeper into the ground. In our effort to control the lateral cutting of the stream, we gradually install rip-rap, sheet piling, and gabions until the entire stream is armored (Figure 5.6). Downstream the channel may even be elevated due to deposition. Soil lost to erosion typically equals 100 tons or so per square mile in woodland but rises to 100,000 tons from areas of heavy development. In addition to water and sediment, the stream today often carries plastic, lumber, concrete, tires, old refrigerators, car bodies, and other rubble. These stream corridors speak volumes about our engineering of runoff and illustrate the consequences of failing to include the conservation of natural hydrologic patterns as a primary

Figure 5.4. The erosive force of runoff regrades the channel of the receiving stream to meet increased volume and velocity.

Figure 5.5. A deeply eroded gully can continuously drain away groundwater.

Figure 5.6. Without protection of historic drainage patterns and without regulation of the small but frequent storm, streams gradually become armored ditches.

regulatory goal. While our attention has been directed toward flood control and other narrow foci, the larger systems around us have collapsed.

One example is the Mississippi River watershed. In the summer of 1988, barges were stranded and forced to wait for water to rise in the Mississippi and Ohio Rivers. More than 100 years of management, directed largely to handle flood stage, had allowed for the loss of the rivers themselves. In fact, most federal funding for water management has a side effect of further decreasing baseflow. People along the upper Mississippi complain of siltation in backwater landscapes, including favorite fishing holes, which is compounded by the navigational dams built throughout the shipping network. Further downstream, however, the complaint is of too little sediment, with extensive land loss in the Mississippi Delta and serious contamination. A sub-marine barren landscape of the continental shelf, described as a "dead zone," has lately stretched to the west of the mouth of the Mississippi well into offshore Texas, a dramatic illustration of pollution and our failure to conserve our water resources.

The trends are already apparent. We can expect to see baseflow, the minimum flow of streams, continue to diminish, a problem aggravated by the climatic variability associated with global warming. Meanwhile, stormflows will continue to be much higher than in the past. There have, for example, been six floods of a magnitude that once were thought to have a statistical likelihood of occurring only once a century in the Mississippi Valley in the past fifty years.

We have infiltrated and "piped" contamination into groundwater too, sometimes literally injecting it. Where stormwater is collected, it is rarely treated and is often dumped directly into downstream surface waters. Where storm and sanitary sewers are combined, even greater contamination results. The river may carry each problem farther downstream, but the supply of problems is ever renewed upstream.

Impacts to the forest occur even where damage is invisible. An interesting example is the phenomenon of "macropores," also called "pipes," created by decay of roots of dead trees. A study by Dobson, Rush, and Peplies (1990) showed that accelerated runoff laden with acid rain was quickly conveyed to lakes through the pipes left by a 1950 regional-scale blowdown in the Adirondacks. Blowdowns are especially common in urban parks and other fragmented forests (Laurance and Yensen 1991) that are often confined to steep slopes. Buildings on the adjacent flatter land channel and intensify wind. Acid rain itself may also predispose a forest to blowdown (Flynn 1994), and forests killed by acid rain also produce extensive pipe systems as do logged forests. Groundwater is always affected.

A whole drainage network forms a fractal pattern extending out like a tree, with the upstream extremities represented by the pipes of rotted roots. Farther downstream we engineer pipe systems of much greater dimensions, drain fields, collect and convey precipitation, channelize streams, and levee rivers. Whatever happens in one part of the system has effects throughout.

For all these reasons, sustaining and restoring the functions and services of the natural drainage system is an integral part of restoring a forest ecosystem. By doing so, we ensure a rich landscape setting and a valuable natural infrastructure, and, in the process, a more sustainable and safer water supply for our own uses.

CHAPTER 6

Terrain Modification

We have damaged soils casually, unaware of the long-term consequences to their structure, soil microorganisms, and potential vegetation. We regrade the landscape, fill lowlands, compact uplands, and strip topsoils. The impacts spill over to adjacent landscapes and urban wildlands. The response of trees to fill or compaction over their roots is not immediately visible on new construction sites, for instance, where large trees may have been spared the chainsaw but not the bulldozer nearby. New homebuyers, unaware of what has occurred, are dismayed several years later when their beautiful trees start to drop large limbs, dying in pieces until they have to be removed at considerable cost. The same problems occur when vehicles are driven in woodlands or trampling is excessive.

The removal of vegetation that accompanies grade changes during construction has continuing adverse effects. Plants improve soil by adding organic matter and by loosening compacted layers. Without vegetative cover, no organic matter is added to the soil and the ground is more vulnerable to erosion. With no plants to break the fall of raindrops, rainsplash action alone is enough to increase erosion and compaction. All compaction inhibits aeration and infiltration and leads to the accumulation of harmful gases in the soil, including carbon dioxide from the respiration of roots, fungi, and other microorganisms. Even relatively light disturbance of the soil may be sufficient to influence the species composition of a woodland. For instance, some species sprout vigorously when their roots are scarified or cut, quickly replacing more diverse assemblages of plants by a thick coppice.

Soil Compaction and Loss of Topsoil

Soil compaction is an especially insidious problem, largely because its effects are often underestimated. Any time there is visible soil compaction, there is also damage to vegetation. Off-trail use may kill herbaceous vegetation outright, even shrubs and small trees. Below ground, the pressure crushes roots and packs the soil too tightly for new roots to grow. A compacted soil typically has too little void space between soil particles for the water and air movement necessary for respiration and growth. Where heavy equipment has been used, soil structure may be permanently damaged, reducing rainwater infiltration, root penetration, and respiration indefinitely.

A corollary problem, stripped soil, is often a consequence of mining topsoil. Stripped soil, where the upper soil layers have been removed by grading, erosion, or other means, exposes subsoil, an impoverished growing medium. Sometimes topsoil is stripped and sold as "landscaping material" or "loam." That is why, in restoring land, we should make every effort to avoid using topsoil stripped from somewhere else to rectify soil problems; instead, we should work to restore the soil itself.

Fill

Many remnant natural areas include some areas of past filling. Development has already consumed most of the flatter, well-drained soils, leaving only steep slopes and wetlands in vegetation. As increasingly marginal lands are built upon, contractors use fill to even out the topography for easier construction, to elevate the surface above some flood level or to construct a flatter or better-drained site.

Despite wetlands legislation and regulations to limit construction on steep slopes, you can find thousands of examples of fill, of both major and minor scale, much of it illegal, throughout a region, often on sites too numerous and too small to demand the attention of enforcement officials. Individual owners casually enlarge lots and widen roads by continuously pushing a little fill over the slope. Even permitted fill projects often exceed their boundaries. In addition to the direct application of fill, there often is a large amount of sedimentation in the landscape from uncontrolled stormwater. Excessive runoff from a storm culvert or road that enters a park or natural area brings large volumes of debris, eroding the banks of a now-overloaded stream channel and later depositing a layer of sediment as fill lower in the valley.

The impacts of fill are severe but may take years to be fully visible. Vegetation that seems to have been spared often slowly succumbs to the presence of fill over root systems. Roots respire, so even a thin added layer of soil inhibits the exchange of atmospheric gases that typically occurs in the upper layers of undamaged soil. New roots can compensate to a degree by extending upward into the fill zone, but that may not happen rapidly enough to save the existing plants and will not occur at all if the fill has been compacted. Gases in the soil may also accumulate to toxic levels beneath the fill. Soil piled against a trunk may rot the bark and damage the root collar. Different tree species are tolerant of differing amounts of fill. Oaks and beech, for example, are notoriously sensitive and may be killed by the addition of only a few inches of soil. Box elder and sycamore are especially tolerant and often survive under conditions that would kill other species.

Where fill has been added to flat sites, the surface is often like concrete because of the use of heavy equipment. Fill on a slope, however, is almost always very loose, having been simply pushed over the edge. It erodes and collapses continuously, making it difficult for new plants to become established on the ever-shifting ground. The new fill is usually poor quality, largely mineral subsoil and rubble, low in organic matter, a poor medium for renewal of the forest and a great medium for less-choosy invasive vegetation. Exotic seeds and bits of exotic plants that can easily take root often arrive with the fill. The term "ruderal," which is generally used to descibe weedy plants, is taken from the latin word for rubble. Even very small pieces of Japanese knotweed, such as a sliver of root or a part of stalk material, have a node that is capable of rooting to establish a new colony.

Sometimes the soil is so poor that there is a fine line between fill and debris or trash (Figure 6.1), which creates even more difficult management problems. Dump sites, if not addressed in a timely manner, may rapidly develop into serious problem sites. The illegal disposal of toxic materials occurs often where

Figure 6.1. Extensive fill, rubble, and rubbish end up in wildlands when people fill, grade, and clear adjacent private lands.

such activities appear to be tolerated or unnoticed. Even where severe contamination, such as a spill of pesticide or other hazardous material, is not an issue, other additions to the soil, such as high pH resulting from concrete rubble or the low pH of acid clays of estuarine origin or mine tailings, may create ongoing management problems. Removing fill is a costly and disturbing event in itself. Sites with these impoverished soils present serious long-term problems for revegetation.

Clearcutting and Agriculture

The filling and grading associated with modern construction overlays a terrain already much modified by previous land uses. Clearcutting results not only in massive soil losses from erosion but also in a kind of rough grading because it usually means that huge logs and equipment such as skids have been dragged across the landscape. The plow that usually followed forest clearance further obliterated the groundscape of pit and mound. Plowing also completely destroyed clonal species—that is, those that spread by runners—and widely separated remaining populations. Insect populations that many species depended upon collapsed when their host plants died, and with them went any hope of regeneration for those plants of the herbaceous ground layer. Although much marginal cropland was soon abandoned, the regrowing forest encountered a drier and more simplified terrain. Many vernal herbs and other species of the forest floor, dependent upon the more mesic and specialized environment of old growth, failed to recover from these impacts.

Soil Degradation

Serious problems result from our tendency to modify terrain with little regard for the life within the soil and the role it plays in the landscape. Beyond erosion or compaction, beyond the loss of spring wildflowers and salamanders, is the long-term degradation of soil organisms, the loss of groundwater recharge, the soil's ability to effect recharge, and the soil's ability to support vegetation. Even a century after severe disturbances like clearcutting, only a fraction of the original soil community is restored. Exotic soil microorganisms and chemical pollutants directly dumped on or borne by water and air onto the soil systems have further impoverished natural soils in the same ways that exotic plants and contamination have diminished native plants and their communities. Lowered infiltration rates have watershed-wide consequences. Reductions in fertility, soil microorganisms, seeds, and root stocks, as well as damages to soil structure, may inhibit the growth of vegetation indefinitely. Unfortunately, we do not yet treat native soil as an irreplaceable resource.

CHAPTER 7

Invasive Exotics

Nonnative species invaded even faster than the European settlers they accompanied. One, the broad-leaved plantain, established itself along trails and at building sites so quickly that it was called "white man's footprint." The introduction and widespread dissemination of alien species, such as Norway maple and Japanese honeysuckle, into environments where they have no natural controls have been devastating to indigenous species. By rapidly increasing their distribution we have set the stage for local and complete extinctions.

The diversity and quality of protected natural areas are deteriorating everywhere in the developed corridor of the East Coast to no small degree because of introduced species. Exotic plants now consume over 4,500 acres of public parkland each year in the continental United States. Invasives have infested over 100 million acres in all, on both public and private land. That area increases by 20 percent, or by an area twice the size of the state of Delaware, each year.

Beyond jeopardizing native habitats, invasive species typically represent a significant maintenance drain on horticultural landscapes and often outcompete desirable noninvading exotics as well as indigenous species simply because they are aggressive generalists well suited to a wide range of conditions. A recently published estimate of the U.S. Office of Technology Assessment, cited by Holloway (1994), indicates that the number of naturalized exotics in the United States is greater than 2,000 and that just 79 of the worst have caused damage of almost 100 billion dollars in this century. As large as this number sounds, because these estimates account for primarily agricultural losses and little else, the actual damage is far greater. Quite possibly the damage to the budgets of parks alone exceeds this number. The true damages to all systems are simply incommensurable.

Mechanisms of Invasion

Though not all introduced exotic species become invasive, as many as a third of all introduced plants become at least temporary colonists. The success of just a few of those species is more than enough to jeopardize virtually every native habitat.

What makes a plant an invasive species that should concern us? It is all a question of degree and context: how fast it spreads, how widespread it is, and, most important, to what extent it replaces whole communities of native species. In the Chicago area, for example, one-third of the plants that now inhabit the region are not native. Out of about 900 or so species, about 150 are so extraordinarily successful as alien invaders that they can now be found in 95 percent of the vegetated landscape (Swink and Wilhelm 1994).

The most problematic invasives replace whole communities, not just a few species. Many are not only reproducing in highly disturbed sites, such as vacant lots, but they are also making significant inroads in less-disturbed areas, such as large woodlands.

What role disturbance plays in a species' potential to be invasive in a landscape is not fully understood. Invasions are not always caused directly by humans. The African cattle egret, for example, first appeared in the New World in the 1950s, blown off course to a new continent that they have rapidly colonized. Animal introductions can happen very quickly, even in relatively undisturbed landscapes. Plants, however, have typically required repeated reintroductions coinciding with large-scale changes in the landscape. At least for animals, disturbance does not appear to be a necessary precursor to invasion, and it may not be for vegetation. Over time we may see that plants may also be able to invade as effectively into undisturbed habitats, albeit more slowly than animals.

At the same time, sudden changes in the landscape tend to destabilize conservative species—that is, those that are highly adapted to a particular place and set of conditions—and in turn benefit opportunistic species, both native and exotic. Many of the exotics are more tolerant of disturbance than native species and can therefore overwhelm native forests and woodlands.

The components of natural habitats found in a region have coevolved over millennia and produced a natural system of checks and balances to regulate the rate and magnitude of change. Ecological communities are in a perpetual state of disequilibrium, and additions and deletions of species are commonplace. The environment's constant adjustment to changing conditions, though it does not eliminate dramatic change, reduces its overall vulnerability to natural stresses.

But the rate and magnitude of recent change has further destabilized systems that were previously, at least on a relative scale, more unchanging. It is true that over time natural systems will adapt to the presence of a new entity or disturbance, but it is also true that such a change can decimate extensive areas of native habitat and limit the capacity for recovery in a system already severely hampered by a wide range of other environmental stresses.

The control of exotic invasive species is one of the most critical management tasks necessary to restore and sustain native plant communities, but there are no simple answers. At present, it is not remotely feasible, or necessarily even desirable, to remove all exotic vegetation. In many areas, local wildlife have become dependent on exotic and mixed exotic-and-native landscapes. Moreover, we also depend on many exotics as cultivated crops. And occurrences are so extensive that, in the woodlands of Central Park, for example, complete removal of exotics would eliminate more than a fourth of all trees.

Our Role in Alien Invasions

Nonnative plants are not invading solely on their own—we help them every step of the way. Our choices of plants and management practices provide ideal conditions for spreading weedy opportunistic species throughout the fabric of our landscape.

Kudzu, for instance, was introduced at the Japanese Pavilion at the U.S. Centennial in Philadelphia and was exhibited at the New Orleans Exposition in 1888. Even though botanist David Fairchild (1938) had observed the invasiveness of this plant as early as 1902, in 1936 the U.S. Department of Agriculture (USDA) began offering assistance payments of up to $8 for each acre of kudzu planted. Fairchild published his findings in 1938; yet two years later Soil Conservation Service nurseries produced 73 million seedlings and employed thousands to plant them throughout the South (Shurtleff and Aoyagi 1985). A single government nursery in Georgia distributed over 34 million seedlings in a single year. Today, this plant as a crop is on the decline, but it once covered over a half million acres in the South, swallowing up forests and farms alike. A single shoot may grow over 60 feet in one season. Once thought to be confined to the South, kudzu has begun a slower but still effective invasion to the north, fueled in part by the suppression of fire, to which it is very vulnerable. It is now a serious problem throughout Pennsylvania (Figure 7.1), New York, and New Jersey.

Railroad companies planted Japanese honeysuckle on a massive scale as an erosion cure-all on steeply sloping embankments. Not only did the honeysuckle fail to stabilize soils as hoped, but it is also highly invasive and largely free of natural controls. Because birds help disseminate many invasive species such as exotic honeysuckles, autumn olive, and multiflora rose, some people defend the use of these plants as valuable to wildlife, which has proven to be a very short-sighted perspective. Plenty of honeysuckle is likely to be with us always, but the continued loss of habitat diversity it causes through its proliferation has been and will continue to be devastating to native birds and other wildlife.

Figure 7.1. The kudzu that ate Philadelphia.

Vine problems are likely to worsen. Growth-chamber experiments have shown that increases of carbon dioxide, such as that resulting from global warming, stimulate the growth of kudzu and Japanese honeysuckle (Sasek and Strain 1990). In urbanized corridors it is difficult to find a fragment of habitat that is uncontaminated by vine growth, and in places nearly all narrow strips can be thoroughly infested. As once-intact forests throughout the larger region are increasingly disturbed and fragmented, they are ever more vulnerable to vine growth. Yet today, other fast-growing exotic plants, such as the invasive Japanese knotweed, are being pushed as cure-alls for erosion problems without adequate consideration of their potential to become invasive in the future or what threats they may pose to native species. There are native honeysuckles that should be encouraged, such as trumpet honeysuckle, a vine, and a shrub of a different genus called bush honeysuckle.

One of the most invasive canopy trees in the temperate forest is the Norway maple, which is gradually taking over many forests in the Northeast, as well as in the Pacific Northwest and even Britain. Because it looks to most people like just another tree in the forest, evidence of the disturbance is less apparent. In late fall, however, when the leaves from other trees have fallen, the butter yellow foliage of the Norway maple reveals a continuous understory of its saplings and reproduction by few other species (Figure 7.2). When the saplings mature, the woodland may be entirely Norway maple, replacing many native trees, shrubs,

Figure 7.2. Norway maple invades the forest with a ground-blanketing wave of seedlings and saplings beneath the existing canopy, dominating reproduction.

and wildflowers. At the Drew University Forest Reserve in New Jersey, for example, Norway maple accounted for almost 98 percent of all woody seedlings in the study plots; the native sugar maple comprised only 2 percent; American beech, 0.05 percent (Wyckoff and Webb 1996). Understory richness overall also was lowest under Norway maple canopy and highest under the beech trees.

While vegetation in general is usually perceived as providing soil stability, Norway maple and another exotic invasive maple, the sycamore maple, actually provide poor erosion control compared to healthy native communities. Norway and sycamore maples, another exotic eurasian maple, have very large and thick dark leaves that emerge early in the spring and typically fall off the tree long after those of the native maples, creating a shade so dark that it severely inhibits ground-layer vegetation. Even more problematic, both Norway and sycamore maples are strongly allelopathic; that is, they suppress the growth of other plant species through the release of toxic substances in the soil. As a result, once Norway or sycamore maple becomes established, reproduction of other species comes to a halt and the ground beneath them is often totally barren. Without the additional protection of ground layer and/or small tree and shrub vegetation, the root systems of the trees do not provide adequate stabilization. On top of everything else, Norway maple grows at nearly twice the rate of our native sugar maple (Kloeppel and Abrams 1995).

We tend to underestimate the impact our plantings have had on the forests of today. You can still make out the lines of planted Norway maple across the Central Park landscape. The progeny of the 70,000 Norway maples planted in Fairmount Park in Philadelphia in the last century are as much a legacy as the Norway maples still being planted today will be in the next.

Despite Norway maple's proven damage to native species, its continued planting faces few obstacles. Two varieties of Norway maple are among the top five sellers nationally. Until very recently, more than half the trees recommended for use on city streets by Philadelphia's Fairmount Park Commission were varieties of Norway maple. Typically, a nursery will have more varieties of Norway maple than all other maples combined. Philadelphia has stopped recommending this tree because of complaints about its invasive tendencies, and some townships discourage its use, but demand for the plant is still large enough that it is cultivated on a massive scale. Because it is widely available, it is also widely recommended. Without better education, both the public and nurseries will remain largely ignorant of the problem.

Many invasives are spread more by management than by planting. Human interaction is the agent for the proliferation of Japanese knotweed, known as "backyardia" in some urban centers for its propensity to show up in human habitats. Like the Norway rat, our most common species of rat, it finds urban conditions ideal. Soil disturbance scatters Japanese knotweed's propagules, and, like mugwort, another nearly ubiquitous invasive of disturbed landscapes, it benefits from brush clearance, irregular mowing, and occasional soil scraping that eliminate all but the toughest herbaceous species.

Japanese knotweed is a failure at site stabilization. It may also be highly allelopathic as its stands are often monospecific; few other plants seem able to co-exist. Although large in size, often up to 8 feet or more, this plant is actually a stout-stemmed herbaceous plant that dies completely back to the ground each winter, leaving no winter cover, setting the stage for erosion. Knotweed also lacks a fibrous root system, which would provide greater soil stability. Spreading quickly by rhizomes or runners, it soon completely displaces other, more stabilizing, vegetation. Yet roadside management practices continue to spread this plant throughout the Northeast, from narrow corridors deep within the Green Mountains of Vermont to alleyways in Boston and gardens in Georgetown.

Aquatic areas, particularly wetlands, have been vulnerable to many invasives, such as the infamous water hyacinth. Purple loosestrife escaped from garden cultivation and has overwhelmed wetlands from New York to Minnesota. Ducks and other waterfowl avoid infested areas. Its spread has been so rapid that a single plant is considered an infestation although, when established, there may be up to 80,000 stalks per acre. This plant is still a popular ornamental. Naturalized plants show a higher degree of seed fecundity when cross-pollinated with cultivated horticultural varieties in the same vicinity.

Some species that are already relatively widespread may have yet to show their full invasive potential. The tree of heaven, one of the most abundant invaders of urban areas and other highly disturbed sites, has for the most part not yet spread in forests extensively except in localized areas. It does tend to spread rapidly along roadsides and may in the future prove to be a greater problem outside of urban areas. The princess tree, first introduced when its seedpods were used as packing material for fine porcelain, sprouts readily even in the most highly disturbed conditions, sending up extraordinarily vigorous and fast-growing young stems from cracks in buildings, sidewalks, in vacant lots, and on rubble piles in the city. These traits have also allowed it to spread widely in the temperate forests of this continent, yet it rarely becomes a dominant element in the forest, at least not so far. Its occurrences in more rural areas often show close association to current or former human settlements, old homesites, and old trash dumps. Interestingly, this tree produces highly valuable logs under the right growing conditions, which has led to a minor tree harvesting industry despite its rather diffuse distribution. The value of the wood has stimulated the interest of large-scale commercial forestry, which means that this tree is becoming increasingly problematic and widespread in the wild. A major home builder has secured global rights to princess tree tissue-culturing technology and is seeking investment in mass production of the quick-growing tree on several continents. The princess tree is a good example of another trait of some of the worst exotics—because its flowers are quite pretty, people transport the plants widely, as they do with honeysuckle, water hyacinth, loosestrife, and other showy plants.

Genetic Consequences

Invasive plants may have more subtle effects—effects that even a student of modern molecular genetics may not fully comprehend. Most invasive plants take up residence only after repeated introductions, not just one contact. Thus, genetic strains from many parts of the range inhabited by the invader in its own native area may be commingled. Some degree of genetic recombination and accumulation of variations in the commingled populations may lead to even more invasive strains. Under such a scenario, an exotic that we currently regard as only weakly invasive could become more of a pest in the future. It is likewise possible that Norway maples, for instance, may reflect new and more vigorous strains as introduced subspecies become intermixed.

Controlling Exotic Introductions

The legislation that might have been used to combat the problem of invasives more effectively, the Federal Noxious Weed Act, largely ignores threats to natural ecosystems and concentrates exclusively on agricultural pests and alternate hosts for insects and diseases, as well as species influencing public health. Even then, agencies at both federal and state levels are slow to act; the USDA has listed only two new plants, both agricultural weeds, since 1974.

Despite the evidence that exotics pose severe threats and cost so much to control, new introductions are still heavily marketed on a grand scale. Promoters today advocate, for example, the widespread use of the sawtooth oak as valuable for wildlife. The plant is already locally naturalized in the Midwest. It seems foolhardy to spread it throughout the landscape in the Northeast, where is it not indigenous, and then sit back to see if it naturalizes here as well, but that is the case today.

There are, by definition, no effective natural controls for exotic plants that are currently invasive. Even though Norway maple, for instance, is subject to numerous diseases and pests that create problems in horticultural settings, such as giant tar spot, as yet none has checked its spread in the wild. Over time, it is inevitable that some organism will take advantage of any widespread host, native or exotic, but that may not happen before extensive areas of native habitat are decimated by invasives.

Exotic pests have come along with exotic plants, either on the living plant or even on raw wood, which is increasingly being imported into this country. The most well known casualty is the American chestnut, which was lost through the introduction of chestnut blight, but most people are less aware of the impact on wildlife of losing this prolific nut producer than its impact on forestry. The tulip poplar and Norway maple that have filled the gaps left by the chestnut are far less valuable to bears and turkeys, for example. The chestnut blight, which was introduced on chestnut nursery stock imported from Asia, was first observed at the Bronx Zoological Gardens in 1904. Less than half a century later it had radically altered the composition of the eastern forest and destroyed the most important hardwood species in the East.

Dutch elm disease, white pine blister rust, balsam woolly adelgid, pear thrips (a serious pest of sugar maple), beech scale, dogwood anthracnose, hemlock woolly adelgid, butternut canker, and many other diseases and pests have taken and continue to take a great toll on the indigenous forest. The gypsy moth alone defoliated over 1 million acres in 1993. Several new invasive pests appear every year.

One of the most recent accidental introductions is the elm spanworm, which began moving on Pennsylvania's forests in the summer of 1994. Unlike the gypsy moth, which is somewhat selective, the spanworm strips every species and leaves behind a winterlike landscape. The USDA Animal and Plant Health and Inspection Service (APHIS), the agency responsible for regulating plant and animal importations, is now considering revisions to its procedures. Tighter and more uniform controls are needed, especially for wood products, although some outbreaks are probably inevitable given the volume of introduced plant material.

If we want to restore the ecosystems we depend upon, we cannot plant species that have demonstrated themselves to be successful invaders at the expense of native habitats. Even following this proscription, which addresses introduced planted material only, would not in any way preclude invasions by accidental means. Because so little is known about many exotics, we strongly urge that plantings in and adjacent to natural areas be confined to locally native species. Where greater diversity is desired, efforts should be concentrated on reintroducing native species that are absent or diminished in abundance and unavailable in the trade.

This policy is conservative, but the consequences of being too optimistic have been costly. In the past, whenever concern about a species has been raised, regulators and producers usually dismissed the problem until it was thoroughly entrenched. State forestry departments and soil conservation agencies throughout the Northeast still have lists of recommended species that include the most vigorous invaders. Only kudzu seems to have demonstrated enough damage potential to be generally viewed as unacceptable. A ban may seem extreme; however, many of the more invasive species are dispersed by wind, birds, and other animals including humans and quickly travel over fairly large distances. All are extremely difficult to eradicate once established, and none is so critical to landscape character that you cannot replace it with another, less-threatening, species. The future of our native forests may depend on immediate action.

The increased globalization of trade and travel is breaking all the barriers of isolation and moving diseases and other pests more rapidly than ever. This biological concoction is hitting the landscape like the conquistadors, slaying even more by disease than outright attack. The great oceans no longer isolate North America and the New World from Eurasia. Globalization has made our continents more like one great landmass. While it is true that larger islands tend to have higher diversity than smaller islands, it is also true that global diversity is diminished by reconnecting once isolated land masses, i.e., by losing isolation. We can expect to see large numbers of species succumb simply because one large interconnected land mass island supports less planetary diversity than many land masses of varying sizes.

Hawaii is a good example of the consequences of losing isolation. Prior to human settlement, new species arrived at an average rate of one every 10,000 years. In the last two centuries, however, people have introduced over 20,000 exotic species to the state (Marinelli 1996). Native flora and fauna are declining just as fast. In comparison to Eurasia, North America has been very vulnerable to invasion because it was for so long isolated from the influences of Europe and Asia despite its similar climate. Eurasia is in direct connection with Africa, and together these form the world's largest land mass with the greatest number of species, many of which are invading the smaller land masses. A major shift in public and private investment to favor indigenous communities is needed, and forest restorationists can start the process.

CHAPTER 8

Opportunistic Natives

The kinds of changes that have occurred in the landscapes around us have significantly altered interrelationships between native plants and animals, even where no exotic species are involved. Today, woodland managers are concerned about the rapid spread of several native species, among them the black cherry. The 1982 tree inventory in Central Park, for example, revealed that almost 20 percent of the trees in the park greater than 6 inches in diameter were black cherry. That ratio was even higher when sapling-size trees were included in the count. This species is clearly proliferating in the park, and some argue that, in the interest of maintaining diversity, it should be treated like an exotic and removed. Black locust engenders even greater concern because it is extending its range at a rapid rate. It is both prolific and invasive. Management plans frequently recommend removing locust.

Cherries and other aggressive native species often take advantage of the opportunities provided when once-tended areas are released from management or woodlands are partially cleared for security reasons. Where disturbance limits the variety of native communities in the vicinity, only a few natives, all easily dispersing propagators, may account for a disproportionate amount of new growth. They also represent a problem in horticultural landscapes, where they grow like weeds. These species of seminatural areas thrive in the kind of simplified environments we make of the landscapes around us. Like whitetailed deer and Canada geese, however, they are not well adapted to the habitat of continuous mature forest, and their reign in the forest may well be shortlived. But a maturing landscape does not necessarily daunt exotics such as Norway or sycamore maple. We need far more research and documentation in order to define effective management strategies. The real conditions and trends over time on each site must be the actual basis for any good proposed management plan.

A general trend is that species native to the midwestern part of the United States, including many common midwestern plants such as the catalpa, osage orange, and western red clover, are spreading into the East. Many of these plants, along with other exotics, spread quickly along the early linear corridors created by trails, roadways, railroads, highways, and seams of disturbance (Brothers 1992). In addition to the original creation of these corridors, humans have further helped spread the species through direct planting and spreading of seed. This trend continues even more intensively today with the popularity of mixed meadow seeds grown in the Midwest or California. The East currently has no major commercial native-seed propagators, so no regionally appropriate mixes of locally indigenous seed are currently available. We have also helped some midwestern natives in their settlement of the East by managing the landscape to replicate the drier native environments of their origins, rather than managing to replicate the actual environment of the East, the result of conventional engineering practices designed to provide drainage and flood protection.

Several considerations are important in evaluating the invasiveness of an exotic species, such as persistence. An important characteristic of an invasive plant is that it replaces indigenous species and often whole native communities, preempting the return to native forest. Aggressive invasive native species are sometimes, but not always, so persistent. Where Norway maples are established, for example, there is no evidence that native communities will gradually colonize the site, even if aggressive natives such as black cherry are kept at bay. Where cherries are established, however, native forest species usually appear within a few years, at least woody species. If proliferation of Norway maple and other exotic invasives could be checked and adequate seed sources for indigenous plants were available, natural succession would gradually replace the cherries with more diverse native vegetation. Nevertheless, the abundant establishment of cherries may retard this process, and management may be warranted in some cases. Black locust, although still expanding its range, may be more easily controlled naturally once more forestlike conditions prevail.

The Supercompetitor

Far more complex questions come into play when we look at the example of the common reed. In this century, common reed has moved from the very fringes of the landscape at the margins of the sea up through the estuaries and river valleys to occupy thousands of acres of once-forested swamp. This invasion includes many acres of marshland formerly occupied by other species. You can also find common reed on landfills, on roadside edges, and in vacant lots, as well as in woodland seeps and along shorelines as it moves farther and farther inland and upland. The pattern of its spread seems much more like that of an invasive exotic than a native and raises questions about what is native and how natural the resultant landscape is (Figure 8.1).

Until recently, two species of common reed were generally recognized, *Phragmites communis* of the New World and *P. australis* of the Old World. Scientists later reclassified common reed and combined these two species under the same scientific name, *P. australis*. The confusion this change created seems quite apt given the already strange behavior of common reed in our landscape. Some people refer to the plant as nonnative, and others recognize it as native. In some ways both are right. The plant that has swept across the Delaware estuary may be less the "native" of the last century and more the "nonnative" of the Nile, the Danube, and the Thames, a new supercompetitor that is a genetic concoction of our own making that combines these international strains. Plants of shorelines are typically very mobile, and their populations are linked rather than isolated by water. This natural mobility, however, has increased exponentially with the transglobal journeys of migrating and trading humans. Seeds carried in grain sacks on wagon wheels, and in ship ballast spread rapidly carried seed from several continents across exceptionally large distances over and over again, and very swiftly. Modern transport is an even faster distributor.

The introduction of distant strains into local plant stock is documented for other species, such as rice cutgrass. Eurasian stock, sold as seed and planted

Figure 8.1. Common reed, growing in monospecific stands, now dominates many once-diverse wetlands and is making inroads on many upland sites as well. (Photo by Ann F. Rhoads/Morris Arboretum)

widely, has invaded natural wetlands and replaced native subspecies. Even introduction of native species in distant locales has impacts. The characteristics of saltwater cordgrass from New England sources planted in mid-Atlantic locales are noticeably distinct from those of local stands, which may be replaced.

Even minor geographic distances may be enough to account for important variations that maintain subspecies identity. Transplants of hairgrass rescued from a construction site that were planted amid local stands were notably different in appearance from the original plants that they gradually expanded to replace (Gehring and Linhart 1992). Some of the behavior of common reed, such as its new affinity for roadside slopes, may, in fact, be due to the same kind of subspecies mixing. The expansion of common reed has been repeated in many eastern estuaries from Maine to Florida. In the Delaware River estuary, this explosion began quite suddenly in the middle of this century for reasons that are not known but may involve many of the factors raised above as well as significant changes in the hydrologic regime (Kraft, Yi, and Khalequzzaman 1992).

There is also some question about when common reed arrived in the New World, because it provides habitat to so few indigenous wildlife species here, while in the vast reed marshes of Europe and Asia complex ecosystems revolve around the plant. Some have suggested that the earliest human travelers from Eurasia, indigenous people, may have introduced the reed, using it like tobacco for rituals well before Columbus and well before the first botanist arrived to find it already established.

Clearly, the horticultural practice of propagating and moving plants to any location where they will survive creates problems for restoration efforts. Among the environmental problems it poses is to threaten subspecies diversity by eliminating the level of isolation between populations. The subspecies level of biodiversity is the least understood and the most endangered. So to retain these often very localized plants, the restorationist will need a new kind of nursery centered on local diversity. Monitoring of changes in naturalizing populations of both native as well as nonnative plants also will need to become an integral part of the growing and transporting of plants. And, last, we must expect the unexpected, especially in this age of supersonic travel. As the world gets smaller, the diversity that was dependent on isolation is lost. In the years to come, new permutations of the problems we are already observing are likely to emerge.

CHAPTER 9

Wildlife Impacts

Few forces so thoroughly obliterate varied natural habitat as urbanization. Paved streets account for almost one-third of the area of New York, rooftops even more. Yet wherever a bit of land is left alone for a while and wildlife is relatively unmolested, an abundance of adapting species may occur.

Because the urban environment is so inhospitable, even to humans, we are surprised by what can be sustained in the developed environment wherever the requirements of natural communities are met, whether by design or accident. In the midst of the busy industrial shipping corridor of Newark Bay, for example, over 29 percent of the entire colonial waterbird population along the Long Island–New York City Atlantic shoreline breeds on three artificial islands used as disposal sites for dredge spoil in the Arthur Kill, the river that separates New York from New Jersey. The waterway was the site of no fewer than ten oil spills between January and September 1990 and yet is home to snowy, great, and cattle egrets; black- and yellow-crowned night herons; little blue herons, and glossy ibises (The Trust for Public Land and New York City Audubon Society 1990). Recent monitoring, to evaluate the impact of an Exxon pipeline leak of millions of gallons of home heating oil, revealed unexpected populations of organisms surviving under conditions that would normally be lethal to individuals of that species accustomed to less-polluted environments. But most of these creatures are living at the very edge, in habitats that are largely unprotected. Without more effective conservation, they are extremely vulnerable. Even a minor added degradation could eliminate entire communities.

Effects of disturbance on wildlife mirror those previously described for vegetation: loss of habitat, changes in the structure and composition of the remaining habitat, changing population proportions, and the introduction of

diseases and competing exotic species. Birds and other wildlife are excellent indicators of environmental degradation no matter how you look at the numbers.
One-sixth of the bird species extant several hundred years ago globally are already lost, as a direct result of human activity (Wilson 1992). More than 1,000
of the world's fewer-than-10,000 species of birds are threatened and 70 percent
are in decline, according to Bird Life International of Cambridge, England.
More than one-third of all frog species in the United States are declining in
numbers. Nearly one-third of the world's known mammal species are threatened with extinction, according to the World Conservation Union Red List. In
only a few years, one scientist alone has collected over 20,000 bird carcasses that
were killed when they struck the windows of a single highrise building in Texas.
Imagine those losses multiplied by all the highrises found in migratory corridors. The Government Accounting Office has reported that activities harmful
to wildlife, such as logging and military bombing, have occurred on 60 percent
of the 91 million acres of the U.S. National Wildlife Refuge system (Chadwick
1996).

Imbalanced Populations

Fragmentation and local extinctions have dramatically changed the species
composition of communities for many wildlife species, to the point that populations are critically altered. Lack of predators can mean overpopulation by a
plant-eating species, and not just by deer, the most obvious example. In the
New York Botanical Garden, for instance, squirrel density was measured at up
to 25 per acre as compared to a forest level of 3 to 5. In Central Park squirrels
seem to consume nearly every acorn and beechnut that is produced and any
that are planted without extra protection. New seedlings are also vulnerable.
The woodland planting crew in Central Park has observed squirrels digging up
newly planted, two-year-old oak seedlings to get at the remaining bit of the
original acorn still attached to the roots of the young tree.

Some species of wildlife are more vulnerable to predation in forest remnants. Fragmented forests, with their continuous and increased edges, provide
greater access to prey than the unconnected gaps of a patchy forest. Introduced
predators, in the form of cats and dogs, can number several per acre. Another
companion of humans, the Norway rat, is also a predator in the landscape.
People also prey on animals, even where hunting per se is not a factor. Reptiles
and amphibians are especially hard hit. There are no snakes, toads, tortoises, or
turtles at all in Central Park, except a few totally aquatic species that shelter
safely under water. Birdwatchers in Central Park who had been following the
progress of a nest of downy woodpeckers were dismayed to find that children
had pulled over the tree stump so they could try to hand-rear the nestlings.
Poaching is common, too, especially in larger protected landscapes. The National Park Service documents over 100 species of animals that are poached and
almost as many plant species that are illegally harvested.

Impacts of Single-Species Management

Recognizing that the whole system represents the greatest value, not just some of the components, is the most important concept in land management. Planting and managing appropriately to benefit native wildlife by necessity means restoring and managing for the full spectrum of native plant communities, not just wetland or rare species. Many efforts to manage for single species of wildlife, especially game populations, have had serious negative impacts on native communities of both plants and animals. People still widely recommend and disseminate many invasive, alien plant species on the grounds that they benefit wildlife when in fact they may actually benefit a few species at the expense of many others. Well-meaning managers planted autumn olive, a highly invasive small tree, throughout wildlife management areas in New Jersey and other mid-Atlantic states to increase the numbers of small gamebirds for hunters, without adequate regard for its effect on other species in the system. More than a third of native streamside birds avoid areas of olive because it supports few insects and has very hard wood with few cavities. The olive can fix nitrogen directly from the atmosphere and compete effectively with native species, establishing in shade under native cottonwoods and willows and growing up to replace them completely. Efforts at exotics control may have similar consequences. To cite just one example, an exotic parasitic wasp deliberately introduced to control the exotic gypsy moth is now preying upon three native moth species that are federally listed as endangered.

Animal Invaders

Some animal species, like invasive plants, have naturalized at the expense of indigenous wildlife. Some introductions have been accidental, such as the escape of a pair of gypsy moths from a professor in Montpelier, Vermont; others have been deliberate, such as the release of starlings into Central Park as part of a larger effort to introduce all the animals and plants that were named in the works of William Shakespeare into the United States. One of the most familiar introductions is the ring-necked pheasant, a bird from China that most of us remember from childhood and think of as native, having never seen the turkey that preceded it. Many states still pen-raise and release pheasants for hunters, such as in New Hampshire's "put and release" program.

Invasions are not just by land. For instance, efforts to stock fish, such as brown trout, for sportfishing have been enormously successful with the license-purchasing public but have had detrimental impacts on native fish, such as the brook trout, which they have displaced. Warmer streamwaters and reduced corridor vegetation already have reduced suitable brook trout habitat; continued stocking where they persist increases pressure on these and other native species.

Modified Native Species

Wildlife management has sometimes affected native wildlife in ways that make their behavior more like that of an introduced species. For example, when the practice of using live decoys was banned, game farms released thousands of Canada geese that had been raised for this purpose and had lost the habit of migration after generations of captivity. The birds readily adapted to farm ponds and pastures and then spread rapidly. Today they are equally at home on turf and in retention ponds and have proliferated with increased development of corporate office parks and golf courses to such an extent that they are often considered pests. Similar changes may have already occurred to the farm-raised native turkeys that are now being released throughout the East. New Jersey, for instance, now boasts over 10,000 wild turkeys. You can easily catch a glimpse of this once rarely seen forest bird seeming very much at home in the hedgerows and cornfields of the rural landscape.

Changing Plant Communities

Remnant natural areas provide habitat conditions for wildlife that are very different from those of the former forest. One aspect is food type and value, which change when vegetation is altered in composition. The nonnative shrub honeysuckles and multiflora rose, spread so widely by the birds and other animals that feed on their sweet fruits, are of less benefit to other species, especially those dependent on the high-lipid, that is, high-fat, fruits of native forest species such as dogwood or viburnum that the invaders have replaced. Migratory songbirds, for example, require lipids to sustain them on their journeys. The native spicebush berry is 35 percent lipids; the multiflora rosehip is 40 percent sugar and only 10 percent lipids. As nonnative species expand throughout a region replacing high-lipid native species, important food sources for migratory birds decrease—yet another reason that restoring indigenous plant species is so important.

Whole Systems Management

Ultimately, the preservation of wildlife diversity will depend on our management of the largest preserves where the whole of the food chain can be represented, from wolves, bears, and bobcats to the smallest soil creatures. The fragmented landscape, no matter how well managed, cannot functionally replace larger sites, which hold out the most opportunities for preservation and survival to the largest number of species.

Nevertheless, fragments are still very important not only as greenways but also as refuges in densely populated areas. For migratory birds, many of whose populations have dropped precipitously in the last century, the distance between sizable fragments is crucial. The loss of stopover sites along their migratory routes may be as serious as loss of breeding or overwintering habitat. A good example is Central Park, which provides at least temporary food and shelter to more than 260 species of birds. This relatively degraded landscape, considered a birdwatching hot spot, is one of the few remaining links on an increasingly tenuous chain of seminatural areas connecting the megalopolitan corridors.

We are not just losing animal species with loss of habitat; we are losing plant species as we lose animal species. Plants and animals have coevolved over millennia, and plant reproduction is inextricably tied to wildlife—for pollination, seed transport, seed scarification, planting, and regulating competition. Where wildlife is impoverished, many plants have no means of effective reproduction and survival. We must include the interdependencies of plants and wildlife in our planning of habitat preserves. We cannot any longer separate one from the other in our thinking and our management practices. This will be no easy task. Managing to improve conditions for native wildlife in landscape fragments will place some of the greatest demands on our skills and ingenuity.

CHAPTER 10

Use, Misuse, and Mismanagement

Our frequent use of wildlands for recreation is testament to the important and exhilarating sense of freedom we experience in them. But inappropriate behavior by humans, both intentional and inadvertent, is all too common in urban and suburban wildlands. Off-trail bicycle use, horses, and vehicles as well as trampling damage habitat and disturb the soil.

Our parks are also damaged by antiquated management practices. Despite current knowledge about environmental effects, the bulk of landscape management practices still consist of actions largely destructive to native landscapes—extensive mowing, application of herbicides and pesticides, poor erosion and sedimentation control—as well as a narrow horticultural perspective on plants. The most common management mistakes are made over and over again and have resulted in direct damage to natural areas as well as degradation to adjacent sites.

Off-Trail Use

Decline of the landscape following off-trail use is a familiar scenario. It begins with damage to herbaceous and small woody vegetation. Some plants are killed when stepped upon; others die from compaction and erosion around their root systems. New plants may not even be able to germinate because the remnant forest exists on steep, easily eroded terrain or in poor, compacted soils.

Before we can address the restoration of plant life, we must ensure that the ground is not subjected to continuing disturbance. Control of off-trail damage from trampling and wheeled vehicles often offers the greatest immediate improvement to the health of woodlands, but it is not always easy to accomplish.

For many people, simply being able to leave the beaten path is at the heart of their sense of contact with nature. It seems so harmless, and we all do it, from the birdwatcher to the dog walker, but the impacts are severe and often underestimated. A survey of the ground-surface condition in Central Park conducted in 1982 revealed that more than one-quarter of the nonturf areas were bare ground, nearly all of it caused by trampling or bicycles. More than two and a half miles of new mountain-bike trails were opened up in the northern wooded section of Central Park in a single season alone.

The rapidly growing use of mountain bikes (Figure 10.1) has created an accelerating problem for managers of natural areas. The Bicycle Institute of America reports that sales of mountain bikes have increased in the United States from 200,000 in 1983 to more than 11 million in 1989, and the figure is still rising. Land managers walk a narrow line between the need to preserve the land and the need to enhance the public's enjoyment of it.

Trails blazed by cyclists often run up and down slopes, so their erosion is not surprising. Additional controversy over multiple trail uses is stirred up by the difference in user speeds. Mountain bicyclists can go quite fast down hills and often startle walkers and equestrians, much to their annoyance and mutual endangerment. Since the widths of new paths may be narrow (some are only 12 inches wide), passing is a constant problem. User conflicts may be serious.

The National Park Service, the U.S. Forest Service, and the Bureau of Land Management have declared that mountain bikes are "a mechanical form of transport," and as such, like motorized vehicles, are banned from use in designated wilderness areas under the Wilderness Act. Cross-country-bike use is permitted on some nonwilderness trails. Enforcement, however, is difficult, since bikers are not easily pursued except by another off-trail vehicle, which, in turn, also damages vegetation. Walkers often complain that enforcement personnel do not take off-trail infractions seriously.

While a ban on mountain bicycles and off-road vehicles in wilderness areas has given some protection over thousands of acres, there is still a great deal of controversy when it comes to urban parks of limited area near dense populations, where damage is rampant and increasing. Off-trail use of bicycles, motorcycles, and all terrain vehicles often goes unaddressed in urban wildlands simply because control is difficult.

One of the most crucial and difficult decisions may simply be determining where mountain bikes and other vehicles can be used appropriately and where they should not be allowed. All users feel they have a place in a park system, whether or not they are aware of the damage they cause. At the very least, sustaining natural landscapes will require that the use of wheeled vehicles be managed to prevent avoidable degradation of the environment. Interestingly, recent studies show that bicycle tires do not cause any more erosion than walkers; so

Figure 10.1. Mountain bikes often cut steep trails that concentrate runoff and damage the ground sur-
face. (Photo by Clare Billett)

the greater issue is to get all users to cooperate by keeping to the paths and con-
forming to trail rules, and for taxpayers and local governments to commit to
adequate maintenance. We need to establish and follow performance standards
to determine when and where use should be discontinued or modified.

All too often, not providing enough amenities to users leads to off-trail use.
Even where the trail system is adequate, users, both pedestrians and riders, may
create new, unofficial "desire-line" trails. These "outlaw," or "rogue," paths may
be hard to distinguish from an official path that is in disrepair, so expecting the
visitor not to use it is unreasonable. Elsewhere, trampling is due to poor design
and poor trail maintenance rather than usage. In all these cases, damage is inevitable because the trail system does not meet the users' needs, even for desirable activities. To keep delicate relic landscapes out of harm's way, the restorationist must carefully evaluate and justify each footstep, and plan trails to take
visitors where they want to go.

Landscape Mismanagement

When we see park maintenance people at work or when we tend the wilder
fringes of our yards, we like to think we are doing something good for the landscape, but often we are not. Chances are the work commenced with clearing out
the landscape—removing briars, poison ivy, and common herbaceous species

such as pokeweed. The problem is that the twigs and natural debris that build the soil disappeared along with the trash, as well as desirable plants of value to wildlife. Soil itself may have been raked to remove objectionable bits from the surface, taking more of what little dirt remained. One of the most destructive but common practices, acquired from horticulture, is overly simplistic ground-plane management.

Much of this misuse and management is attitudinal, rooted in our failure to recognize that landscapes are living systems with their own patterns and organizational principles. Forests are composed of complex interrelationships that take time to develop. They are not static objects to be manipulated at will. Nor will restoration bring about an instant transformation. It took time to mess up the landscape, and it will take time to restore it. A living system is not like a garden or a capital project; you can't do it all at once—it will require ongoing, sustained effort.

Conventional management tends to focus on isolated tasks such as litter removal or lawn mowing. As long as these activities are completed on schedule, maintenance is considered done, sometimes regardless of the level of continuing degradation of resources, both natural and cultural.

We have institutionalized bad landscape management habits that are costly and damaging, such as deferred maintenance. Landscapes under public management, such as edges of roadsides and rights-of-way, are bushwhacked until they support invasives almost exclusively. Site impacts that degrade adjacent natural habitats, such as erosion, are still routinely ignored. Because management is usually started too late, instead of when problems first become evident, too much ground is lost. Our efforts should go toward sustaining what we still have now, rather than trying to re-create it later. A habitat should not become degraded before we turn our attention to it.

As land managers, we have often failed to confront the most difficult problems, the ones that drive changes occurring in the landscape. We have allowed the details to divert us from the big issues. The landscape has been degraded and maintenance has been deferred to the degree that we now fail to uphold reasonable standards. Most construction standards, for example, mandate only a temporary cover of quick-germinating grass to meet regulatory requirements where there may formerly have been forest. A corollary problem is the use of standardized specifications for a wide range of environmental conditions. Our failure to customize our techniques and approach to each region produces unnecessary damage to native communities and preempts opportunities to work with, rather than against, natural processes. This is as much a problem in professional practice as it is in construction and maintenance.

Overuse of Heavy Equipment

There is also an unfortunate pro-equipment bias in our culture. A good example is the general myth that the use of heavy equipment always saves time and money. It's true that it can take manual workers days to move dirt that a bulldozer can move in hours. But it is also true that equipment use has many other, greater costs, not all of which are acknowledged, including the debt payment on the equipment, maintenance, mobilization expenses, and higher labor costs.

More rarely recognized, the use of heavy equipment results in added costs for access roads and from the direct damage done by vehicles. The larger the equipment, the more likely that a larger-than-necessary area will be affected, which, in turn, creates the need to manage a far larger area than would otherwise have been undertaken. This is one of the reasons there are ever fewer leftover places in the landscape. Both the landscape and our budgets are victim to our heavy equipment mentality.

Costs in health and safety, like the costs of deferred maintenance, are usually overlooked as well. A host of environmental changes, from soil compaction to air pollution, are intimately associated with our tendency to motorize, electrify, and oversize most tasks that were once accomplished with manual labor. A gasoline-powered mower spews out the same load of some air pollutants as thirty average cars for every hour of operation. Southern California's Air Quality Board has called for replacing nearly 2 million pieces of gasoline-powered lawn and garden equipment in Los Angeles. Lawnmowers alone account for 25,000 serious injuries and seventy-five deaths yearly; one in five victims is a child.

Our management of all landscapes, from turf to woodlands, tends to be quite consistent: we too frequently apply the same principles no matter what the situation. Let's take a look at some of our conventional landscape management practices and compare them to the perspective of sustainable restoration.

Woodlands and Forests

Often, conventional recommendations made in the past by foresters and horticulturists were inappropriate for natural forest and woodland areas. Take the use of exotics, which they routinely recommended for wildlife enhancement. Such concepts as "create more edge" or "thin the canopy to stimulate regeneration" have little relevance today in fragmented woodlands. Even though some individual trees may be a century old or more in urban woodlands, the prevalence of invasives and native species characteristic of early stages of forest regrowth, as well as the absence of old-growth species, attest to the relative youth of these landscapes. However, these woodlands also may be rightly termed senescent when there is a canopy of older trees and no reproduction of canopy replacement species because of suppression by vines, exotics, trampling, or any number of other chronic disturbances.

The directive to thin canopy to stimulate understory regeneration is almost always inappropriate in remnant woodlands because it tends to favor the wrong species. Woodlands in the megalopolis are fragmented and disturbed, often with a discontinuous canopy. Many older woodlands are increasingly losing canopy and are subject to blowdowns, in part because of increased storm intensity and frequency resulting from higher global temperatures, and in part because of damage from chronic disturbance. Successional species such as black cherry, black locust, and mulberry, as well as even more problematic invasives such as Norway maple, are favored by thinning—not the more desirable mast species of oak, hickory, and beech.

In the early part of this century it was fashionable to thin remnant woodlands to "beautify" them. Middlesex Fells in Boston, cleared of understory to "open it up" around 1920, showed a dramatic loss of many once-common native species in a recent recensus. The impacts of thinning practices in Prospect, Central, and countless other parks in the fifties and sixties to increase visibility and perceived safety are still felt today.

A forest consists primarily of long-lived plants whose reproductive sites develop aboveground. These plants are not "rejuvenated" by cutting, as turf is. The actual damage may go unobserved because something usually regrows. It is the shift from native wildflowers to more vines and exotics that often goes unnoticed. But every time the forest is cut, it is further simplified.

A corollary problem is the too-tidy forest that has not necessarily been brush-hogged but where too much of the understory has been weeded away. With it wildlife shelter and food vanish, as well as the next generation of forest plants. Overvigorous removal of exotics has the same result. The moral of the story is: Do not clear more than will be replaced naturally by adjacent native vegetation or what you are able to replant and tend.

We must remember that the ground is the cradle of the landscape, the place of regeneration. We must make every effort to avoid unnecessary soil disturbance, such as grubbing (digging out roots of undesired plants) and rototilling. These methods may be appropriate for renovating a horticultural bed planting but are not suited to woodlands. If the surface is presently stable, even if it is supporting only exotic invasives, beneath the soil will be roots of numerous different plants, often originating a great distance away. Tree roots, for example, are completely opportunistic, seeking any favorable ground, and easily extend 30 or more feet beyond the farthest reach of their branches.

Grubbing and rototilling also disrupt fragile microorganisms and may damage fungi associated with plants, mycorrhizae, on which good forest growth depends. Such disturbance can also stimulate many root sprouts of certain trees that can strongly alter the character of a site.

Lawns and Meadows

The greatest mismanagement of lawns is simply that they are used far too extensively. More land in the United States is maintained as turf than is managed under wilderness designation. Overlarge turf areas are damaging to adjacent environments and preempt other landscape types, including indigenous communities. All remnant natural landscapes would benefit if we could drop our attachment to lawn.

Turf is, in fact, so widespread that much of it is simply undermaintained, especially in areas where turf may have been inappropriate or difficult to maintain in the first place (Figure 10.2). Unfortunately, where turf is well maintained, the off-site impacts—high levels of pesticides and water usage, air pollution—are often even more severe. Well-watered and fertilized lawns grow faster and require more mowing. "Cost-saving" techniques such as close mowing, often 1 1/2 inches or less, like that employed on a golf course—have grown in popularity, but, unfortunately, the grass soon suffers from desiccation and may need more fertilizer and irrigation. Cropping too closely also results in early dormancy of desirable cool-season grasses and early appearance of crabgrass and other turf weeds, which then are likely treated with pesticides. When modern mowers cut very close, they frequently scrape the ground where the terrain is uneven, resulting in a skinned soil surface with only patches of remaining vegetation.

Even when we relinquish lawn, the resulting released landscapes often bear little resemblance to the waving grasslands and wildflower meadows of the past. Thinking that a meadow means no management at all is a common fallacy. In a climate that supports forest, a meadow is by definition a managed landscape. In a disturbed landscape, an unmaintained meadow overwhelmed

Figure 10.2. Many lawn areas are riddled with bare places and poorly maintained, poor quality turf, especially under trees.

Figure 10.3. Honeysuckle, miscanthus, and other weedy exotics have completely taken over this aban-doned farmland in a protected preserve.

with heaping Japanese honeysuckle (Figure 10.3) or Japanese knotweed serves as a continuous source of infestation to other natural areas and will not help in changing public perceptions of the messiness of "natural" landscapes.

In the end, most mismanagement is based on misinformation and, in particular, a failing to see long-term trends. Sustainable landscape management is a holistic approach that seeks to complement and work with natural systems. It is primarily centered on places rather than tasks. This larger ecosystem focus informs site-specific management and provides a context for evaluating the success of management. It is time that we confront the wide-ranging consequences of conventional landscape maintenance. It is time to give up myths about the natural landscape that have dominated horticulture, landscape architecture, and forestry. We must turn to newly acquired skills to restore ecosystems.

CHAPTER 11

Atmospheric Change

Some of the most pervasive and least understood impacts on forests result from human-induced changes in the Earth's atmosphere, globally, regionally, and locally. Long-term consequences of these changes may affect forest biodiversity as severely as fragmentation and loss of habitat do.

Local and Global Impacts

The effects of the atmosphere and air quality on vegetation are both local and large scale. Plants growing in the city are the most obviously stressed. The urban environment creates a heat island that, when combined with reflected heat, glare, and desiccating winds, creates a high water demand on plants, whose roots may be under paving or compacted soil and, hence, experiencing drought. Even when uncompacted, urban soils often exhibit restricted levels of water infiltration that may be attributed to hydrophobic (water-repelling) materials, like motor oil, in the soil. High levels of grime and particulate matter clog breathing pores on leaves. Ozone, hydrocarbons, and airborne pollutants, such as lead and other heavy metals, damage leaves, roots, and tissues and may increase opportunities for parasites and pathogens, such as tree fungi, to invade plants.

Although atmospheric changes, such as pollution or ozone thinning, have been known for some time as causes of serious health problems in humans, public awareness came slowly and change even more so. Despite our awareness, fine particulates, tailpipe exhaust, coal-fired power, and other forms of combustion still kill more people than car accidents or guns each year (Schwartz 1994). And we have consistently underestimated air pollution's pervasive impacts on indigenous landscapes. If we are slow to act when it involves our own lives, is it any surprise that we are slow to recognize the overall impacts of atmospheric changes on forest systems—impacts that are unfamiliar to most people and regulatory agencies? The scenario is not unlike that of invasive exotics, whose impacts were denied until the evidence was overwhelming and the damage so excessive that some claim it is now too late to take action.

Changing Types of Impacts

About the turn of the century, signs of air pollution began to show up in tree rings, the result of the advent of coal- and oil-fired engines and the emergence of large-scale industry. The enormous increase of carbon dioxide, sulfur, nitrogen, and many other contaminants released into the atmosphere since then has been accompanied by several other changes that have had profound effects on the environment.

First, the proliferation of trucks and autos meant that pollution sources were no longer confined to fixed locations, such as factories, where the affected area could be readily distinguished from more pristine landscapes. As greater and greater amounts of traffic moved far and wide across the landscape, the impacts became pervasive. By the 1950s trees began to die noticeably faster than they had in the first half of the century. By 1969 there was one mile of road for every square mile of land in the United States. By the mid-1980s more land was disturbed by roads than by all mining activities. Today, no area that we would have previously referred to as pristine still remains.

Another basic change came with a shift in the by-products of combustion, especially oxides of nitrogen, the so-called NO_x pollutants. The use of wood, coal, and oil as fuel always produced some oxides of nitrogen, but the higher-temperature combustion processes that developed in industry created a huge new source of NO_x pollutants from the nitrogen in the air itself. Little progress has been made in reducing the amount of NO_x pollutants, and they are still largely unregulated.

Unfortunately, we still know far too little about such problems. The National Acid Precipitation Assessment Program (NAPAP 1991), carried out between 1981 and 1990 to evaluate acid rain–related problems, actually served to obscure the real issues by its assertion that the forest was healthy (Likens 1992). By limiting studies primarily to sulfur-related acidity and ignoring nitrogen-related acidity and other problems (Aber 1993), NAPAP deflected public concern as well as research efforts, especially in this country, except by a few persistent investigators.

Acid Rain and More

The NO_x constituents of air pollution tend to become fine nitric acid droplets or aerosols that wash out of the atmosphere as acid rain or fall directly onto the landscape as dry deposition, both of which add to overall acid problems that originate from other sources, such as the huge amounts of sulfur oxides also generated in fuel combustion. This complex problem, referred to by various names, such as acid precipitation, acid fog, dry acid deposition, and occult deposition, is one of the more important ways that we have modified forest conditions. Like other acid precipitants and ozone, NO_x contaminants are directly injurious to humans, other animals, and plants alike.

Excess acidity creates many problems, such as the leaching away of important mineral nutrients like phosphorus, calcium, magnesium, and potassium, which may become factors limiting growth. High acidity also releases toxic elements such as aluminum and heavy metals. Long-term studies by Likens, Driscoll, and Buso (1996) document the loss of over half the pool of calcium in the soil in the past thirty years at the U.S. Forest Service's Hubbard Brook Experimental Forest in New Hampshire's White Mountains. The trees have nearly stopped growing, and the stream waters are still very acid because the soils have lost the ability to buffer these impacts.

William Sharpe of Penn State University has concluded that depleted and leached soils have weakened root systems in the Allegheny National Forest and that liming, fertilizing, and fencing will be necessary to regenerate them (Reidel 1995). In the Netherlands, calcium became so scarce in some woodlands affected by acid deposition that snail populations sharply declined. Without this source of calcium in the diets of birds, consequent eggshell thinning lowered reproductive success (Graveland et al. 1994).

Reduced Lignin

Another phenomenon associated with high levels of nitrogen is a reduced proportion of lignin, which is the woody tissue in leaves and litter (McNulty, Aber, and Boone 1991). The lignin component of forest litter tends to decompose more slowly than cellulose and other components, so a decline in the lignin ratio increases the rate of decomposition that is already accelerated by higher nitrogen, resulting in reduced surface litter and increased soil exposure to erosion.

Some protection from plant-feeding insects is also lost with reduced lignin levels because leaves become more succulent. This nitrogen-induced succulence, which in turn cycles nitrogen more rapidly back to the forest floor, increases feeding by gypsy moths and other exotic and native herbivores. Experimental nitrogen fertilization of hemlock stands affected by the hemlock woolly adelgid, an exotic pest that is currently devastating hemlocks over large areas of their range, hastens their death (McClure 1991). Similarly, tests simulating acid rain worsened the rate and severity of fungus infection on dogwoods. Air pollution from industry and vehicles is in fact fertilizing the forest by adding nitrates and thereby accelerating its decline.

Altered Competition

Despite the problems created for hemlock and other species, some plants benefit from increased nitrogen and are proliferating rapidly. An alien plant with a name like mile-a-minute vine, which is also known as dog-strangling vine, is not nutrient starved; rather, many such ruderals (plants of disturbed landscapes) and weed species are documented to be "nitrophilous," or nitrogen-loving. Scientists in Nancy, France, demonstrated a consistent pattern of greater eutrophication (that is, increased nutrient levels) at the edge of a small woodland (Thimonier and Dupouey, 1992; Thimonier, Dupouey, and Becker 1994) on the upwind side, where pollutant deposition is greater. This "edge effect" was measured in the vigor of nitrophilous and acidophilous species and increased nitrogen and acidity in the soils. In northern Europe, a species of hairgrass, *Deschampsia flexuosa*, is a bioindicator of nitrogen deposition, especially in heathlands, where its replacement of the typical heather shrub layers during this century has been notable (Steuben 1992). Recent studies in the southern Appalachians (Wilson and Shure 1993) suggest that *Rubus*, a genus that includes many brambles, briars, and berries, may be a good bioindicator of nitrogen enrichment in eastern forests. In general the ground-layer vegetation in a forest is a good indicator of changes in the nutrient balance in the ecosystem, exerting significant pressure on species composition over time (Eichhorn and Hutterman 1994).

Nitrogen deposition in Europe has increased the production of forage plants for roe deer and, therefore, the population of roe deer in forests (Ellenberg 1987). We tend to think of overpopulations of deer as caused by lack of predators and change, in forest age and structure, but part of the increase is the result of added nitrogen, causing increased succulence and abundance of some food sources.

Soil Changes

This rain of fertilizer and other once-uncommon substances has changed the character of soils as much as the vegetation. Some of the primary impacts of the overenrichment process now under way are due to changes in soil biology, including its most fundamental aspects, such as total soil respiration. Changes in soil respiration affect the oxygen available to root systems and in turn the health and vigor of the forest.

One of the most important studies being conducted today, at the Institute for Ecological Studies of the Carey Arboretum in Millbrook, New York, is examining the types of changes to soil systems occurring in the most urban to more natural landscapes. Scientists are monitoring the similarities and differences of oak woodlands growing on similar soils on sites extending from Central Park in Manhattan to a distance of about 200 miles north of New York, where the institute is located. This ongoing study has already shown several important urban-to-rural gradients in and around the New York City area (McDonnell, Pickett, and Pouyat 1993). Heavy metals in the soil were higher in urban areas and declined outwards. Salt in the soil and soil waters trended in the same way. Urban soils tended to be more hydrophobic, that is, more resistant to wetting, as well.

Urban soils also had higher rates of mineralization of nitrogen, which is the conversion of organic forms of nitrogen, such as soil organic matter, to inorganic forms such as nitrate. The authors suggested that the principal changes brought about by nitrogen deposition may be an increase in the abundance of bacteria in the soil and litter combined with reductions in fungal and invertebrate populations. Urban forest-floor litter also decomposes more quickly than that of countryside forests. The authors further suggest that higher nitrogen deposition on the urban end of the gradient leads to the faster release of organic nitrogen from soil and litter that they observed.

Soil structure is affected by these changes, too, particularly by a shift from fungal dominance to bacterial dominance. The webby mycelia that comprise the bulk of soil's fungal component serves to knit together soil particles and bits of organic matter, while the substances secreted by bacteria (exopolysaccharides) are slippery and cause soil to slump when it is exposed to rain (Harris, Birch, and Short 1993).

Unfortunately, conventional soil tests do not measure such factors as total biomass, that is, the living component of the soil, nor do they identify levels of specific fungi or invertebrates. Conventional soils analyses do not reflect the high levels of nutrients cycling through ecosystems today because the analyses reflect what is present at the time rather than what is flowing through the system. Thus we have remarkably little information on the soil's functional character.

Eutrophication of the Landscape

Acidity is not the only problem occasioned by nitrogen as an air pollutant. Nitrate and ammonium ions, the important forms of nitrogen in deposition, are powerful stimulants to many biological processes.

Nitrogen levels have increased substantially, not just from automobiles and trucks, but from other sources as well: intensive agriculture, wastewater treatment, landfills, water pollution, and other activities that involve large quantities of organic waste. The levels of nitrogen loading on natural systems today are in many cases more than three times the natural level of loading.

This shift is an example of a general phenomenon that ecologists call "eutrophication," an increase in levels of one or more nutrients. Although in the minds of regulators and the public, eutrophication is usually associated with aquatic or wetland habitats, eutrophication of terrestrial systems has occurred as well and will continue to severely affect forests as well as aquatic systems:

> The global nitrogen cycle has been altered by human activity to such an extent that more nitrogen is fixed annually by humanity (primarily by nitrogen fertilizer, also by legume crops and as a byproduct of fossil fuel combustion) than by all natural pathways combined. This added nitrogen alters the chemistry of the atmosphere and of aquatic ecosystems, contributes to eutrophication of the biosphere, and has effects on biological diversity in the most affected areas (Vitousek 1994, 1861–2).

Unfortunately, ecosystems are generally adapted to relatively low rates of nitrogen input. Nitrogen amounts above certain levels will drive a system to change. Prior to extensive human generation of nitrogen by industry, agriculture, transportation, and other activities, nitrogen from precipitation and deposition from the atmosphere was relatively low, and forested ecosystems were necessarily adapted to the efficient utilization of a limited flux of nitrogen in the system. Widespread association with mycorrhizal fungi by different kinds of forest trees is one of the adaptations that allowed forests to develop under natural conditions that were typically low in nitrogen. Certain trees, such as alder and locust, also have associations with nitrogen-fixing microorganisms. From the trunks and canopy branches of forest trees, lichen communities and their nitrogen-fixing symbionts shed a slow rain of nitrogen to the forest floor. At one time the total amount of nitrogen added to a system came from these and other small sources—such as minor amounts generated by lightning that then fall in rain. The shift to high nitrogen loading has been relatively swift, and the consequences are pervasive.

According to Orie Loucks, Director of the Lucy Braun Association for the Mixed-Mesophytic Forest, which monitors the impacts of air pollution:

> In my view, the Mixed-Mesophytic has lost its immunity to diseases and insects from the impacts of these pollutants. In this sense, the death can be compared to the AIDS epidemic afflicting the human species. One only has to walk in the forest and see the premature leaf drop in early July to understand that ozone is dramatically impacting a predominance of species. As for nitrogen, we know that the forest is receiving three times the amount of its estimated tolerance level (quoted in Flynn 1994, 34).

Philip Wargo, a U.S. Forest Service plant pathologist, uses the term "full-force decline" to describe the combination of severe mortality and failing reproduction. European forests are even more strikingly affected. "The Dobris

Assessment," a comprehensive report on the environment of greater Europe, indicates that one-quarter of the forest trees in thirty-four countries are defoliated. More than 50 percent of the forest in the Czech Republic, for exmaple, may be irreversibly damaged.

Nitrogen Loads in Wetlands

Although discussion of the eutrophication of terrestrial systems is somewhat new, its effects on aquatic systems are all too familiar. Relationships between upland and wetland eutrophication are not, however. Much of the nitrogen that is causing eutrophication of rivers, lakes, and coastal water bodies derives from atmospheric deposition over the watersheds draining into them.

The bulk of the nitrogen in a forest soil is tied up in the soil's organic matter and surface litter. Biological decomposition typically releases nitrogen in the form of the ammonium ion, in a process referred to as "ammonification." But when excess nitrogen is available, certain microbes convert the ammonium to nitrate in a process referred to as "nitrification." Plants or microbial populations may take up both the ammonium and nitrate forms of nitrogen; however, under more natural conditions, ammonium in limited quantities rather than high levels of nitrate would be typical. Ammonium, although less available, persists in the soil while large amounts of nitrate seep away in runoff in groundwater and contribute to downstream eutrophication. Presently, atmospheric deposition provides 10 to 50 percent of the nitrogen from human activities that is added to coastal waters, contributing to eutrophication and ecosystem changes (Paerl 1993). In the Chesapeake estuary, for instance, 30 percent of its nitrogen originates from atmospheric deposition in the watershed (EPA 1994). A study in southeast Australia found that deteriorating riparian eucalypt woodlands had high levels of nitrogen, especially as nitrate, and nitrophilous plants beneath the eucalypts had high foliar nitrogen and high levels of nitrate reductase activity (Granger, Kasel, and Adams 1994). Riparian woodlands, including swamplands, can be especially affected in this way since they receive burdens of nitrogen from agricultural runoff, wastewater, and atmospheric deposition, including not only that which falls directly on them but also that which washes down from upstream in the watershed.

Global Warming and Ozone Depletion

Problems of nitrogen deposition at the global level go well beyond regional effects. Oxides of nitrogen in the atmosphere, especially where human-caused nitrogen deposition is high, contribute to global warming as well as to destruction of the ozone layer. High levels of nitrous oxide further increase incident ultraviolet radiation levels on forests and the biosphere owing to its role in the destruction of ozone in the stratosphere. Nitrous oxide is more or less stable in the lower atmosphere, but in the stratosphere it converts into reactive forms that become nitric acid aerosols. These droplets act as catalysts in the reactions of the various forms of chlorine with ozone and thus are implicated in the stratospheric ozone loss (Sander, Friedl, and Yung 1989).

Because coal and oil are fossil fuels, unlike firewood or other biomass fuels, their use releases carbon fixed long ago very rapidly back into the atmosphere as carbon dioxide, which is linked to the problem of global warming. The accelerated release of nitrous oxide from increased denitrification by soils receiving excess of nitrogen amplifies this greenhouse effect when high levels of nitrate are transformed by anaerobic soil microbes into gaseous nitrogen or gaseous oxides of nitrogen, especially N_2O (nitrous oxide), which escape to the atmosphere.

Nitrous oxide is an important greenhouse gas itself, especially problematic because it is more persistent in the atmosphere than the other two prominent greenhouse gases, carbon dioxide and methane. Impacts of the greenhouse effect also have great relevance to the survival of woodlands. Temperate forest areas will possibly experience increased storminess, changed soil moisture conditions, changed hydrology, and increased growth of certain species over others, including many exotics. Associated problems will preoccupy managers of parks and other lands for many years to come. Blowdowns, for example, already a major problem in many parks and woodlands, are partly a consequence of fragmentation because forested areas have greater exposure to windstorms, especially when accompanied by soil saturating heavy rainfall. Should storms increase in frequency and intensity with global warming, the problem of blowdowns will worsen, especially if fragmentation continues.

Researchers have chronicled the plight of beech, an indigenous species that has shown a high degree of sensitivity to the climatic changes already under way. Early responses include reduced growth and seed production, increased insect and pathogen attacks, and, in older trees, direct physiological stress. Compounding these effects, beech also competes poorly with increased disturbance, such as trampling or windstorms. Their prognosis is very bleak:

> With increased disturbance and the temperature scenarios we are using here, persistence of beech for longer than 50 years in reserves in the southeastern United States seems unlikely. To compound the difficulties of management, it may become difficult to protect forest reserves. Moribund beech forests will have little aesthetic appeal and may also be viewed as fire hazards or centers for disease that endanger surrounding vegetation (Davis and Zabinski 1992, 303).

As disturbing as some future concerns are, damage visible today in the landscape underscores the need to expand roadless areas, especially in national park and forest lands. Direct impacts along every road corridor amplify the consequences of regional air pollution and add to the complex of edge effects. The following description of forests in West Virginia should be specter enough to make everyone concerned with forests seriously reconsider our use of the automobile and of fossil fuels in general:

> Aliff began noticing the decline in the 1950s—in native red mulberry and butternut walnut growing on the lower slopes, in the fields, and along the creek banks. Then some twenty-five years ago, it climbed to the ridges where it hit yellow locust at about the same time it struck the hickories and oaks. From there it invaded the coves and in the past five to eight years it has spread like wildfire until I don't think there's a single species unaffected. I used to think of survivor trees, like yellow locust and red maple. But that's not the story anymore. The death is system-wide. Wherever I go, here at home and into other states, I walk among tombstones (Flynn 1994, 35).

Recent reports on the impact of atmospheric change on the growth of white pine in New England are at first glance encouraging. One study (Bennett et al. 1994) suggests that some of what we see is individual tree sensitivity, rather than the vulnerability of a whole species. This thorough review of field surveys of damage to white pine since 1900 suggests that many reports are flawed and that current inventories show that growth of white pine throughout its range is vigorous. It appears that not more than 10 percent of the population is hypersensitive and that these individuals in particular declined in the years between 1963 and 1973. Fewer losses are occurring now as this stock is gradually eliminated from the population. Still, is this good enough news? Is 10 percent too great a stress at this time? Is the sensitivity of other species much higher? Isn't a more natural range of air and soil conditions more likely to sustain greater diversity? How much farther do we think we can deviate from historical conditions?

For many of us watching the treasured landscapes of our youth collapse before our eyes, it is hard to think what it might take to restore the landscapes of the ancient forests. As Bill McKibben (1996) noted in his afterword to a book about old-growth forest in the East,

> When we look at a hemlock on a slope above Cold River in the Berkshires, or a towering white pine south of Cranberry Lake in the Adirondacks, or a massive tulip poplar in a cove in the Smokies, we must not imagine that its glory devalues the second- and third-growth birch and beech a quarter-mile distant. Instead the majesty of the ancient forest makes this tentative wildness all the more valuable, for it shows what it might become someday. Old growth is not simply a marker of past glory, an elegy for all that once was. It is a promise of the future, a glimpse of the systemic soundness we will not see completed in our lifetimes but that can fire our hopes for the timeliness to come (363).

PART II

The Restoration Process

This guidebook is rooted in the idea that restoration is a heuristic process, one in which the participants learn by doing. Restoration is best accomplished by those who live in the place, guided by their shared mission because they are naturally more invested in the community. There is no step-by-step method. The following chapters are devoted to an adaptable approach for landscape restoration that reflects the way change really happens and how people really work together. Instead of rigidly specifying everything down to the letter, it establishes a structure in which concepts are agreed upon and the strategies are fleshed out over time as opportunities unfold. The details will change with time as the participants learn about one another and their shared environment. At the same time, real knowledge gathered through the monitoring program and compiled in management logs and databases will lead to far greater understanding of the site's systems. Restoration is a kind of purposeful research. The questions asked in the course of attempting to restore a system expand the scientific and social dialogue and require us to integrate higher levels of information more effectively.

CHAPTER 12

Restoration
in Theory and Practice

What is restoration? At a literal level, the term implies that we are returning the landscape to some former state. The past is often our reference ecosystem for the natural potential of the landscape, since more species and ecosystems existed in the past than today.

When we aspire to restore a landscape, however, we usually mean to the pre-settlement condition when that landscape was most likely fairly intensively used and managed by indigenous people. Our concept of "wilderness" is actually the landscape formed by the practices of indigenous peoples. Their burning, foraging, and planting methods over thousands of years influenced the composition and character of the forests. Kat Anderson, an ethnoecologist with the Natural Resources Conservation Service, describes the landscape management by indigenous peoples in this way:

> For most [Native American tribes], hunting and gathering were but two components of a much more comprehensive land-management system. Simply put, acorns get wormy, old berry bushes produce less fruit, fire-dependent mushrooms don't grow every year, bunchgrasses decline in productivity as they accumulate dead material, and meadows shrink as trees encroach upon them—all processes that can be reversed by active management to maintain the abundance and diversity needed to support human populations. In fact, most of the plants tribal people value are shade-intolerant and depend on burning or other forms of disturbance to maintain the early successional communities they inhabit (1996, 158).

Ecosystem restoration today likewise requires active management to reestablish the historic environmental conditions on which indigenous plants and animals depend. We cannot sit back and wait to see what pops out of the soil. We are just as much a part of the land and just as responsible for it as our predecessors.

But we must also recognize that historic conditions cannot necessarily be re-created. This understanding is as important as trying to re-create historic con-ditons. Every condition that has been forever altered represents a potential evo-lutionary dead end for some species. This was once a rare and episodic occurrence that is now all too common. As Dennis Martinez, director of the In-digenous People's Restoration Network of the Society for Ecological Restora-tion, has noted:

> We recognize that we cannot recover the unrecoverable—the ecological details and minor idiosyncrasies of the pre-contact landscape. We are only saying here that the key or basic ecosystem features, keystone and conservative species and communi-ties, and critical natural processes within historic ranges of variability are probably necessary for full ecosystem function, integrity, resiliency and stability. Restoration or protection of biodiversity alone is not enough. Biodiversity per se is not the un-derpinning of ecological stability. Ecosystem integrity and function set the necessary conditions for biodiversity to flourish by achieving stability. Biodiversity then is a function of the relationship between ecosystem structure and dynamics and processes. But it is a stability precariously balanced—constantly ebbing and flowing with each wave of ecosystem change. The critical natural keystone species in main-taining this delicate and ever-shifting balance is Homo sapiens (1995, 26).

In fact, we must recognize that a true forest restoration is *not* possible. There is no landscape where we can say we have reestablished a forest system as com-plex and diverse as that displaced. The degree of recovery is usually more de-pendent upon how damaged the site was in the first place or by how much time has elapsed than on our deliberate restorative actions. This is in large part be-cause until very recently the restoration and conservation of ecosystems were not our objectives. We did not minimize damage to ecosystems nor factor nat-ural systems into the construction of public works or subsidies related to land use. Until now, sustaining indigenous communities has not been a regulatory, aesthetic, or economic goal.

Although the indigenous forest will not be restored to its former condition, we can hope to sustain enough of its patterns and components to support much of the biotic richness that has persisted until today. The object is to re-create and sustain historic natural processes to the extent *feasible*. We can work to bring about solutions that mimic or are analogous to local natural condi-tions as closely as possible. Therefore, the objective of restorationist should also be to shift conventions sufficiently to favor restoration over degradation, grad-ually tipping the balance in favor of sustaining natural systems. It will take decades of planting indigenous species to overcome decades of planting exotics. It will take thousands of projects favoring groundwater recharge to offset cen-turies of increasing runoff. We can change our management of roadsides and other rights-of-way to gradually foster rather than eliminate native communities simply by altering mowing methods and investing a proportion of the maintenance budget in strategic exotics control. There is ample evidence to de-omonstrate the economic feasibility of altering our practices.

This necessary change of attitude will not happen until enough agencies, institutions, and landholders make a commitment to integrate an ecosystem approach into all aspects of landscape management. For restoration to occur, it cannot be limited to isolated instances or to a few high-profile projects or to work completed by volunteers. Significant change would have to occur at every level of our use and care of the landscape, on both public and private land. But as overwhelming a challenge as that is, each of us, in whatever sphere we work, can make significant contributions. Restoration is everyone's business, from the roadside maintenance crew to the local zoning board; the home gardener to the forester, the farmer, and all the other members of the community.

Even when this process starts only as lip service, it often leads to real change because a new dialogue is started. When we ask new questions, fresh ideas emerge. Ecological restoration as a larger goal also helps to get past the immediate constraints of special interests, jurisdictions, and the project of the moment to the more enduring aspects of the landscape and the community.

New strategies for restoration will become clear as we establish networks and recognize overlapping interests. What might be seen now as a capital project in need of funding, such as a woodland restoration, might become an ongoing monitoring and research project undertaken as a cooperative venture with area schools, integral to the curriculum and supported with outside funding and grants. The need for large quantities of propagated native plant material might lead to the creation of a for-profit nursery coordinated with a parks department to augment its management budget. The object is to create an environment where such connections can be made and where all actions reflect an understanding of the larger context, of the whole system.

The most important conditions for a restoration project are that it be community-based and that it be science-based. To be community-based, it must represent a consensus, which in turn requires that it be participatory. To be science-based, it must be documented and monitored. While we cannot necessarily know enough at the outset, we can establish a process for learning, for restoration that proceeds developmentally.

"Developmental" Restoration

Restoration is, by definition, an incremental and developmental process. Restoration proceeds in phases representing the levels of health and complexity of the system, instead of area-by-area project completion. A typical scenario might be to stabilize all bare soil areas and to initiate exotics removal while starting planting, and then to evaluate the success of plantings and natural regeneration before developing a more detailed planting plan. An important key to this approach to restoration is the principle of "minimal intervention," that is, taking only those actions that are necessary to counteract disturbance, but also taking no actions that may inhibit the natural processes of restoration. Actions that are reversible are usually preferable to those that make irreversible changes or commitments. Where complete restoration is not feasible immediately, the best course is to make no decisions that foreclose the possibility of instituting future restoration options.

Restoration is not a one-time thing, any more than raising a child is. Grandiose projects that involve extensive grading and replanting are appropriate only in the most drastically disturbed conditions where there are no remnants of indigenous communities to protect and renew. Restoration is also not a simple procedure of planting followed by maintenance.

Sometimes our knowledge of and experience in gardening, horticulture, and landscape architecture create very false perceptions about forested landscapes because we presume a high level of control over the landscape and tend to focus on the individual plant, rather than the community. We change "unsuitable" conditions as needed, concentrating a lot of effort in a relatively small area, attempting to accomplish as much as possible all at once. Thousands of laborers, for example, were brought into Central Park to implement Frederick Law Olmsted and Calvert Vaux's Greensward Plan, draining swamps and filling over barren bedrock to create rolling hills with idyllic pools and cascades along meandering streams. The Ramble, at the center of the Park, planted as an idealized, romanticized forestlike landscape, is crafted with the highest-level skills of Victorian design. It was a triumph, and it was a significant habitat alteration, but it was not an intentional habitat restoration. The goal to restore disappearing natural landscapes is relatively modern. The evolving approach is to learn how to recognize and optimize the natural regenerative processes of each place; to learn from nature, not to change it.

When we view forest restoration as if it resembled a garden or landscape installation that can be completed in one phase, we preclude opportunities for the site to design itself and impose our own preconception of what the landscape can become. Moreover, the sheer amount of activity all at once may be in itself a severe stress to fragile remnant systems. A large park project that includes architectural reconstruction may span several seasons, during which time nearly all valued wildlife is displaced from the site. If the area is a temporary resting site for migratory warblers, for instance, the wildlife may have no other site to use in the interim.

It follows, then, that we should consider undertaking large-scale grading operations, extensive soil reworking, and massive planting efforts only where the landscape is in collapse, overwhelmed by invasives, or extensively eroded. Similarly, where the vegetation is a mix of desirable species and pests, complete elimination of all invasives at once may actually open up the landscape so much that a reinvasion, perhaps even greater in scale, ensues. Such sweeping actions are simply not effective or even acceptable in most forests or other remnant natural systems.

We cannot know enough at any one point about a site to accurately predict the future or to fully specify what actions are appropriate to take. Such a level of uncertainty may be difficult for site managers who feel obliged to provide fixed budgets and timeframes for landscape projects to accept. On capital projects, where scheduling and immediate results are priorities and adequate maintenance is unlikely, we sometimes try to compensate by overplanting a site, thereby losing the opportunity to benefit from incremental management over time. It's no wonder that conventional solutions prevail over more innovative but realistic approaches in the regulatory arena, where projects are often expected to be completed in eighteen months or less. The timeframes are too short for what is really needed to restore or establish a landscape and do not allow for a more developmental and monitored approach. Wetlands regulation is replete with projects that were approved with no follow-up monitoring and were poorly constructed; they failed to function as promised. Others were never built because of lax enforcement. Even "successful" projects rarely have a regulatory life beyond the date of project acceptance and may slowly degrade over time.

A common concern is "Who will care for this restoration after it is complete?" This is a false question, presuming that a restoration is accomplished and then maintained. It is also a difficult question to resolve, since the participants must always keep in sight that restoration is a long-term, continuing process. The more appropriate question is, "How is the community going to be involved in the continuous monitoring and management of this site over time?" By bringing science to the community and the community into the decision-making process we will be better equipped to face the challenge of restoring the forests we live in, despite the ecological uncertainties that lie ahead.

> Confront uncertainty. Once we free ourselves from the illusion that science or technology (if lavishly funded) can provide a solution to resource or conservation problems, appropriate action becomes possible. Effective policies are possible under conditions of uncertainty, but they must take uncertainty into account. There is a well developed theory of decision-making under uncertainty. In the present context, theoretical niceties are not required. Most principles of decision making under uncertainty are simply common-sense. We must consider a variety of plausible hypotheses about the world: consider a variety of possible strategies; favor actions that are robust to uncertainties; hedge; favor actions that are informative; probe and experiment; monitor results; update assessments and modify policy accordingly; and favor actions that are reversible (Ludwig, Hilborn, and Waters 1993, 36).

CHAPTER 13

Community-Based Education, Planning, and Monitoring

The parks belong to those who use them.
—*John Muir*

One of the most important concepts about restoration is that all members of the community have a role to play. Unfortunately, all too often the natural landscape has been left to experts or regulators. In only a few generations we have moved from a culture in which most people knew the names of most plants as well as many of their medicinal uses and other properties to one in which the public has little knowledge of the natural world. We have accepted the idea that wild plants and animals are better left to the concern of someone else, such as a scientist or a lawyer.

The degradation of the environments around us is ultimately due to a breakdown in the relationship between the local community and the landscape. Without real knowledge of the details and patterns of a place observed and monitored over time we will not be able to restore indigenous landscapes that are used and enjoyed by the community. Sustaining the values of the landscape over time will depend upon reestablishing positive interactions with each place within the communities who use them most. Those who use and care for a landscape are responsible for sustaining its value over time, but they cannot do that if they are not involved, informed, and empowered. For restoration to have a chance, we must encourage open, direct communication and a broad level of participation at every opportunity to empower users and managers alike with both responsibility and accountability for the restoration of their local landscapes.

It is interesting that this approach presents new opportunities and at a time when land managers are all too aware of fiscal pressures to downsize or enact hiring freezes. The most difficult item to get in a budget these days is a new full-time employee. In this climate, asking what roles are appropriate for volunteers is important. The use of volunteers to facilitate the loss of jobs is not desirable. Ideally, it is not the role of a volunteer to pick up trash, work that is usually performed by a paid person. Trash removal is important enough to justify paying for it. It may, however, initially take a group of volunteers removing trash from a site to demonstrate the value of the landscape and get agency and media attention. It may also take volunteers to raise public awareness sufficiently to diminish the volume of trash in the first place. The most important role a volunteer can play is to break new ground, to open our eyes to what needs to be done by doing it. If successful, the new management practices and innovations will in time create jobs and contribute greatly to long-term economic and environmental sustainability.

Consensus Planning

Restoring and sustaining natural systems in the fabric of a developed landscape require us to learn to support many interrelated values, rather than favoring one over the other. Disparate individuals and groups, even when in agreement on larger goals, often have difficulty coming together on policy and implementation, in part because their positions are restricted by a narrow set of interests and experience. The mountain biker and the birdwatcher often can see only conflict despite their mutual affection for and use of the landscape. We must find the shared vision of the landscape that brings the community together.

Determining that shared vision and the means to achieve it is, in the best situations, "planning by consensus." Consensus planning means that all constituents participate in the decision-making process and in the process sign on to the collective decision making that occurs. Anyone who has worked with groups knows how difficult that can be. All too often each person coming to a project or initiative has a personal agenda. Setting aside preconceived ideas and opinions and learning to work as a group is a real challenge. Achieving real consensus among all constituents in the community means that each person has a say. By necessity it means we must go well beyond conventional planning and implementation procedures. It depends upon ongoing education and participation, and also high levels of communication. Most people are in fact unfamiliar with consensus planning because it rarely happens.

Consensus planning is not the same as majority rule. It is not about win-lose-compromise. In the process of achieving real consensus, each participant must grow and change. Each participant must come to a larger understanding of the site context and an awareness of how other people's concerns relate to their own goals before they will "buy into" the larger vision and set aside their own more specialized interests. If the process goes well, it will be disconcerting at the outset as preconceived ideas are shaken. However, that period of uneasiness is followed by a time when we recognize new opportunities and new resonances brought to light among different objectives. As Stephanie Mills, ecologist and bioregionalist, has noted:

> The guiding assumption is that each member of the group possesses an important piece of the truth, and that the work of the meeting is to elicit these truths and create the best decision possible at that time. It does not proceed by voting but by gathering agreement (1993, 6).

Two essential rules must be followed for the process to be successful:

- *Everyone is welcome.*

Consensus is an open process. It is desirable to have a broad array of users and managers in the process to help reconcile conflicts. There is often an objection to some group or individual, but the bottom line is that if anyone who has impact on the landscape is left out of the process, vital information will be missing as well as important support for the effort. Site disturbance will continue to increase if all the users are not effectively involved.

- *Make no decision and take no action until real consensus is achieved.*

Typically, a few individuals will resist early on, testing the commitment. It is at that moment that the group finds out whether consensus is the real goal. Consensus will require that the team acknowledge the lack of a good resolution and get back to work. All objections need to be taken seriously, discussed, and resolved. As tiresome and as difficult as that might seem, the reward is trust in the process. Then decision making and action taking begin to happen with increasing speed and focus.

Decisions based on real consensus tend to get implemented because they meet multiple goals. Agreements are more durable because they are not bad compromises but rather represent real community values. This is the real advantage of this approach for those who seek to effect change in what seems like a bureaucratic or political stalemate. Stepping outside those boundaries and working with an outside group that operates under very different rules can be most effective in situations that seem paralyzed by red tape and misinformation. A consensus-based group can break a deadlock and often is acceptable to those who must give up power because it achieves results.

You can bring together a consensus planning team from the bottom up, with volunteers who seek to influence and work with local land managers, or from the top down, through an agency that wants to confront the problem of a deteriorating landscape. Regardless of how the effort is initiated, an effective team will include all who have a stake in the landscape. There may be some distrust of an agency-initiated participatory planning process at the outset, but trust will begin to build among participants if real communication is established. An institution or agency can play a crucial role by simply providing meeting space or start-up funding for a newsletter, one of the most important ways to build support, gain membership, and coordinate with other organizations.

A common argument against consensus-based planning is that it is very difficult and that few decisions get made. Actually, one of the most important virtues of consensus is just that, since, in hindsight, we see that much of what gets done is wrong anyway. Managers will often admit that if they had done everything they thought at the outset should be done, they perhaps would have done more damage than good. Restraint in landscape management is not always a bad thing. If the focus is on the most important restoration issues and there is widespread commitment to the solution by the community, real progress toward restoration is possible with only a few vital decisions.

Another argument against consensus planning is that dissension can stall community groups just as much as any other group. This is true but not valid. We have all experienced a civic or volunteer setting that was rendered ineffective by a belligerent participant or undermined by entrenched power. The win-lose mentality encourages people to dig in their heels and hold fast to original positions. Consensus actually offers a way out of the conflict by relying on real agreement instead of decision making based on personalities or politics.

Many agencies still perceive community-based planning to be a giveaway of their power. This is emphatically not true. Those agencies are ultimately the decision makers for lands under their jurisdiction by law. If given a chance, consensus-based community planning teams usually make a broader-based decision than the agencies would on their own. If the process fails, the agencies still have final responsibility and can step in. That alone is a powerful incentive for a community group to reach a meaningful consensus.

It is also important for agencies to realize that they have as much power as any community member does in a consensus-based process. That is the whole point: everyone is empowered to be part of a solution. This is a much more satisfying role for each player than being one of two polarized sides of an argument.

Communication and education are needed before participants will be able to grasp the larger community issues. The success of the process depends in large measure on how well the participants know their community and the extent to which information is shared. Innovative solutions that are not merely extensions of existing applications will require a sophisticated understanding of the site's natural and cultural systems and the ways in which they have changed over time. In order to not settle for the narrow interests that individuals may represent, those who participate will need to appreciate the length of time required to learn and understand a larger vision. Part of the understanding will happen with the sharing of information. Then it will become evident that the site, and the ecosystem it is embedded in, are participants also—as important in the process as any other.

One of the most popular strategies for consensus planning is called "envisioning," or "revisioning," which brings together diverse groups of people to craft new visions of the community's future. Community members participate in workshops that focus on hopes and dreams rather than on immediate crises and past failures. The object is to describe a new reality that is compelling enough and positive enough to entice people to see freshly and change their habits. The process is always open to a great breadth of participation and a lot of visionary thinking.

Professional facilitators are available both to set up and implement consensus-planning processes as well as to teach a group how to do these tasks for themselves. There are also two excellent guides that can assist any group in the shift to consensus planning:

Getting Together: Building Relationships As We Negotiate by Roger Fisher and Scott Brown of the Harvard Negotiation Project, New York, Penguin Books: 1988.

Involving Citizens in Community Decision Making: A Guidebook and *Pulling Together: A Planning and Development Consensus Manual,* by Program for Community Problem Solving, Urban Land Institute. To obtain copies, contact the institute at 915 Fifteenth Street, Suite 601, Washington, DC, 202-783-2961.

Regional Environmental Education

Empowerment is not sufficient in itself for a restoration project to receive community support. Knowledge, too, is an important element in gaining consensus. Often, what seems like apathy among community members masks a simple lack of understanding of the landscape and their role in it. Most people are unlikely to know the difference between a severely degraded woodland of exotic trees and a native oak forest; nor would they be aware of the differences in habitat value for the plants and animals that might inhabit the place. Our lack of familiarity with natural landscapes is compounded by the frequency with which people change their places of residence today, so that noticing what is being lost, and how rapidly loss is occurring, is even harder. Our actions might be very different if we knew our immediate environs well and comprehended the long-term consequences of our actions.

A premise of living sustainably is that those who live in a place should be the ones who know the most about it. While impacts to the landscape may have their roots all over the globe, restoration is fundamentally a local act. Its success depends to a large measure on how well those who live in a place understand the landscape they use. But many people today have learned more about animals on the African savanna than about wildlife in their neighborhood parks. Indeed, we are inclined to assume that only an expert could identify the trees and wildflowers of the local woodland, yet this is knowledge that all of us should come to possess once again. We will not be able to meet the environmental challenges of conserving biodiversity and sustaining viable habitats for ourselves as well as other species if we are not skilled observers of trends in our landscapes. Dramatic change is occurring at faster rates than at any time in recent history, making it increasingly important to record conditions in the local landscape, close at hand and continuously.

Environmental education has been evolving. The focus in the past was often very general, avoiding local issues that actually determine what kind of place we are creating for ourselves. What began as nature study now includes conservation and environmental awareness. Current school curricula typically include local natural and cultural history as well as general environmental topics. There is also a growing trend to study the regional ecosystem in the core curricula of schools and to incorporate real-world experiences and research into academic programming.

Students can be excellent scientists, and, through their schools, can bring a high degree of continuity to a restoration and monitoring program. Across the country, many schools are already carrying out important scientific research and participate actively in restoration. A project in Seaside, Oregon, that began as an effort to develop teacher training programs, has become a major nonprofit institution employing students and teachers alike to do research on local ecosystems. The research is funded by grants from the Environmental Protection Agency and other agencies. In one high school in Maine, the curriculum is centered on the Kennebec River and watershed as the primary vehicle to integrate the different disciplines of the academic program. In Florida, the Tampa Bay Watch High School Wetland Nursery Program operates five saltmarsh nurseries in area high schools, each of which presently produces over 5,000 planting propagules every six months for marsh restoration.

There is also a growing interest in creating schoolyard habitats that provide a tangible and integrated learning experience that can be easily coordinated with school curricula. Every school is an environment for learning, and every school environment gives us a lesson, whether intended or not. Daily contact with natural landscapes may provide even more important lessons about the way we live our lives than any curriculum can. Studies in England, for example, have demonstrated that violence and fighting in the schoolyard were reduced when the environment became more natural and complexly organized (Kaplan and Kaplan 1989). As E. D. Cheskey states in *Habitat Restoration: A Guide for Proactive Schools* (1993):

> Restoration is perhaps one of the most important actions that we can do as educators to prepare the future decision makers in our society to face the environmental and ecological challenges that await us. Habitat restoration at a manageable scale can contribute to a generation of nurturers and healers that have the knowledge, skills and values to address greater societal issues (3).

The entire community can become an environmental classroom. For example, the state of New Jersey, under past-Governor Tom Kean, developed a model Master Plan for Environmental Education that recognizes the need to address environmental literacy at every level of society. The task force developed recommendations for areas of basic competence at all grades and at every level of government and outlined the role that institutions and agencies could play as well.

In order to conduct restorative efforts that show long-term success, we must ensure that the participants/residents/users/caretakers come to understand the real nature of the landscape and its scientific underpinnings and have available a means to join with their fellow citizens to pursue new patterns of behavior. Studying ourselves and the places in which we live, forming a team of local and regional scientists from many disciplines to make the restoration of the landscape comprehensible to its managers, and involving community participants in monitoring projects through which strategies and policies are refined—all of these elements are essential.

Local Knowledge

Several researchers at Massachusetts Institute of Technology undertook a statistical survey to quantify environmental values in American culture, interviewing all kinds of people from all walks of life, from members of Earth First! to operators of dry-cleaning establishments. Perhaps not too surprisingly, all of those they interviewed valued the environment very highly, with little difference between people of different backgrounds. It was in the second half of the survey, which examined perceptions about the actual state of the environment, that the differences between people became apparent. Although all of us support the environment and want clean air and water and unspoiled places, we have very different information sources and perceptions about the environment that relate strongly to factors like politics, occupation, and income. We have little real knowledge or access to it and are exposed to a lot of pseudo-science (Kempton, Boster, and Hartley 1995).

There is much to unlearn. Outdated information, such as promotional material on plants that are invasive, still abounds in the agencies and literature. The media typically reduce environmental issues to pro and con controversies, often without full consideration of scientific data, often devaluing real information by implying that both sides are equally valid and that everyone and everything involved is in one or the other opposing court. By studying the real landscape ourselves, we can escape the landscape of preconceived notions and conflicting truths created by the media and conventional educational curricula.

The best way to convey real information to a community is to have them gather that information themselves, through monitoring both before and during restoration. One of the main reasons that monitoring by staff and community members themselves can be so important is that it gives credibility not only to the information but also to the actions taken in response. We are more likely to believe and understand our first-hand observations. Like students, volunteers in environmental groups and parks departments, if effectively coordinated and adequately trained, can perform accurate and valuable monitoring. John Reiger, a California restorationist, asked if he really thought that volunteers could do monitoring, simply replied, "Can they count?" Not only can community members provide an important, often overlooked, service to a project, but the in-depth knowledge that monitoring brings to those who live and work in the community also builds pride of place, ensuring in a real way that restoration efforts will continue long after the seed is sown.

The Role of Monitoring

No one yet knows how to achieve the goal of sustaining ecological health over time. Our potential for success is restricted by the extent of the damage already done as well as our lack of understanding about how landscape systems function. Some information, although at various scales, is generally available on geology, soils, watershed boundaries, climate, extent of tree cover, and general land use and topography. Often, however, there is no systematic information on plant and animal communities and those aspects of the landscape that are most likely to change over time. Restoration must have a scientific basis. We need good science, not simply anecdotal experience, to monitor conditions and the effects of our efforts, and we need good scientists to assist us.

Monitoring could be defined as simply obtaining accurate information and maintaining a long-term accessible record of it. Monitoring entails a purposeful and systematic observation and documentation of the landscape. It is an essential part of restoration and often the most neglected. Without monitoring we are simply making policy decisions and implementing management based on oversimplifications such as "remove exotics." We do not monitor our landscapes well, and this is a great loss because landscape histories, whenever they can be found, are very informative.

Monitoring is invaluable both before and after beginning restoration. The objective at any stage is usually to describe the ecosystem as fully as possible. Before beginning the project, we need to observe and document the living components of the landscape; we need to know what we have, and we need to know what we are losing. It is also important to document human actions on the landscape, both the causes and effects of misuse and our restorative efforts.

Monitoring is the only way that we can tailor the work we do to the specific conditions of the site. While the broader issues are defined by the larger landscape context, the details and particularities can be discovered only in the real place. The monitoring program serves as the voice of the landscape, telling its story, ensuring that our actions are ultimately justified or invalidated by what happens on the ground. Monitoring, because it also helps to overcome the hurdle of time, compensates for the difficulty of observing the landscape accurately in our own lifespans. When we are able to capture the image of the landscape through time, it is always compelling.

The success of our approach to restoration will depend on our ability to recognize the patterns that are inherent in the landscape rather than imposed by us. The goal is to learn how to recognize and optimize the natural regenerative processes of each place; to emulate nature, and to understand the dangers of trying to improve upon it. By documenting the patterns observable in a landscape we can recognize the patterns by which it designs itself when it is allowed to do so. Implicit in this concept of self-design is the process of continuous assessment of what is happening and continuous adaptation of the management program as trends are observed. At the Crosby Arboretum in Picayune, Mississippi, for example, ongoing vegetation mapping has revealed more subtle site distinctions than the topographic mapping in that very low relief terrain, informing the managers about microhabitats and special opportunities for displays.

Despite efforts to make it so, monitoring is not a truly objective process; our goals and preconceptions shape the outcome from the outset when we decide what to monitor. On actual construction projects, monitoring is often limited to the initial phases and does not continue through the design and construction phases. In reality, successful monitoring does not end with the collection of baseline data. At the same time, monitoring must be accessible so that we can use it and integrate the information.

Accessibility of information is now more possible than in the past. Previous monitoring efforts were hampered by the lack of an infrastructure for record-keeping or an archive for data and reports. Now, however, the ability to keep very large archives of information, usually referred to as "databases," is easier and cheaper. In the last decade, many county and local planning agencies have acquired computerized geographic information systems (GIS) and other hardware and software that can provide an unparalleled level of shared data and community education. It should be possible for any individual to go to a local planning office or work from a home computer to access up-to-date information on natural and cultural resources, planning and regulatory issues, current proposals and projects (both public and private) for review, and related individuals and organizations.

A necessary step in any restoration or management program is to establish an ongoing site database that continuously records and informs our actions. The database becomes the vehicle by which the site speaks to us. It helps to overcome preconceptions about the landscape and to compensate for the difficulty of observing the landscape accurately. All scales of observation are not only relevant in a database but crucial to our understanding of the ecosystems affecting our small corner of the world. It is quite all right for the site database to record different scales of observations, from effects of climatic and atmospheric changes at the global scale and the watershed to indigenous plant and animal communities at a more local scale. The ecosystem view goes beyond a site or watershed or biome and is related to both larger- and smaller-scale systems than the site in question, but it will also inform the activities you use in your restoration project. The site database is discussed in detail in Chapter 21, "Monitoring and Management."

A key objective is to prioritize management activities to ensure that the participants' efforts are effective. The most severe problems are readily identifiable, but we have avoided them for years. We must address them now despite how difficult it seems at the outset. If we document the magnitude and rate of deterioration, we can ensure that a chronic problem is recognized amid the daily crises that divert attention and funding in public agencies and institutions.

In many situations, you won't find detailed background information about your site, but a lack of existing baseline data should not be an excuse for inaction. Not taking action in a landscape in the face of ongoing disturbance is, in fact, a very powerful management decision that supports accelerating landscape decline. At the very least, we can work to reduce the worst and most chronic impacts on a landscape while we gather more detailed data. We can organize user groups and involve them in monitoring. If you engage people in observing how a landscape is used, issues such as trampling and off-trail use of wheeled vehicles will raise themselves. We can also start by controlling excess runoff from adjacent sites or restricting the planting of exotics. There is always somewhere to begin, a set of priority tasks, that a group of concerned individuals can acknowledge as important despite other areas of conflict or inadequate information.

Documentation has been crucial to the effort to restore a 90-acre woodland in Central Park known as the North Woods. The wake-up call came with mapping undertaken in the 1980s revealing that 25 percent of the ground was bare soil while another 25 percent was dominated by Japanese knotweed. This information convinced those who had previously thought the site would restore itself naturally; it underscored the need for active management and stimulated action by park managers. In 1992 the staff updated the tree canopy inventory of the North Woods undertaken in 1988 to evaluate how quickly the landscape was changing. During that four-year interval, more than 100 new exotic maple saplings had attained a size of 4 inches in diameter or greater in the 70-acre woodland, where 30 percent of the more-than-400 canopy trees greater than 6 inches in diameter already consisted of Norway and sycamore maple. The survey also revealed that seedlings of exotics were numerous and that native seedlings were absent altogether over much of the area, presenting a disturbing picture of what this forest would likely be like in the near future.

The Central Park Woodlands Advisory Board clearly recognized the potential threat of invasive tree species with this evidence. Despite their initial reluctance to remove any vegetation at all, they decided to initiate management with exotics control and to get staff and volunteers to monitor the effectiveness of the procedures and any changes in ground-layer vegetation. They began to remove sycamore maples up to 4 inches in diameter in the understory, and a year later young seedlings of native ash, sassafras, and tulip poplar appeared in their place. Alien maple seedlings still showed up, but at least the natives were present as well, which encouraged further intervention.

Pilot Projects and Field Trials

Even where unity of vision exists, agreeing upon or enforcing a plan may be impossible at the outset because there is no clear sense of what would result from any proposed action. Reaching agreement on the larger goal while disagreement remains about how to get there is common. Where information is inadequate, designing and implementing a pilot project or field trial that includes monitoring to determine actual impacts is appropriate and commonly employed where issues are controversial.

For example, if your group is disagreeing on trail management, and cyclists contend that their activities can be compatible with sustaining environmental values while others insist that degradation is an inevitable result, a field trial produces tangible evidence to support the decision-making process. A typical pilot project might initially span several years or more to assess trail erosion, use, and abuse, and to complete the "learning curve" of users so they comply with rules of the trail. Signage, trail condition monitoring, and trail repair are key components of any such program. Trail users should also be involved in trail maintenance as well as habitat restoration activities, such as removal of invasive exotic vegetation, closing outlaw trails, stabilizing the ground layer, repairing gullies, and undertaking woodland planting projects. By bringing together all the users and caretakers of the landscape, we can come to agree on performance standards and cultivate and communicate a true stewardship ethic to everyone in the community.

Before You Begin Your Project

Planning for Safety

An essential part of any planning process is to take adequate precautions to ensure the safety of project workers on the site. Before a near-miss or accident forces you to take it seriously, devote adequate time and resources to institute safety procedures and make sure everyone adheres to them.

- Precede every action on-site with a formal review of possible hazards and appropriate procedures and equipment.
- Insist on adherence to appropriate safety standards and precautions as well as protective clothing from eyewear to chain-saw aprons. Carefully read all labels and instructions. Take special precautions with older, modified, or oversized equipment lacking safety features.
- Be sure all project participants are aware of any potential site hazards and can recognize them. The most effective protection from poison ivy, for example, is avoidance, which requires identification.

Just as important, be sure to take precautions against Lyme disease, which is very common in the Northeast. Make every effort to avoid infections in the first place. Not everyone gets the telltale bull's-eye rash or notices a tick bite; other symptoms include fatigue, headache, joint stiffness, irritability, and forgetfulness. The peak months for deer ticks are May and June in the Northeast although ticks can be found at almost any time.

To protect against Lyme disease, make sure that all field workers wear protection whenever working in infested areas and follow these guidelines.

- Tuck your pant legs into your socks and tuck in your shirt to limit access to the skin.
- In areas of very high tick infestation, spray your outer layer of clothing with a tick repellent outside or in a well-ventilated area. Let it dry before putting it on.

Pyrethrum repellents can be used only on clothing that does not come into contact with the skin. If you use products with DEET, be sure you are aware of health concerns and precautions. Read all labels before using any such product.

- At the end of the day, put clothing into a dryer so the heat will kill any clinging ticks.
- Take a shower and check your body for ticks.

Ticks typically must feed on you for several hours before transmitting the Lyme disease bacteria.

Liability Concerns

Be sure you also address issues of liability before initiating any fieldwork. Consider putting together a team specifically to address issues of risk management and instituting regular safety reviews on the site. Addressing safety in a comprehensive manner reduces risk as well as the likelihood of being held responsible for something that is genuinely accidental, rather than due to carelessness, a lack of knowledge, or inadequate procedures. This effort will not only reduce actual risk but may also reduce your insurance premiums. An attorney can advise you on liability concerns and help you determine if you need to obtain waivers or releases from liability from project participants in case of accident or injury.

As you plan your project, also consider emergency preparedness, determining how you would communicate with emergency services when necessary. It is also advisable to have people on-site who are qualified in first-aid procedures.

The need to keep accurate written records cannot be overemphasized. Be sure you document your periodic safety reviews on-site, your follow-through to correct the condition, and any incidents. Thorough documentation is vital, especially if problems arise.

Complying with Legislation and Other Requirements

Some restoration projects that are carried out on private land and do not involve environmentally sensitive areas or wetlands and watercourses can be undertaken without any involvement with regulatory agencies and the permitting process. But most projects will sooner or later involve a regulated activity or place. Even where regulation is not a factor, a great deal of coordination with agencies and landowners is usually necessary except on the smallest sites.

Although the permitting process seems daunting, it is actually an excellent time to better integrate the agencies' role in regional restoration. The opportunities for shared learning also are important to your future success. Look upon this effort as a vital part of the restoration. Be sure to assess required licenses and construction-type permits as well as environmental policy concerns and compliance requirements. Any project using federal monies must comply with the 1970 National Environmental Policy Act (NEPA), which requires an Environmental Impact Statement (EIS). Start talking to state and local officials and familiarizing yourself with the overall regulatory requirements in your area. As you learn more about the impacts that are occurring to the site, you may also become involved in addressing gaps in the legislation and areas where the intent of the law could be better met than in current practice.

Get written permission for access to any land you will be using, even incidentally, from any and all landowners, institutions, and agencies. Obviously, clear agreements are all important. Be sure to spell out what is proposed and who is responsible for what, and any restrictions on use and access in as much detail as possible.

Associating Your Group with Another Organization

If yours is a fledgling organization, consider associating with another, more established one. In Philadelphia, for example, a group that wanted to start trail repairs became the Wissahickon Conservation Committee of the Friends of the Wissahickon, an existing nonprofit associated with the park. The group did not have to acquire separate nonprofit status and was covered under the Friend's liability policy, which allowed them to devote all their resources to the work on the ground. In return, the partnering organization, the Friends of the Wissahickon, received an infusion of new members and energy, thereby broadening and deepening its scope. One group of high school students solved their liability problems by affiliating with the Boy Scouts of America. They formed a group of Eagle Scouts who received badges for restoration projects.

Learning More

To learn more about restoration, consider joining the following organizations:

The Society for Ecological Restoration (SER). SER is an international organization of professionals and others committed to the repair and ecologically sensitive management of ecosystems. SER facilitates communication among restorationists; encourages research; promotes awareness of the value of restoration; contributes to public policy discussions; develops public support for restoration and restorative management; and recognizes those who have made outstanding contributions in the field of restoration. With your membership you can receive either or both of two journals: *Restoration and Management Notes,* a forum for the exchange of news, views, and information among ecologists, land reclamationists, managers of parks, preserves, and rights-of-way, naturalists, engineers, landscape architects, and other committed to the restoration and wise stewardship of plant and animal communities," and *Restoration Ecology: The Journal of the Society for Ecological Restoration,* in which local chapters focus on regional collaboration in restoration efforts. Contact: The Society for Ecological Restoration, 1207 Seminole Highway, Madison WI 53711; 608-262-9547.

The Natural Areas Association. The Natural Areas Association is a national nonprofit organization whose mission is to advance the preservation of natural diversity. The association works to inform, unite, and support persons engaged in identifying, protecting, managing, and studying natural areas and biological diversity. Membership includes the *Natural Areas Journal,* a quarterly publication offering articles relating to research or management of natural areas, parks, and rare species, land preservation, and theoretical approaches to natural areas work. Contact: The Natural Areas Association, P.O. Box 900, Chesterfield, MO 63006-0900.

The Society for Conservation Biology. The goal of the society is to help develop the scientific and technical means for the protection, maintenance, and restoration of life on this planet—its species, its ecological and evolutionary processes, and its particular and total environment. *Conservation Biology* is a joint publication of the Society for Conservation Biology and Blackwell Science, Inc. The society's offices are at the Department of Wildlife Ecology, University of Wisconsin, Madison, WI 53706, 608-263-9547. For subscription information contact Journals Subscription Department, Blackwell Science, Inc., 238 Main Street, Cambridge, MA 02142.

The Eastern Native Plant Alliance (ENPA). Membership is open to individuals who promote or demonstrate native plant conservation in the eastern United States or southeastern Canada and to individuals committed to serving as liaisons to an appropriate organization or audience. P.O. Box 6101, McLean, VA 12206.

CHAPTER 14

Restoration at the Macro Level

The smallest ecosystem known for certain is the biosphere.
—L. Margulis and D. Sagan

In order to identify strategies that will serve the interests of restoration, we need to understand and promote in our own landscapes the large-scale, or macro-level, processes that underlie the health of the landscape we seek to restore. With these as a foundation, we can begin to incorporate the restoration of the function, structure, and composition of natural systems into all aspects of management in our landscapes, whether it is a backyard or national park.

Ecosystem Health

Our approach to restoring landscapes is in part directed by how we define the health of an ecosystem. But what *is* ecosystem "health"? It depends. In its broadest sense, Aldo Leopold (1949) defined ecosystem health as "the capacity of the land for self-renewal." It is the exact opposite of ecosystem degradation, defined by James Karr of the University of Washington as "biotic impoverishment," or the "systematic reduction in the capacity of the earth to support living systems" (Karr 1992). For commercial purposes, forest managers define it as productivity: the vigor of individual trees; their resistance to stress, disease, and pests; and their rate of growth expressed in yields of logs ultimately. Those interested in sustaining old-growth forest look at a different set of factors, such as an uneven-aged and multilayered forest with many gaps, abundant ancient trees, and large amounts of dead wood—conditions that support many rare and specialist species.

Hammish Kimmins, with the Department of Forest Sciences at the University of British Columbia, proposes a definition that centers on sustaining all the components of the landscape and the processes that drive the system. He sets two conditions for ecosystem health in a forest landscape:

- The pattern of forest ages, ecological conditions, and seral stages is within, or close to, the range of these variables that is characteristic for that landscape.
- The scale, severity, pattern, and frequency of disturbance do not impair the landscape-level processes that are responsible for providing, at the overall landscape scale, a sustained supply of all the values that are desired from that landscape, or for the recovery of that landscape following a disturbance to a condition that once again provides these values (1996, 69).

This definition of health includes all stages of forest succession, including its early stages. Much concern today focuses on species of the old-growth forest, but many species of early successional landscapes are now in decline as well. Between 1880 and 1980, for example, the amount of forest nearly doubled in New Hampshire as almost 200,000 acres of farmland were abandoned, creating abundant but rapidly changing habitat and affecting proportions of species with each successional shift. During that time, twenty-six species of migratory birds suffered declines because of maturation of forest rather than loss of interior. When cottontails declined with the loss of open fields, so did the bobcats. In the early fifties, more than 300 bobcats were killed by hunters annually; by 1969, only 36 (Litvaitis 1994).

Kimmins's definition also recognizes that some external disturbances are inevitable and normal for all ecosystems. It is not stress per se that is the problem, but rather stresses that deviate substantially from the historic range of disturbances to which systems are adapted and upon which natural processes and species depend. The components of natural habitats found in a region have co-evolved over millennia and produced an intricate system of checks and balances. While there is no protection against dramatic change, the overall vulnerability of a healthy complex community to natural stresses is reduced in such a situation, simply because it is complex. In the face of serious disruption of historic patterns, these interrelational bonds are rigid and thus brittle. The systems collape under severe stress.

Sustaining natural processes includes restoring natural hydrologic regimes, fire patterns, predator-prey relationships, migratory routes, crucial levels of both isolation and connectivity, and scales of large natural areas adequate to support diverse communities. Many small-scale sites cannot replace forest expanses or support the same species that a single large landscape can. There is no substitute for a migratory path and no alternative to a few missing links in the food chain. Only in large natural areas where natural phenomena prevail to the maximum extent feasible can we even attempt to sustain the full breadth of regional biodiversity.

Pattern and Scale

A basic principle of modern design is that form follows function. In forest restoration, pattern, an aspect of form, determines function. The continuing disintegration of the larger, complex patterns in the landscape is accompanied by an associated loss of function, only some of which can ever be replaced.

The largest-scale patterns are being lost with the greatest finality. The breaking up of the last unbroken expanses is occurring at every scale, from inroads into the Amazon and Congo to the proposed development of a one-of-a-kind 17,500-acre tract just 40 miles from midtown Manhattan, where you can see bobcat and bear as well as the tops of the World Trade Towers.

It is not the forest that exists today that we seek to preserve, but its underlying health—the potential of the landscape to sustain that forest over time in all its diversity. To sustain the richness of our forests we must address the full spectrum of landscapes, from the most pristine remaining wildlands to the developed urban corridor. Instead of saving isolated sanctuaries, our restoration projects need to play a role in promoting an interrelated, healthy ecosystem, one that includes at least these components:

- **Bioreserves**—regional-scale ecosystem reserves where biodiversity is sustained under largely natural conditions
- **Managed reserves**—smaller-scale sites, where the full range of indigenous species and subspecies can be researched, propagated, established, and managed under field conditions and in natural landscapes
- **Bioregional landscaping**—native plant communities in unprotected areas that are the matrix in which natural reserves occur and regional-scale nurseries that propagate and distribute plant species characteristic of the native communities of each area

If we have learned anything about the environment, it is that all landscapes are connected and interdependent. Each is affected by the management of all the others. And we have also learned that the survival of each level in the system depends on all the others.

Bioreserves

The richness of all landscapes is ultimately dependent on the quality of wilderness. There is no substitute for wilderness as a reserve for biological diversity. All actions that are applied to landscapes must be built on the foundation of preserving diversity, since it is the necessary premise for every other objective of restoration. No amount of management within a reserve can make up for a site that is simply too small or too intruded upon to support the conservative species, large-scale systems, or the full range of local diversity. The overriding goal of all restoration activities is to work to expand our largest reserves, all of which are experiencing growing losses of indigenous species. We must also re-assess and consolidate the more isolated pockets we now call nature preserves.

Our concept of a wilderness reserve must also include the highest trophic levels, that is, the reintroduction of the largest predators in order to reestablish the processes that shaped those communities. At the same time, primary intrusions into wildlands, such as roads and other infrastructure, should be reevaluated and removed wherever possible. "Roadlessness" has become one of the most important attributes of wilderness.

The Great Smoky Mountains National Park, for example, in the center of biodiversity for the eastern deciduous forest, is only just over a half million acres and is laced with roads. By McCloskey and Spalding's (1990) definition, this forest does not qualify as wilderness. In the Adirondacks area is an even larger area of protected forest, but that also is even more fragmented by inhold-ings of private land, roads, and settlements. Therefore, that area does not qualify as wilderness either, since it is not all one piece. The only other sizable expanse of forest in the East is in the state of Maine, but this forest is almost en-tirely private fragmented parcels, unprotected in any way from further depre-dation. Recently, developers acquired well over 100,000 acres of large forest tracts in New York and New England despite efforts of local environmentalists to protect the lands. Re-creation of conditions suitable for sustaining the wilderness may still be possible in these areas and yet they are not receiving the protection they so critically need.

One major hurdle to restoring wilderness in the eastern United States is our failure to recognize the importance of forest. Some people dismiss the eastern forest systems as already gone or too fragmented and modified to merit major acquisition and preservation. We must think very creatively about reestab-lishing wilderness. At the same time that sprawl is consuming tens of thousands of acres, much of Appalachia and New England is experiencing a loss of popu-lation. Recreation, rather than resource extraction, is the growing industry, and open space has different values in such a context. Because large scale is integral to protecting biological richness and scenic quality, regional-scale, or eco-system-scale, planning and management is required. Many people who testi-fied in favor of open space protection at hearings held by the U.S. Forest Service concerning forest protection in New York and New England wanted to establish a new national park in the Northeast. Others wanted to expand state, federal, or private programs like the Pinelands in New Jersey and the Adirondacks, in New York.

To sustain landscape-scale processes, then, we need to both protect far more land and reclaim the wasted ground of our conventional landscapes. The con-figuration of open space has rarely been determined by what is needed to sus-tain natural processes; yet without adequate land devoted to forests and fields, the conditions necessary to sustain native plants and wildlife cannot be met.

One problem is that most zoning plans provide little room for open space. Once buildout has occurred, our parks and protected areas are likely to be far too small in area to sustain much of what we hope can survive there. In the state of Rhode Island, for example, 95.5 percent of all land under local government jurisdiction is zoned for development.

In addition to size, we must also protect age and old growth, especially primary forest that has never been clearcut and soil that has never been plowed. Significant primary or original forest still remains, even in the Northeast, and it should be at the core of the bioreserve system.

Until recently, we greatly underestimated the amount of old forest left. Mary Byrd Davis (1993), cofounder of Wild Earth, inventoried up to 1.5 million acres of old growth in the Northeast, including 36,000 acres in Maine, 27,000 in Pennsylvania, and 15,000 in Maryland. New York State supports the most old growth in the Northeast, with over 260,000 acres, predominantly in Adirondack State Park, where the Five Ponds Wilderness alone accounts for nearly 50,000 contiguous acres. Michigan's Boundary Waters area supports the greatest expanse of old growth in the East, over 376,000 acres.

Today, nearly 2 million acres of old growth are estimated to remain east of the Mississippi. That is only 5 percent of the total of 380,330,000 acres of remaining eastern forest, and these places represent the greatest opportunity to preserve the most historically important forests. They are especially important in light of the failure of other landscapes to completely recover ground-layer species of both plants and animals even centuries after logging and subsequent replanting (Duffy and Meier 1992). Further surveys under way include Will Blozan's effort to update Stupka's big tree register. Blozan has found an abundance of huge and ancient trees, such as a red maple 135 feet in height and 23 feet in circumference and a hemlock over 500 years old but only 20 inches in diameter. Already the Forest Service has entered into a cooperative agreement with The Nature Conservancy in the southern and eastern regions to identify likely old growth that will be evaluated and left uncut if it meets old-growth criteria (Seaton 1996). These evaluations and inventories will help identify landscapes that may harbor the most threatened of landscape types as well as serve as models for forest restoration.

What will it take to have true, successful bioreserves? Ecologists Reed Noss and Allen Cooperrider (1994) suggest that 25 to 75 percent of the total land area in each bioregion, including connecting corridors and buffer zones, will be needed. Twenty-five years ago, Eugene Odum estimated that 40 percent of the state of Georgia would need to be kept as natural or seminatural area to sustain the full range of indigenous biodiversity (cited in Zahner 1996). These numbers may seem large until you look at some encouraging facts. The city of New York, for example, contains 27,000 acres of city-owned parkland, over 13 percent of the land area of this densely populated city. Much of the parkland was acquired in this century, including 1,500 acres in 1994 and 1995, with thousands more proposed for acquisition. While only a third of this land is devoted to natural areas, it makes clear what we can accomplish even in an area perceived to be already developed. Imagine what is possible if we take action now in places that are still largely rural or in areas where population is decreasing.

It has been 500-plus years since Europeans discovered the continents that they named the Americas and began to introduce novel land-use practices and rapidly increasing populations. The landscapes of that time have been replaced. The trackless wilderness was virtually eliminated, and today, only 3 percent of the United States is protected under statutory wilderness designation, not enough to support even half the nation's species over time.

Natural Systems within the Developed Landscape

There is no wilderness in the heart of the city, but the nature of our cities affects the heart of the wilderness. The management of regional- and community-scale open space, both public and private, is as important to the regional effort to sustain biodiversity as are larger bioreserves. Conservationists sometimes discuss the merits of a single large preserve or several smaller ones but in fact both scales and configurations are necessary. The successful management of parks, greenways, and other fragmented natural landscapes both affects and is affected by the quality of wilderness. Through restorative efforts we add and sustain vital linkages, including the system of natural drainage corridors and shorelines, the connectivity of forest systems, flyways and spawning journeys, ridgelines, and a host of other physical landscape connections. By connecting landscapes, these corridors and middle-scale reserves can also serve as buffer areas for larger core reserves, provide places for species to disperse after a catastrophe, and allow for migrations.

Looking at the larger context of a site is very helpful when it is seriously disturbed and a return to the historic community is not at all feasible. The Fresh Kills Landfill, for example, now rises hundreds of feet on top of what was once a tidal marsh. Short of moving all the garbage from the world's largest landfill, restoration of the tidal marsh is not likely here. Yet restoration is a very important factor to those who hope to manage the site as city parkland one day. The 2,500 acres of the Fresh Kills Landfill are strategically located between the area's largest tidal marsh preserve and the rest of the Staten Island Greenbelt. Coupled with the landfill, the greenbelt will eventually consist of over 8,500 acres of protected land connecting the forest to the sea in a rich mosaic of natural and managed seminatural landscapes across a fabric of dense residential and commercial development.

The Smallest Habitats
No site is too small to merit restoration even though its possibilities as a habitat are limited. For example, even the garden, which expresses our most intimate

relationship with nature and mirrors our attitudes toward the larger planet, can be a place that contributes toward sustaining and restoring indigenous communities. Indeed, some of the smallest habitats may be the most influential in changing our perspective on how we deal with all landscapes. To note just one example, many of the native trees planted in Central Park's North Woods restoration project were grown on a Manhattan balcony.

The Wildness Continuum

Landscapes occur on a continuum of relative disturbance, from the nearly pristine to those entirely modified from their presettlement condition. A gradient of human impacts on the landscape also existed well before European settlement from the activities of indigenous people. When land uses at the most intense end of the gradient occur adjacent to the wildest places, there are usually severe conflicts. Forests near New York City are typically under greater stress than those located at greater distance from the city on the urban-to-rural gradient. Similarly, a woodland next to a farm is usually less disturbed than one next to a housing development. Natural areas typically benefit from a gradual transition to mitigate the impacts of development. A meadow strip adjacent to a path or lawn, for instance, minimizes the disjunct quality of new barriers and edges in the landscape. A buffer area around a protected natural area also creates a gradient of wildness as well as lower intensity of use that helps reduce outside impacts.

This principle applies at every scale. That means the quality of the transitional area is important to the health of the natural area at the core. For example, an exotics-ridden woodland surrounding a patch of old-growth forest may cause more problems than a belt of agricultural land that does not harbor forest pests.

Corridor Connections

If we make some room for wildness in all places, we can ultimately reconnect a larger, continuous natural network. Restoring continuity to forest systems is a difficult challenge in the Northeast, but it is also certainly true that we cannot sustain biodiversity in the fragmented forest in any other way. As long as natural areas are reduced to smaller and smaller islands, they will be about progressive extinctions. Restoration at the landscape scale begins with recognizing, preserving, and connecting the remaining natural lands. The first step toward this goal is a network of corridors of less-disturbed areas connecting forest remnants. But this is just an interim, not a long-term, solution. The resulting forest landscape would be a network of threads connecting a patchwork of remnants, one consisting of many edges. We must make the journey from natural areas that are islands in a sea of urbanization to patches of urban development in a matrix of forest. To do so, we must have a continuous forest matrix where more intensive land uses are managed as gaps in the forest.

Increasing connectivity in the landscape has inherent risks, but the health of the ecosystem is impossible without it. A newly planted forest overwhelmed by weedy exotics and connecting two natural areas could serve to introduce invaders to them, but without linkages to other natural areas each of the forests will be diminished over time. Certain regional sites, almost always the largest, typically account for most surplus populations, but every local habitat will vary in populations from time to time. Even a site that is frequently a sink for some species may occasionally become the source for recovery. No single site can adequately protect local biodiversity; rather, it takes many sites to provide sufficient backup populations. This redundancy is an important strategy in the real world of cyclical and sudden change, and is a necessary strategy for sustaining native plants and animals into the future. It is the persistent regional population that is the most important to the long-term survival of a species. Many sites, small as well as large, are necessary to ensure adequate overlap.

Guidelines for Protecting and Reestablishing Landscape-Scale Linkages

All the concepts described previously are reflected in the following guidelines. Although they are simple and somewhat obvious goals, they will not be easy to implement because they represent such a significant shift from current patterns and trends.

• *Do not displace or significantly modify any relatively healthy natural system.*
Restoration should focus on redevelopment more than new development. In planning, give every consideration to locating facilities for human activities such as roads, parking lots, or buildings outside the boundaries of remnant natural areas. Where such disturbances already exist as intrusions to otherwise-intact natural areas, the goal should be to relocate them to already-disturbed and less-sensitive lands. A good rule of thumb is to not build on raw land.

• *Minimize disturbance to any remnant natural area.*
In addition to road building and facilities development, other disturbances, such as the clearance of vegetation, should be carefully avoided. Where disturbance does occur, it should be confined to the edges, which are already subject to increased disturbance and more accessible to management.

• *Do not compromise natural and cultural resources that cannot be re-created elsewhere, such as rock formations, mature forests, stream corridors, and historic sites, by activities that threaten their character and preservation.*
These places are defined by their locations. Even if disturbed, they represent opportunities for restoration that cannot be duplicated in other places.

• *Protect and expand remaining natural wetlands, streams, and recharge or storage areas. Reestablish natural drainage patterns and hydrologic regimes where they have been disrupted.*

Hydrologic corridors comprise a vital natural infrastructure. Natural patterns and processes are the most efficient and should be relied upon to the greatest extent feasible rather than preempted by costly built infrastructure for stormwater management and control of pollutants. That may sound obvious, but it is remarkable how often proposals are made to relocate a stream, for example, as if its location was not related to its function.

• *Reassess and modify the management of existing and potential connecting corridors such as rights-of-way to favor indigenous species.*

At present, rights-of-way are typically seams of disturbance when in fact we have the opportunity to manage them as corridors in a network of habitat. Because these sites are already disturbed and require active management, they offer ideal opportunities to experiment with new techniques, such as made soils and innovative exotics control, without putting more intact areas at risk.

• *Protect critical forest linkages and reestablish missing links.*

Continuity and configuration are as important to smaller-scale forests as to larger areas. Islands of habitat isolated from surrounding natural systems experience a decline in native species diversity and are less adaptable to stress over time. Management and proposed alterations in current forest patterns should always enhance, rather than reduce, the continuity of natural habitats. Projects should effect a decrease in habitat fragmentation by including provisions to link and expand existing natural areas, especially along hydrologic corridors, steep slopes, and ridges. This is equally important at all scales of the landscape.

• *Manage corridors and buffer zones to favor indigenous species.*

Without management, continuous edges can become seams of disturbance rather than vital linkages. These edges are the landscapes where we live and work; they comprise our gardens, parks, greenways, cemeteries, and golf courses. These lands create a landscape matrix that can either support or diminish native plant and animal communities.

CHAPTER 15

Managing with Succession

Much research in succession has focused on the journey of secondary succession from field to forest. For obvious reasons, researching phenomena that might be occurring over centuries rather than decades has been more difficult. But now that more than three centuries of intensive forest modification have passed, the longer-term consequences—such as the loss of the ground layer and plant and animal species—are becoming more apparent. Even more profound changes are occurring over even longer timeframes.

Many questions arise from this line of thinking, such as: Will the precise set of conditions that are needed to regenerate a majestic white cedar or tamarack swamp occur while their propagules persist? Will deer convert once-forested landscapes to brushland? Will the consequences of changing fire patterns further diminish the prevalence of oaks? Oaks have disappeared from many areas they once dominated, replaced by maple and beech where fire is fully suppressed. In the coastal-plain pine barrens landscapes, however, where fire was often too frequent for oaks, oaks are benefiting, at the expense of pine, from fire suppression.

For these reasons many in the restoration field view the reestablishment of historic successional patterns as the primary prerequisite for sustaining fragile landscapes. Indeed, restoration is sometimes referred to as "directed succession." For some perspective on using succession as the context for management and managing succession as a primary strategy, let's look at its principles and processes in more detail.

Mechanisms of Succession

Frederic Clements (1916) is usually referred to as the father of succession theory, and his description, though outdated, still prevails as the popular perception of succession. He described succession as occurring in sequences, called "seres," of discrete phases, each creating conditions necessary for the succeeding stage of the sere. In his view, succession is a kind of ordered unfolding of the potential landscape based on climate. He made little reference, for example, to the influence of the management by indigenous peoples or the impacts of current human uses.

Frank Egler (1954), also a successional theorist, recognized that the patterns of community succession are influenced by the presence of plants that have a head start because their seeds and other propagules are already in place and develop more quickly than those that must travel to that place. His theories of initial floristic composition and relay floristics held that the determining factor driving the unique patterns of succession in each place was the serial expression of those species that were established at the very outset because they determine much of the site context from that point on—by casting shade, by preempting space that would otherwise support other plants.

These patterns are easily observed in nature and have shifted with the growing prevalence of exotics. A meadow abandoned in central New Jersey today shows very different successional patterns than meadows in the same vicinity in 1940. Multiflora rose, exotic honeysuckles, and a variety of exotic invaders unfortunately make meadow management today more difficult than in the past.

Any change in the structure of the landscape will influence to some degree the future patterns of the landscape. Something as simple as planting a pole in the middle of a field will have consequences that will shape the future structure of the landscape. By providing a perch, a pole attracts birds and the plant propagules they bring with them. Succession is given a jump-start in such a place. Decades later, the largest and oldest trees in the forest will be found there, visible evidence of the early development of woody species in that spot.

Less well understood is the mechanism of "allelopathy," chemical inhibition between species and individuals, in successional change. Virtually all plants exude chemicals from their roots, in their decaying leaves, and in the moisture conveyed across their leaves. These substances may affect other plants. The difficulty of growing other plants with black walnut and in the barren zone within the dripline of many eucalypts is legendary, but allelopathic processes also affect forest succession as well as the compatibility of agricultural crops, determining which plants can be grown together. Allelopathy influences which species can become established and persist in any given landscape. Perennial herbaceous plants such as goldenrod and aster actually inhibit the germination of tulip poplar and pine, thereby prolonging the open phase of a meadow and postponing their eventual demise under forest cover (Brown and Roti 1963). The Norway maple has been observed to be strongly allelopathic. Its fallen leaves can retard growth of many species, so that the ground beneath its thick canopy is often markedly bare. The tree-of-heaven may be even more potent. Extracts from its leaves and twigs have been shown experimentally to act as an effective herbicide; commercial applications are currently being evaluated (Heisey 1996).

Jack McCormick, one of the most important plant ecologists in the field of allelopathy, noted in 1968:

> [S]ufficient evidence now is available to suggest that allelopathy may be of great significance in succession or lack of succession. In addition, autotoxicity may . . . be involved. Autotoxicity may result in inhibition of seedlings in the vicinity of mature plants of the parent species and, thus, may play a major role in population density control. It may also produce reduction of stature of annual plants growing on a site formerly occupied by dense stands of the same species, fairy ring configurations of perennial clones, and other effects (32–33).

Managing allelopathy may offer intriguing opportunities to influence succession but will require many field trials. For example, mowing and mulching of perennial grasses with their own freshly cut stalks may stunt their growth and be a component of an effective control strategy for some grasses such as common reed and miscanthus. Cutting and removing the end-of-season vegetative growth of goldenrod and aster may reduce their ability to retard the germination of woody species. Removing the herbaceous cover from a mown meadow may accelerate the return of woody species.

Arresting, Accelerating, or Altering the Course of Succession

All management is an effort to direct the process of natural change, to arrest it by mowing and applying herbicides or to accelerate it by planting and irrigating. Here in the East, where rainfall is more abundant and evenly distributed seasonally than in the great open plains in the central portions of the United States, almost all grasslands, whether they contain native prairie species or exotics, require management to keep trees from developing and overwhelming the herbaceous species, eventually returning the landscape to forest.

Human uses more frequently arrest the speed of succession than accelerate or promote it; consequently, the landscapes around us are getting younger at the same time that they become more fragmented. Cropland is held forever in a juvenile state, sustained primarily by resown annual species. Wildflower meadows need to be mown or weeded of woody volunteers at least periodically or a forest will grow. Lawn is perhaps the most extreme example of this pattern. In a landscape that has the potential to support a great and multilayered forest it takes a lot of labor, energy, fertilizer, and pesticide to prevent that forest from growing. The heavy watering and chemical use that lawn requires in turn means excessive runoff, much of which is highly enriched or contaminated, with adverse impacts on remnant natural areas. The resulting landscape performs none of the functions of forest and supports almost none of its creatures.

All the landscapes around us represent some moment in time on the successional continuum, whether we are performing conventional or ecological management, and it is from the perspective of that particular moment that we want to manage each landscape. The stage of succession will determine where the landscape is trending. An important and simple distinction is between those landscapes that are still primarily herbaceous and those that are predominantly woody.

The most common activities directed toward managing succession in an effort to control the direction of change include adding species or taking them away as well as manipulating environmental conditions by burning or flooding by impoundment. In the case of forest restoration projects currently under way in Central Park, all of these actions are taken simultaneously. Propagules of all native species mentioned in the historic record are added to the site because there are no longer any natural sources of these propagules. At the same time, exotics are removed to reduce competition with natives.

Successional Landscape Types

In order to manage with succession, you will often find it useful to identify the seres, or sequences of successional stages, that occur on your site as well as determine as clearly as you can what your successional goals are. For example, are you trying to arrest succession at a meadow stage or revert to a grassland from a vine scrubland? Are you trying to accelerate succession to create conditions more favorable to species of the primary forest?

Although the phases associated with succession are more gradational than discrete, simplifying your management review by generally summarizing the characteristic stages in your landscape can be helpful. The following description is simply a guideline for what is typical of many temperate second-growth woodlands. In your own region, you should describe its special and characteristic successional changes, specific plants, and plant–soil associations. In most environments there are numerous gradients—for example, the gradients between moist and dry soils or between highly disturbed and more natural sites.

Herbaceous Landscapes

Herbaceous landscapes are typically small in scale, ranging from short turf underfoot to knee-high or even shoulder-high and taller grasses and wildflowers. When herbaceous landscapes are effectively stabilized, you cannot see the ground because the herbaceous cover is so dense. Natural herbaceous landscapes typically occur after a disturbance such as fire or vegetation clearance and are relatively temporary. They may persist where the land is too wet or too rocky to support trees. Modified herbaceous landscapes such as flower gardens and lawns require maintenance to arrest natural succession, or they too will succeed to woody species.

We have many opportunities to enrich the range of herbaceous landscapes around us by taking a cue from natural succession. By confining lawn to the smallest area necessary, we can then explore alternatives to turf such as a "greensward," a taller grass lawn in which grasses and short wildflowers are intermixed, which was typical in the last century. Tall-grass and wildflower meadows, prairies as they are also called, are another possibility. In addition to reducing the amount of lawn, we can restore many waste places and weedy landscapes that are merely the result of poor maintenance but that with management could become quite rich. Today there is renewed interest in meadows, both as scenery and as a cost-effective alternative to turf. Maintaining a meadow costs on average only $50 per acre a year, less than one-tenth the cost of lawn. But we have lost the historic art of meadow management in our recent attachment to lawn and must relearn the techniques for tending meadows.

Savannas and Woody Oldfields

Along the continuum from grassland to forest, there is a long period of transition as the maturing woody plants gradually assume predominance over the herbaceous vegetation. Woody plants, of course, have been there since the outset, less visible amid the tall grasses and wildflowers, but they take on a dominant character as they mature. Such landscapes are known as "woody oldfields" or "savannas."

John Curtis of the University of Wisconsin Arboretum, where he initiated the first prairie restoration, defined "savanna" as grassland with up to 50 percent forest canopy (Curtis 1959). In the Midwest the savanna landscape occurs as a transition between the limits of the eastern forest's range and the drier grasslands of the western prairies and the plains. Indigenous peoples historically burned the midwestern savanna in the fall, creating and sustaining the mix of grassland and woodland of the savanna landscape type.

In the East, the term "savanna" is less commonly used, but a savanna-like condition of about half trees and half herbaceous species frequently occurs for a period of time on the successional continuum. Another name for this landscape is "woody oldfield" (literally "old field," that is, abandoned pasture or cropland). In an oldfield abandoned from agriculture, for example, this transitional semiwoody condition might last up to several decades.

At the outset a woody oldfield is dotted with small trees in an herbaceous matrix, but soon the growing size and number of trees creates a closed canopy in patches at first and then more generally. Although historically less extensive than those of the Midwest, savanna or woody oldfield landscapes in the Eastern forest support species that are adapted to this temporal niche, including many now-rare species as well as the once far less abundant white-tailed deer. With management, you can sustain such a landscape indefinitely, providing an alternative to turf and specimen trees where an open parkland character is desired.

Woodlands and Forests

Woodlands and forests are landscapes that are predominantly trees and other woody plants where the canopy cover ranges from greater than 50 percent to completely closed. They have a structure composed of several layers of woody plants, from canopy and understory trees to shrubs and a ground layer. Herbaceous cover is typically relatively limited. In a forest you can often see the ground, which is usually covered by a layer of leaf litter. If, however, the soil is bare, that is usually an indication of disturbance.

The term "forest" usually refers to multilayered, multiaged, and multispecied wooded landscape; "woodland" generally is more open and younger in age and structure than a forest. Forests and woodlands are stabilized by the multilayered system of roots of multilayered vegetation as well as a multilayered ground topped with a litter of leaves and other forest debris. The ground-layer vegetation may be patchy or only seasonally apparent. Soil stability does not depend on very dense vegetation, stem to stem, as in an herbaceous landscape, but rather on overlapping systems of stories, canopies, and a complex root structure linked with ectomycorrhizal fungi.

Restoration of woodlands and forests may entail the addition or removal of either selected layers of the woodland or specific species, as well as changes to environmental conditions, to accelerate or retard the processes of natural succession. The management team may focus on reestablishing a more continuous canopy where forest cover is too spotty or on enhancing the diversity of patches depending on the current condition of the landscape and their overall goals.

Within the forest matrix are a variety of patches that reflect different environmental conditions such as soil chemistry and available moisture as well as past history such as blowdowns or active browsing. Through management, you can mimic a variety of different environmental conditions to favor selected species or meet specific functional requirements of the landscape. Forest glade, forest edge, and light forest are special management options that may be suitable to places on a site.

The term "glade" is often used to describe an area within a forest where vegetation is less dense and thereby provides an opportunity for a special plant species to thrive and creates a place of pause for the walker. The conditions favoring a glade may occur naturally in the landscape for a variety of reasons. The light forest or glade character often results when a layer of the forest landscape is missing—for example, where a large canopy gap has fostered a grove of locusts growing in a ring or where the largest and oldest tree in the forest has limited the other vegetation beneath its sway to an ancient carpet of woodland ephemerals. These places draw the eye, and hence the walker, to stop for a moment and observe the interplay of light and shade, texture and pattern.

Public concern for personal security is an issue in many woodlands where public use is high, such as in urban areas, but you can increase visibility in the landscape in selected areas without resorting to wholesale clearance of all low-level vegetation. Limited reduction of vegetation to create a light forest condition is one way. Management for this purpose consists of a careful editing of selected plants in the forest layers to enhance a sense of an opening within the enclosed setting. By keeping a managed glade small in area, rarely more than the width of a single tree canopy, you can retain the sense of forest and canopy closure above. Or you can extend light forest over a slightly larger area, such as the margins of a path or at the juncture of two or more paths, without creating a large gap. Forest edges are generally the most managed places in the landscape because so many sources of disturbance stem from them. A good rule of thumb is to "seal the edges" by developing a dense, multilayered edge to minimize the distance that edge impacts extend into the forest. This strategy is especially important where new edges have been created and where control of exotics is necessary. In many other cases, the goal will be to accelerate succession to re-create conditions that are more similar to the ancient forest of dead wood, pitted terrain, and deep humus.

Working with Succession

Keeping a perspective on change is vital in any management program. Each moment in time on the successional continuum presents different opportunities and constraints. Management affects the range of successional landscapes in each place and the potential of future landscapes.

Roadside landscape maintenance exemplifies how management can be completely in conflict with the processes of succession in native communities. For instance, wherever a roadside is not managed as turf, the usual practice is to periodically clear away the woody vegetation. This habit of intermittent brush clearance along roadsides guarantees that the development of woody species cannot proceed naturally. At the same time, the landscape is mown too infrequently to sustain grasses and wildflowers. Weedy species tolerant of frequent perturbation, such as Japanese knotweed, are the inevitable result.

A key difference between maintenance and management is that maintenance, like housework, must be repeated endlessly. Mowing, for instance, is continually needed: we don't expect a lawn to mature and then need less mowing. However, we can often expect management to diminish or at least shift focus as the restored landscape becomes more established and as disturbance is better controlled. A wildflower meadow may always require some mowing or burning or other woody control, but with time our efforts may be directed more to species enhancement than to elimination of exotics such as multiflora rose. In Central Park, after only a few years of invasives management, workers were able to shift focus from removing Norway and sycamore maple to restoring microorganisms necessary to soil quality.

In conventional management, the assumption is that if you are diligent at maintenance, you can be sure of the results. We can expect a green golf-course-type lawn if we mimic the management practices recommended for such turf. The assumptions for restoration management are somewhat different, even though restoration does involve a lot of repetitive activity. The major distinction is that a restoration program should change continuously in response to changing site conditions. The more rote the activities, the more you should look for negative impacts that have been taken for granted, such as overriding the seasonal and longer-term natural variations that are necessary for the recruitment of highly specialized species.

Changing Management Direction

Frequent changes in management direction can be stressful to a landscape and almost always favor opportunistic species. The more frequent the disturbance, the more simplified living systems are likely to become. Deciding to release a meadow—for example, letting it go to woody species such as multiflora rose without any follow-up management—will achieve a more degraded meadow. Similarly, clearing exotics in a forest and failing to restabilize, replant, and monitor as necessary may only invite further invasion in the new openings. We need to learn to examine any proposed changes in management direction closely and be able to defend them based upon monitoring. It is axiomatic that it is easier to sustain existing native communities than it is to reestablish them. It is also easier to achieve a minor shift in the composition of a native community than it is to go from an exotic-dominated to a largely native landscape.

More active and timely intervention is a goal of restoration, but also keep in mind that it is easy to overreact. Removing exotics too aggressively can be as damaging to relic native populations as it is to the target invasives. Planting more than is necessary for restabilization may overwhelm natural seeding and can obscure natural patterns of reproduction. Restoration is a form of research that requires an openness to the new and previously unseen. Observation of interrelationships and influences in the landscape is the manager's most important skill. All too often we ignore the mechanisms that are driving change in the landscape rather than focusing intently upon them.

The Role of Wildfire and Prescribed Burning

All forests have a fire history, whether the cycles are long or short, catastrophic or light. Land-use and settlement patterns have dramatically altered successional fire-related patterns that persisted for millennia. For example, infrequent fires in forest expanses sometimes created large gaps that provided opportunities for species like white cedar. These conditions necessary for regeneration do not occur in fragments. Some areas, such as sandy coastal landscapes with mixed oak and pine, often burned more frequently in the past.

Today, more mesic woodlands typically burn either more frequently or not at all. Urban grasslands may burn yearly or even more often, while more rural ones may never burn but are regularly mowed instead. Without these historic patterns of fire, succession has been dramatically altered.

Attitudes toward fire illustrate how management perspectives are shifting in the direction of restoration. We now recognize that fire suppression alters plant succession and results in far more devastating conflagrations over time because excessive amounts of accumulated fuels build up. Practices that foster more natural fire and hydrologic cycles also make natural habitats more resistant to invasion by some diseases and pests, both introduced and naturally occurring. The restoration of native shrub and understory layers and the reestablishment of a natural fire regime appear to be more effective in controlling pine bark beetle, for example, than vigorous eradication and clearance efforts. Maintenance of natural fire cycles in the Pine Barrens of New Jersey is required to sustain the fire-dependent or fire-tolerant species in this landscape.

In addition to providing beneficial effects that are difficult to replicate in other ways, fire is a relatively low cost management tool. Meadow managers generally acknowledge that fire is the fastest route to a more diverse wildflower meadow or a denser stand of native grasses. Management approaches range from virtually natural wildfire cycles on large reserves to highly controlled burns. Even on small sites we can try to replicate processes that without management would not occur except in very large reserves. Ultimately, the goal of all fire management in restoration should be to restore as close to a natural fire cycle as possible given the landscape context of the site, whether it is rural, natural, or bordering a residential area.

In larger forested tracts, the reestablishment of more natural fire regimes should be given high priority. A pulsed cycle, with varying intervals of burn that closely mimic natural fire patterns, rather than regular intervals, is preferable. Even on smaller sites, management can seek to reestablish a more natural and varied pattern.

At the Connecticut College Arboretum, William Niering, an ecologist and former director of the arboretum, conducted long-term studies on prescribed burning in both woodlands and grasslands. In one forest tract, annual burning was undertaken for the first few years to reduce fuel accumulation. That also served to renew many herbaceous species and resulted in increased germination of oaks and other species with previously poor recruitment. At this point, Niering withheld fire to allow the new saplings to develop to a sufficient size that they would not be killed by a light ground fire. Spotted wintergreen and other woodland ephemerals spread notably across the newly opened ground.

In most built-up areas, light controlled burns limited to restricted sites will be all that is feasible, but you can still mimic a natural fire. Where smoke must be severely curtailed, burning in very small scale patches can help to create greater variability in landscape patterns and to replicate the conditions of greater fire intensity.

The restorationist should consider using fire more extensively as a management tool for many tasks that are now accomplished by more destructive methods, while maintaining stable cover and reducing long-term management costs. For example, where visibility in woodland areas is currently managed by mowing, consider using fire as an alternative to encourage herbaceous species and compact, low shrub growth while sustaining sight lines in limited areas. Native grasses, such as broomsedge, are long-lived perennials invigorated by periodic burning, which at the same time controls new tree growth that would suppress the grasses. Prescribed burning may also reduce the impacts of allelopathy because the aerial parts of the plants are consumed, rather than left behind as a mulch, which is usually the case with mowing. In turn, it may also extend the reign of the plants already established on a site by reducing the rate at which their litter accumulates and the autotoxicity resulting from allelopathy. In this way, fire also may limit the introduction of some exotics into the landscape.

Use of controlled burning entails a sometimes-lengthy permit process because of air-quality concerns. Local air management regulations vary and must be reviewed carefully. Of particular concern are possible hazards of smoke reducing visibility on roadways, for example. Requesting the assistance of local fire departments gradually develops a wider network of personnel trained in controlled-burning techniques who can help you ensure adequate safety and compliance with insurance requirements.

Adequate control of a prescribed burn is dependent on a system of firebreaks, which may include natural features, such as streams and wetlands, or built features, such as roadways and lawn areas. Where new firebreaks are required, careful review is mandatory to ensure that the firebreak does not serve as a route for disturbance, disrupt natural drainage, or otherwise adversely impact the landscape.

The Nature Conservancy has made a major commitment to the restoration of fire in the landscape and conducts training workshops on prescribed burning and maintains burning programs in many of its reserves. The National Park Service also recognizes the importance of fire as a management tool and has developed a manual on wildland fire management for the control of wildfires and the management of prescribed and research burns (National Fire Policy Review Team 1989). Major topics included in this manual are the identification of roles and responsibilities of governmental agencies, procedures for fire analysis, documentation, and staff training and distribution, as well as guidelines for wildfire control and management objectives for prescribed burning.

Controlled burning has optimal results when it reflects seasonality and historic patterns of wildfire as well as fire management practices by indigenous people, which in the past were complementary to and shaped wildfire patterns. As a practical matter, though, managers today schedule most prescribed burning for the winter and early spring for safety reasons, especially where fire frequency has been suppressed and fuel loads on the ground are excessive. In general, spring burning favors woody species that sprout. Fall burning tends to favor herbaceous plants, as well as nonsprouting species, like many conifers, that require a specialized seedbed. Fall fire is also more typical under natural regimens, but it may not be feasible because of fuel loads. Where the prolonged absence of fire has resulted in overgrowth, conservative burning practices are often necessary, at least initially, because of excessive fuel accumulations. The prolonged absence of fire in some areas affects the kinds of techniques that can be utilized. Burning downslope with firebreaks and backburns is often necessary until fuel loads no longer pose unnecessary risks to more aggressive burning techniques.

Burning upslope is swifter and more characteristic of wildfire. Headfires that move with the wind are cooler at the surface but hotter at about 18 inches and above and therefore more effective for killing shrubs and trees while burning deadwood and woody debris. Backfires—that is, fires that move against the wind—are slower and hotter at the surface and reduce damage to the overstory while consuming large fuel loads. Fires that burn sideways to the wind are safer than headfire but not as slow as backfire. Ultimately, the objective is to re-create historic cycles and scales of wildfire as well as burning like that of indigenous people. Much more research is needed to document historic people's management of fire in the landscape as well as their other practices, since the habitats of the lands they once managed are those we seek to sustain and restore.

Initiating a fire management program is not easy. Agency personnel and government guidelines are often so cautious that a natural pattern of fire is difficult to achieve. Volunteers have often had to reinvent the techniques of fire management as they go along. Nonetheless, the effort is well worth pursuing.

Monitoring Succession

A key component of the effort to sustain and restore the natural landscape is the monitoring of population trends in order to evaluate succession in that landscape. By monitoring population change you can also evaluate the success of your efforts to reintroduce or protect selected species or to foster specific communities. Even a decision to take minimal action, such as to allow for natural recovery after a disturbance, should be supported by ongoing monitoring. As McCormick (1968) noted, "The most valuable evidence from which to develop successional theory is obtained from direct long-term studies of vegetation on specific areas." The forest of the future is growing today, and change will be expressed over decades and centuries.

CHAPTER 16

Restoring Natural Water Systems

One of the most changed processes in the landscape is the water cycle. Watersheds in the East that once absorbed over 85 percent of the precipitation they received to support aquifers and stream flow now shed that amount or more as runoff because watersheds have been deforested and cleared for fields and farms or converted to pavement, turf, and rooftop. In parts of the arid West, where water rights are bought and sold as property, water tables have dropped as much as 300 feet or more in this century alone. Yet we continue to render the landscape less permeable as well as increase consumption and waste of water resources.

Our greatest mistake in managing water in the past has been to focus on single issues like flood control or erosion rather than the larger system. We now understand, however, that every action in the natural world has many ramifications throughout the whole system. This is nowhere so immediately evident as when we are dealing with water. One property owner fills his land to protect it from flooding, only to divert those floodwaters to a neighbor's property that then becomes more vulnerable to flooding. Another bulkheads his land to save a shoreline and dramatically increases the shoreline losses of an unarmored adjacent site. All these actions as well as their consequences have severe negative impacts to the forest, from subtle shifts in population composition because of altered hydrologic regimes in the landscape or the sudden blowout of a stream channel and the vegetation along its banks from uncontrolled stormwater.

Therefore, we must restore the patterns of water that have shaped the landscape if we wish to restore the forest. The first step is to look at water as a system. The second, which follows from the first, is to recognize that we do not know how and cannot afford to replace the functions and services of natural water systems. A great flaw in the regulatory system as well as conventional planning and construction is the failure to recognize and protect the whole spectrum of values of a natural system.

When we recognize the way water works, we will no longer separate aspects about the *quantity* of water, such as flooding and erosion, from those related to the *quality* of water, such as biological pollutant reduction. They are ultimately related. The riparian, or floodplain, forest, for example, not only slows water's velocity, but it also reduces and attenuates the quantity of runoff by holding water in the floodplain on its vast surface area of leaves and limbs as well as through transpiration. The forest also treats runoff by filtering pollutants, using nutrients, breaking down petrochemicals, and reducing oxygen demand. When that forest is cleared, we lose the biological treatment system that was in the vegetation. When a stream corridor is exchanged for a pipe under fill, the worth of what we are losing—habitat, flood control, and water purification—often far exceeds the value of what we are trading for it, a few additional square feet for construction in a landscape that is already overbuilt.

The biggest problem is almost always that too little land has been left to natural patterns. Stream courses, wetlands, and floodplains in many urban and suburban areas have been constricted and buried by filling, paving, and structures until all that remains of them are steep-sided stream channels with no real floodplain corridor left at all. Because the conventional engineering perspective has defined stream function almost solely by the capacity to convey water, our regulations are gradually producing streams that are more like pipes or flumes. There is inadequate regulatory protection for a natural stream corridor, which has intrinsic value beyond the functions of flood storage and drainage: as plant and animal habitat and for pollutant reduction through biologic treatment.

Wetlands protection has been under continuous assault since its inception, no less so now. With over half our nation's wetlands gone, there are still strong incentives for landholders to encroach upon, fill, and drain them. Forested wetlands are especially vulnerable, and nearly all would have been removed from even the most minimal protection under the recently considered definitions of wetlands at the federal level.

Opportunities to save wetlands are frequently overlooked, even when we can do so at minimal cost. Little more than one-quarter of the nation's wetlands are publicly owned, yet virtually none of the wetlands acquired by the Resolution Trust Corporation after the savings and loan failures were given long-term protection. All, including extensive forested wetlands, were instead turned back over to private entities, unprotected, often at a fraction of market value.

Floodplains are no less imperiled. Despite public awareness about the problems of building in floodplains and along shorelines and despite demonstrable cost savings to municipalities and agencies in limiting development, construction funded by both public and private investment continues in flood-prone areas, most of which should be returned to forested conditions. Ultimately, the public bears most of the costs of floodplain mismanagement. At the same time, no actions are taken to reduce actual risk, making it all the more likely that property and casualty losses will be great, whether the insurer, the taxpayer, or the individual pays the bill.

The problems of flooding, tornados, and other climate-related catastrophes will only worsen with continued global warming. Because there is more heat, there is more energy driving global climate patterns, and the consequence is increased storminess. The effects of the extent of warming already have been evident for years and take a severe toll on natural environments as well as built infrastructures and property. Remnant forests along stream corridors that already experience severe erosion from runoff will be further stressed by more extreme storm events.

Resource protection benefits the whole community, but not directly the owners of natural resources such as wetlands, for example, who have little or no financial incentive to preserve land. We have invested a great deal of public money to build an engineered infrastructure at the same time we have been dismantling the natural infrastructure. This approach has failed miserably, and at great cost. The built infrastructure is already crumbling from lack of maintenance while consuming ever more tax dollars for expansion. The shifting hydrologic conditions created have degraded natural environments in uplands as well as lowlands. It is now time to turn the tables, to provide for water retention and infiltration within the fabric of development, rather than shunting everything to the stream. Rather than investing in trapezoidal ditches and costly concrete bunkers that provide no infiltrational reduction in contamination, we can meet those goals while protecting forest and reconnecting stream corridors and wetlands. We can reestablish a landscape rich in intermittently wet places. Such an approach recognizes the fundamental irreplaceability of the natural water infrastructure, its biotic richness, and its cost effectiveness over an engineered alternative. We cannot sustain natural ecosystems without restoring the historic hydrologic conditions that depend on this natural infrastructure.

Strategies for Restoring Hydrologic Patterns

We must take every opportunity to enlarge natural stream corridors wherever possible, and certainly in any restoration project. There is no doubt about the value of protecting a stretch of stream corridor and nothing but rewards to reap. Unlike the removal of exotics, which may in some places prove futile or even counterproductive, sustaining the basic patterns of natural drainage in the landscape is vital to every endeavor related to ecosystem restoration.

A substantial proportion of the stream corridor must be kept in a relatively natural state to retain all its functions: pollutant reduction, flood retention, fish hatchery, plant and animal habitat. Ideally, the entire floodplain as well as a buffer strip should be protected in addition to a sufficiently wide corridor to protect the most demanding local species, which will vary with context and scale. In an urban area, the amount sought often depends upon opportunity: how much is left unprotected, how much could be acquired, and how much is restorable. On a smaller scale, restoration may include taking a stream out of a pipe, reestablishing a natural meandering channel, and planting trees along the stream corridor as well as optimizing the opportunities for infiltration and storage in the uplands (Figures 16.1 and 16.2).

Reestablishing more natural patterns of drainage ultimately requires modifications in the management of landscapes throughout whole watersheds, relying on multiple solutions at many different points, rather than a single cure-all at the point of discharge. Once the problem has reached the stream, it is too late. The solutions lie throughout the watershed, and entail actions that range from protecting and reforesting stream corridors to eliminating lawn and replacing it with meadow or woodland.

This guidebook recommends six strategies to restore and manage natural water cycles in the landscape. An essential part of any restoration project—before work begins and throughout the project—is to plan and evaluate any actions in the context of these strategies:

- **The water budget as a management tool**—Monitoring the whole system
- **Restoring historic drainage patterns**—Sustaining and reestablishing natural drainage patterns
- **Stream restoration**—Sustaining and reestablishing natural water courses
- **Changing landscape type**—Changing vegetation to reduce runoff
- **Alternatives to turf**—Creating open landscapes that have fewer negative impacts
- **Dispersed, small-scale storage and filtration**—Restoring and creating new wetlands

You can apply these approaches at a variety of scales throughout the watershed in and beyond your project area. All of the strategies described in the remainder of the chapter respond to the consequences of development patterns in the temperate region and the resulting need to foster more natural hydrologic regimes by reducing the impermeability of the land surface. While correcting the problems of water management goes well beyond the scope of this guidebook, following the recommendations in this chapter will work to mimic natural patterns of rainfall and runoff. If the goals of retaining runoff and maximizing recharge are vigorously pursued in the uplands, the problems of

Figure 16.1. The swales along this roadside edge had been "stabilized" with dumped stone although the arrowhead plants of the earlier wetlands still showed through.

Figure 16.2. The wetland swales reestablished along the roadside treat a significant amount of contamination in stormwater before it reaches natural wetlands.

erosion, sedimentation, flooding, and falling water tables and base flows in lowland areas will become more manageable.

The Water Budget

Our efforts to sustainably manage water resources over time will depend to a great degree on how well we understand the workings of the larger system. Many of our worst decisions, in hindsight, reveal that we really did not know, or ask, where the water would go. Ignoring the small but high-frequency storm has resulted in a landscape of gouged stream channels that now run like torrents after even a minor storm. Any decision about drainage, flood control, or water treatment will have implications for other components of the system, implications that can only be appreciated when we have a model of the whole system.

Anyone concerned with water management must understand the comprehensive water budget of the environment, not just specific factors such as design floods. Instead of single aspects of water management, we must seek to quantify the interacting components of the system.

A "water budget" for the landscape attempts to model the whole system over time and includes all major water inputs and outputs. Typical inputs include surface flow from streams and creeks plus other runoff onto the watershed or site, subsurface inflow from groundwater moving into the landscape, and precipitation (rain or snow). Typical outputs include surface outflow, which includes water moving in streams, creeks, ditches, and pipes; subsurface outflow (which is moving as groundwater); and evaporation and transpiration (from plants). In most landscapes, this natural pattern has been overlaid and modified by an engineered system of channels, pipes, and other outflows that must also be included in the model. Irrigation may be another input that also has to be incorporated into the model.

You can model a site or a region. The best models relate multiple scales of information for cross-comparisons. Do, for example, the quantifications used for local-scale basin design and stormwater management regulatory criteria confirm or correspond well to regional-scale data and models? If performed on an ongoing basis, such an evaluation allows us to integrate the monitoring that is undertaken for individual sites as well as data gathered by agencies and nonprofit organizations to create a regional database or model of how water really works in that area and that grows more accurate over time. The process of refining the water budget also helps to identify the work that is accomplished by natural systems.

Historical Drainage Patterns

Sustaining and restoring landscapes depends on restoring and protecting historic patterns of drainage. One reason is that the reproductive cycles of organisms are timed to take advantage of seasonal variations in drainage regimes, such as flooding cycles. For example, the great galleries of cottonwood trees that overhang the bluffs of a flashy prairie stream germinate in the mineral soil exposed by rushing floodwaters. With effective flood control, however, this natural disturbance has been tamed, so all too often there are no young streamside cottonwood forests to replace the few ancient ones that remain. Here in the East, on the other hand, established floodplain vegetation has been disturbed by erosion and collapsing streambanks. In both cases, the conditions that fostered the regeneration of the historic landscape no longer occur.

Everywhere in our urban and suburban communities, the landscape infrastructure has been left out of our codes and ordinances. We need "green" site standards as well as "green" building standards to achieve real sustainability. Effective watershed protection requires a more comprehensive approach than largely piecemeal development projects and a few big and costly government projects to implement a water management plan. In a more environmental approach, design standards for engineers would seek levels of soil recharge and rainfall infiltration comparable to those that occurred under historic forest conditions as well as effective pollutant reductions. Implicit in this approach is a recognition that the pattern of the natural stream network is an irreplaceable ecosystem infrastructure, that sustaining natural systems within the fabric of development will depend upon our preservation of the natural infrastructure that supports them.

This change in perspective need not be deferred until some future time. Today, using computer modeling and new software, we can tailor an engineering design to the specific conditions of a particular stream and watershed, rather than to a more simplistic regulatory standard. By monitoring and modeling historic and future patterns of drainage we can use the capacity and requirements of the receiving stream, not the anticipated volumes of runoff generated by a site plan, to define engineering standards for a site.

To recover natural hydrologic conditions, at whatever scale we are working, means learning how the land once looked and functioned. Each site has an inherent geometry that reflects its role in the historic water patterns of the regional natural system. Each site has its place in the larger context. It is not a microcosm or a statistic of the larger region; it is an element of the region, and its role in that region needs to be recognized and preserved.

Stream Restoration

A primary rule in stream restoration is not to destabilize the banks in the name of stabilizing them. Vegetation is often cleared along a stream bank for any number of reasons: to reduce the weight on unstable slopes (by removing heavy trees, for instance), to allow easier visual inspection of stream bank conditions, to allow for unobstructed views, or to make it neat.

Trees are often removed in the belief that their roots create de facto "pipes" that convey water and undermine the stream embankment. The trees, however, may be all that is holding the slope. In fact, dead roots of cleared trees serve as pipes far more effectively than living roots, an effect that persists for decades, so clearing a slope or embankment that is wooded generally creates more problems than it solves.

The desire to improve trout habitat is often the primary incentive for stream restoration efforts. Trout, in addition to a steady flow of water, require food, shelter, and cover, as well as areas suitable for reproduction. Sustaining a forested stream corridor is an important component toward meeting these requirements. A forested stream corridor improves water quality and helps retain stormwater and control sediment. Sediment control is vital because many of the insects that are trout's preferred food, as well as the trout themselves, need well-aerated, silt-free areas to feed and reproduce. Shaded streams are also cooler and provide better cover. The forest's leaf fall is an important source of detritus, the base of the stream's food chain, as well as the insects and other creatures that fall from branches overhanging the water.

Most of our problems with streams stem from fighting their inherently dynamic quality, of resisting their tendencies to meander, to flood, and to change course. George Parmentier, a noted stream restorer, pioneered an approach to stream restoration that utilizes those very processes to harness the stream's own energy to accelerate its return to its former condition: he re-creates natural stream meanders in channelized and/or stormwater-gouged streams by using the large mass of felled tree trunks to deflect the flow of water. By placing a sequence of trunks and root masses to veer the water more deliberately back and forth, he re-creates a longer stream channel and with it many of the stream's former characteristics. His approach mimics natural processes of the past that shaped the character of streams, the major storms that toppled trees as well as the actions of beavers.

This same kind of approach is used extensively for large-scale stream restoration on previously logged land as well as in urban areas. Clearcutting creates many of the same problems for a stream that development does, leaving banks gouged by increased runoff and a sharply incised channel. The use of large tree trunks, large rocks, and other natural debris in the channel, rather than old tires and rebars, can help check flows and allow sediment deposition to gradually bring the stream back in touch with its banks. These debris dams shift over time under natural conditions and should not, in general, be anchored in place except temporarily.

Deflectors are structures that jut into the channel and change the direction of stream flow. They can be made of stone, logs, or reinforced soil. A live boom is a bioengineered deflector made of soil and living plants set in the stream's path. Deflectors may be used in a variety of situations. For example, where there are two channels, neither with adequate flow for good fish habitat, a deflector system may be used to block one of them to increase flow in the other. Deflectors of logs, half-logs, or boulders may be used to improve cover for fish.

Low (12- to 18-inch) dams that create a plunge pool below them also serve as good cover and aerate the water.

Stream corridor protection overlaps many different programs and jurisdictions. Restoration as a primary objective can serve to integrate existing efforts better than the current, more fragmented regulatory approach. Already-existing agricultural policies and farm programs can assist in the effort. Over 30 million acres have been enrolled in the federal Conservation Reserve Program since 1985, which takes marginal land out of cultivation. Researchers estimate that this land left fallow or reforested may be able to remove from the atmosphere nearly half of the carbon released by agriculture annually in the United States. In addition, this program can reduce erosion by one-quarter billion tons (Gebhart et al. 1994) by taking highly erodible land out of cultivation.

Simply using grass to filter sediment and break down nutrients is an important landscape strategy for managing non-point sources of pollution, which could provide added habitat opportunities as well. Wide swaths of natural forest are even more effective than grasses at reducing pollutant loads. Although ineffective at reducing or compensating for wetland loss, the Swamp-buster Program, another federal effort similar to the Conservation Reserve Program, has enormous potential, with some modification, to be a major vehicle for habitat restoration. Both programs need to be reinforced with acquisition through purchase and increased attention to habitat restoration and management, not simply set-asides of acreage (McElfish and Adler 1990).

Changing Landscape Type

The most straightforward way to restore natural hydrologic patterns is to reforest as much of the watershed as possible. When you look for the areas that might be returned to forest, you will find many and varied opportunities, from roadside margins and vacant lands to closed landfills and mines and waste sites to cemeteries, schoolyards, and front or backyards. While beginning to work on marginally useful sites, though, we should also concentrate on enlarging and connecting existing areas of forest and forest corridors along stream channels.

Reforesting efforts like these make a significant difference. The Natural Resources Conservation Service recently evaluated about 18 miles of the Winooski River in Vermont that was stabilized and revegetated through soil bioengineering techniques from 1937 to 1941. Although the original plantings have long since disappeared, a forest has grown up along these once-barren stream banks where techniques that are underutilized today (and described in Part III) were used to reforest the corridor.

All scales are important and add to the cumulative effect. Boy Scouts, as part of the Cache River Restoration in Illinois, planted over 2,000 pounds of acorns in a single season. Along the Middle Run Creek, volunteers from the Delaware Nature Education Society planted 700 trees in a single day as part of an ongoing reforestation program.

These projects offer the opportunity to expand habitat for many creatures favoring successional habitats, which all too often have been diminished by the modern wave of suburban development. When there was more forest and logging was more common, a region almost always contained a patch at every stage of succession, providing habitat, for example, for vireos and redstarts. Today, large-scale sites such as corporate office parks, golf courses, and historic estates could convert substantial areas of lawn to tall-grass meadow and savanna and young woodland to provide successional landscapes for species that depend on them.

We still need to develop effective installation techniques for native woodland communities. When erosion- and sediment-control regulations were enacted, rapid establishment of vegetative cover was required. The emphasis in research, as well as practice, shifted from trees to grasses (Smith 1980). This trend is changing once again with greater interest in and funding for reforestation. The next step is not to plant exotics, such as Norway maple, but rather native species.

Alternatives to Turf

It is remarkable how ubiquitous turf has become, considering how expensive it is, not only environmentally but also economically. On average, suburban landowners use substantially more herbicides and pesticides than farmers. In fact, ten times the herbicide is used in landscape maintenance than by all agriculture, most of it unregulated. The average acre of well-tended lawn costs approximately $500 to $5,000 a year to maintain. It is cheap to install, however, and the technology is widespread and nearly universally accepted. The lion's share of the grasses-related research dollar, both public and private, has gone to turf varieties while far less has been spent on experimentation with native grasses.

Simply altering the management of landscapes, such as replanting forest and reducing the area of turf, substantially reduces runoff. Turf areas, when even only gently sloped, shed water nearly as rapidly as pavement. A lawn is not only a preempted natural area but often means a lot of fertilizer, lime, herbicides, and oil and gasoline from mowers polluting nearby natural landscapes by way of runoff. Reducing the extent of lawn is one of the easiest and most effective ways of addressing stormwater management while increasing the area of potential wildlife habitat (Figures 16.3 and 16.4).

Changing our attitudes toward lawn may be one of the most important things we can do to save the eastern forest. Not only could we clear far less land for development purposes and save more forest in the first place, but we could also help protect the quality of those areas of forest that are preserved.

It is very common in parks and other landscapes for lawn to extend right up to the edge of a steep wooded hillside. By maintaining a simple margin of

Figure 16.3. The seeded lawn and substantial amounts of soil washed away completely after one heavy rain on this steep, newly deforested lot.

Figure 16.4. Wildflowers, ferns, and many seedling-size trees and shrubs now create a successional meadow that retards runoff and will gradually replace the lost understory layers of the landscape.

meadow as a buffer between turf and forest, you can reduce runoff, and, in turn, reduce erosion and sedimentation. A strip of meadow paralleling the forest edge also helps reduce the likelihood of off-trail trampling, which turf often invites. Turf trails through the meadow edge can be used to direct pedestrian movement to the legitimate trails instead of desire-line or outlaw trails that further concentrate stormwater. By avoiding the use of lawn grasses, you can more readily maintain the areas under groves of trees or large specimens in a parkland setting as tall grass and greatly reduce the amount of hand labor required for trimming and the damage done to mature trees by mowers. Too many patches of turf end up being mown over and over simply because no one challenged that notion or made the effort to establish some other landscape in their place.

The primary obstacle to the use of native grasslands and meadows is the lack of simple, effective establishment and maintenance methods. There are two major problems. First, the Northeast has no local, large-scale seed production nurseries; and second, in most instances regulatory agencies do not require or recommend native species. Part of the problem stems from the regulatory requirement to provide erosion control rapidly. Because native grasses, like woody species, typically develop slowly at the outset, agencies have often overlooked them in favor of faster-germinating, early season exotics. Energy and money went into methods that promoted quick stabilization, not longer-term solutions.

It is certainly time to master new techniques. The establishment of native grasslands, tolerant of extreme soil conditions, is a more appropriate objective for many sites. We do not suggest that short-term erosion control should be sacrificed to these longer-range goals, but rather that we need to develop more comprehensive and cost-effective techniques for establishing native plant communities that also meet the needs of immediate erosion and sedimentation control.

Dispersed, Small-Scale Storage

Remarkably many opportunities exist for very small scale storage of water. Intermittently wet places that provided special habitat were once common in the natural landscape, but we have graded away many of these low spots and wet places, which now persist only as puddles in asphalt and water in backed-up storm drains. Small swales (shallow drainage channels) and ephemeral wetlands provide a high degree of storage as well as pollutant reduction, and they can be repeated throughout the landscape. Small impoundments are probably the simplest to construct and can be designed as richly vegetated amenities. Replicating the condition where a fallen log or large branch, for example, slows the flow of water from a small section of hillside helps stabilize the landscape. Each such structure contributes to re-creating more historic drainage patterns.

The value of isolated wetlands that are not necessarily connected by a visible stream network but are rather supported by groundwater or seasonal surface water is generally not recognized today. These include vernal pools, which are usually small (less than 2,000 square feet), shallow (less than 1 foot), and typically contain standing water, usually in the spring, for at least two months but are dry at least part of the year. And because they are small and temporary, they support important breeding populations of amphibians and invertebrates, including fairy shrimp, the burrowing mole salamanders, and wood frogs, that do not survive in larger, permanent pools that must be shared with predators such as fish and bullfrogs. We have lost countless of these ephemeral pools, but they can often be partially re-created where woodland habitat still remains as long as there is a relatively steady supply of groundwater or sediment-free surface water. In addition, restoration efforts need to direct attention to the creation of new wetland and lowland habitats—not just as mitigation, but to compensate for past losses (Figures 16.5 and 16.6). Unfortunately, landowners often avoid wetlands establishment for stormwater management purposes in the fear that the land would then be subject to wetlands regulations. New ordinances that encourage wetland creation as well as the retrofit of existing stormwater facilities to create more natural habitat are obviously needed.

New wetlands can also help provide adequate levels of water treatment to allow for groundwater recharge or direct discharge without further contamination or stress to existing systems and remnant habitats. This allows water to recharge ground and surface aquifers to restore historic patterns rather than requiring transport for treatment and nonlocal discharge. Created wetlands are being used successfully to treat acid mine drainage and sewage effluent and are increasingly incorporated into stormwater detention facilities as concern about non-point-source pollution grows. The National Academy of Science has set a goal of a net increase of 10 million acres in wetlands nationally as part of a broad strategy to restore surface-water quality (National Research Council 1992). The restoration of these wetlands will contribute greatly toward restoring historic drainage regimes on which many plant and animal communities depend.

Using created wetlands for biological treatment of contaminated water, whether wastewater or stormwater, is often perceived to be too land-intensive and may be ruled out where available land is in short supply. The reality, however, is that large amounts of land are simply wasted virtually everywhere in the landscape. Lawn is but one obvious example; the typical highway illustrates other overlooked opportunities. The roadside landscape, both upland and disturbed lowlands, can and should be the first treatment system for road runoff. Instead it is typically turf, often mown well beyond what might be required for visibility, further contributing to the rate of runoff. Runoff is collected and conveyed by grading and pipes, rather than retained for use in vegetated landscapes (Figure 16.7).

Figure 16.5. After regrading, this once steep-sided pond now includes a smaller, upstream pond for sediment collection and shallow banks providing ample habitat for replanted native communities.

Figure 16.6. Rootstocks of wetland species grew profusely the first season beneath the thin canopies of the future swamp forest that was planted at the water's edge.

Figure 16.7. Broad shallow swales with native grasses between parking bays allow for stormwater infiltration and pollutant reduction.

Rather than conserving and treating water at roadsides, we often turn them into deserts. For example, here in the East, where rainfall is abundant, xeriscaping has recently become popular for coping with roadsides and other desiccated landscapes created by modern construction approaches. Xeriscaping is a planting technique best applied in areas of low rainfall because it employs species adapted to those conditions to minimize or eliminate irrigation. Sometimes it takes an extreme form where bare gravel and no plants are used with repeated herbiciding to maintain the barren state. In the East, a more appropriate method than using plants adapted to dry conditions or gravel would be to stop creating dry conditions and to see water, even road-contaminated stormwater, as a potential resource to be conserved and recycled in the larger system (Figures 16.8–16.10).

Figure 16.8. The retopping of this parking lot provided the opportunity to remove all unnecessary paving with no reduction in the total number of spaces.

Figure 16.9. Infiltration trenches, dug at the margin of the parking lot and filled with the uniform-size stones, collect and filter runoff from this lot.

Figure 16.10. The larger soil area affords the opportunity for a substantially improved environment for street trees.

CHAPTER 17

Soil as a Living System

What most struck the woodland manager of Central Park on a visit to the Adirondacks was a forest floor so soft he could plunge his hand into it. The ground was visibly alive and completely different from the dead concretized soil of the urban forest in Central Park. Soil wears its problems on the surface. Where trampling or high rates of decomposition prevail, the litter layer and topsoil are entirely absent. Until recently, the annual leaf fall in the woodlands of Central Park typically did not accumulate or even persist from one year to the next. With no litter layer, there was no nursery for the next generation of the forest.

Nearly a decade of woodland management is rebuilding the ground layer in Central Park's woodlands at the north end of the park. The site is becoming increasingly stabilized as erosion is controlled and bare areas are replanted. The many small saplings and seedlings that were planted or that volunteered after exotics removal help to hold the ground. During the icebound 1993–94 winter season, some remains of autumn's leaves persisted under the blanket of ice until spring. That was a turning point for the woodlands. The following winter was unusually mild, and by spring 1995 there was a relatively continuous litter layer.

In time, the organic litter on the forest floor will create humus, an organic soil horizon. Within it, most of the life of soil occurs. As organic matter is continually broken down into humus, it becomes incorporated into the mineral layers of the ground surface to build topsoil.

Soils are forming all the time, and like vegetation, integrate and express all of the ecosystem's processes. Soil is a reflection of climate, parent material, topography, vegetation, and time. The layers of soil tell a more recent history than the rocks beneath.

The soil's abiotic, or nonliving, factors are generally the primary focus of conventional soil assessment. Much of our thinking in the past was oriented toward an "ideal" soil model that balanced sand, silt, clay, pore space, moisture, minerals, and organic matter. These standards determined whether a native soil was judged poor or good, and where soils did not conform to the ideal, soil amendments were used to modify texture, acidity, fertility, or other characteristics. Many early mitigation, stabilization, and restoration projects suffered from this agricultural/horticultural approach. Standard soil specifications, for example, call for routine topsoil stripping, fertilizing, and liming even though many disturbed or made soils are already less acid than in their native condition because of the repeated addition of lime by means of concrete rubble and urban dust. Most regulations related to development sites, highways, landfills, and abandoned mines require from 3 to 6 inches of topsoil spread over new soil surfaces before revegetating. That topsoil comes from somewhere, so the restoration of one site frequently means the destruction of another. We need more research on alternatives to topsoil, especially those that reuse waste materials appropriately to amend local soils and that avoid environmentally costly products such as fertilizers and peat. Even where topsoil has been stockpiled on a site before construction, the living organisms it contains die within days.

The Soil Food Web

A food web is the structure of relations among the organisms within an ecosystem based on what each consumes. Primary producers consume water, minerals, carbon dioxide, and a few other things to produce organic matter, which is consumed by most of the rest of the creatures that are, in turn, consumed by still others. Some organisms have very specialized food requirements while others feed quite omnivorously.

Both soil and water are media in which plants and animals live and grow. And in a very real way, both are living systems. One of the most important contributions to the history of water management occurred with a shift in perspective that originated with Ruth Patrick and others. When one views water as a living system, its quality is measured by the richness of its biota, instead of physical and chemical factors such as flood levels or biological oxygen demand. Its biological components are a defining measure of health that reflect a more complex array of factors. This same kind of revolution is happening in our perception of soils.

In 1968 Ruth Patrick wrote about aquatic food webs:

> The various pathways in the food web and the various types of interrelationships of species to each other are two of the most promising avenues of research.
>
> Most food webs are composed of at least four stages[;] . . . the stages tend to be few because so much energy is lost between stages. . . . Since the stages of the food web are few, . . . diversity is expressed by many species forming each stage or level in the food web.

This strategy of many species at each trophic level has developed a food web of many pathways which seems to give stability to the system . . . [W]e see there are many food webs within systematic groups as well as between groups. It should also be pointed out that the size and the rate of reproduction vary considerably in each of the major systematic groups. These types of variability in food chain, size of organisms, and reproductive rates help to ensure the maintenance of the various systematic groups, and in turn preserve the trophic stages of the food web of the whole community (1968, 37).

Soil ecosystems are strikingly similar. Like aquatic systems, they have a great deal of redundancy. Very simple systems with simple food webs can be drastically altered by the appearance or disappearance of one or a few species. In more complex systems there may be multiple ways in which energy flows through the food web. Thus the more complex systems are said to have redundancy and are not so dramatically changed when a few species change. Many soil components even lie dormant until favorable conditions occur. The full soil structure is not required for most basic soil functions.

Rather than focusing simply on the nonliving aspects of soil, restoration should enhance its living components, primarily bacteria, fungi, and microfauna. Most of the work of forming humus is done by plant roots and by animal life in the soil, which depend on a permeable soil crust, stratified soil layers, and appropriate amounts of organic matter. There are up to 3,000 arthropods per cubic inch of productive soil. A litter layer of leaves 1 1/2 inches thick and a yard square might contain 5,000 miles of fungal filaments.

Plants are the primary producers of organic matter in the forest soil system. Ants and other invertebrates initiate the breakdown of ground-layer litter. Soil microorganisms including fungi, bacteria, protozoa, and actinomycetes continue this process of converting organic matter into soil minerals that in turn become available as nutrients to plants. In food-web nomenclature, these organisms are "consumers." Primary consumers (herbivores) feed directly on the "producers," which are the plants; secondary and tertiary consumers are predators and parasites, which feed upon each other as well as upon herbivores. Food webs also contain other decomposers and detritivores that feed on litter, such as mites, woodlice, and earthworms. Woodlands typically support more diverse assemblages of soil organisms than grasslands. If soil organisms are included in the species count, temperate rain forests are richer in biodiversity than tropical rain forests.

The soil food web performs the primary function of the soil, which is to cycle energy and nutrients, including nitrogen, sulfur, and phosphorus. Native soil systems are very efficient and succeed in recycling, for example, upwards of 80 percent of the nitrogen in the system. The cycling of nitrogen is intimately associated with the cycling of carbon, which is tied up largely in organic matter. Nitrogen, in part, determines the rate at which carbon is broken down. Bacteria and fungi take up the nitrogen as they decompose soil organic matter, and some fix atmospheric nitrogen. This nitrogen too is released into the soil to be again available to plants. Nitrogen's slow release from an organic to an inorganic form, which is available to plants, is called "mineralization."

The microbial community performs three major functions: as discussed above, conversion of organic nitrogen to a plant-available form such as ammonia; nitrification when ammonia is converted to nitrates; and denitrification when nitrogen is recycled into the atmosphere as a gas. The soil microbial community also contributes to soil stability, another vital function. Fungal hyphae knit bits of organic matter together to create a denser, stronger litter layer and upper soil horizon.

Not all soil food webs are the same. Fungi appear to dominate in forest soils, bacteria in agriculture soils. Thus, soil communities change over time as the landscape succeeds to forest. The nature of the vegetation determines the nature of the fuel/food available for soil organisms. Grasslands litter, a relatively easily decomposed herbaceous material, does not typically contribute all of the soil's organic matter. The extensive root systems of grasslands are also a major source of the soil's organic matter. The roots of grasses exude carbon directly into the soil as sugar, amino acids, and other forms to feed soil fungal associates and activate bacteria and other microbes.

As the landscape matures, the litter becomes more difficult to break down. While herbaceous litter is primarily cellulose, the litter of the forest becomes increasingly higher in lignin, the woody component of plants. Tree leaves have more lignin than grasses, and the leaves of late successional species, like beech and oak, typically have more lignin than ash, tulip poplar, and other early successional species. In woodlands an important shift occurs as leaf fall and other litter become the most important sources of organic matter, rather than the direct contribution of carbon by the roots as in the grasslands. There are also larger volumes of wood on the ground in the form of fallen twigs and limbs, which directly foster fungi because bacteria are unable to decompose lignin. The mycorrhizal filaments from tree roots reach up into the old wood to extract the valuable nutrients. Insects such as beetles and ants are also able to break down wood. Wood in contact with the soil and standing dead trunks, "snags," create many opportunities for various wood and soil invertebrates of the forest.

The soil communities continue to change along with the vegetation communities. Over time, the cycling becomes less rapid. In a humus-rich forest soil, the organic matter that remains the longest is rather stable organic compounds that degrade much more slowly. By then the humus is important more as a site for important chemical processes and for the physical qualities it gives the soil than as a stockpile of nutrients. The humus, for instance, increases the water holding capacity of the soil.

Another important role of dead wood is to serve as a water reservoir for the forest in times of drought. Dead wood, especially larger logs approaching a foot or more in diameter, soaks up water like a sponge and retains it for long periods. Old logs or stumps make great nursery sites by carrying vulnerable seedlings through dry spells. Salamander populations also depend on large logs for needed moisture, which is, in part, why they are absent so long after clearcuts and timbering, although they may number one or two per square yard in old-growth forests. Logs increase local stormwater retention as well by inhibiting overland flow and by absorbing water in place.

Fungi in general foster acid soil conditions, whereas bacteria can increase alkalinity. The bacteria and their predators in grasslands help maintain the soil's pH and the form in which nitrogen is made available, as well as nutrient cycling rates that work to the advantage of grasses. Where fungi are more abundant, as in natural forests, the nitrogen is converted to ammonium, which is strongly retained in the soil system. In bacteria-dominated systems, the bacteria convert nitrogen to nitrate instead of ammonium. Nitrate leaches more easily from soils than ammonium; however, the growing patterns of grasses tolerate this condition. But when woodland soils become bacteria dominated, rapid leaching may leave most native old-growth species poorly nourished while invasive exotics and some early successional natives are flush with nutrients. Some species are more sensitive than others to soil nutrition. Conifers do not grow in bacteria-dominated soils whereas agricultural crops cannot be grown in fungi-dominated soils. Indeed, in woodlands, a high ratio of bacteria to total biomass is an indicator of disturbance (McDonnell, Pickett, and Pouyat 1993). These factors, which seem to be dependent upon soil organisms, play a greater role in succession than previously recognized (Ingham 1995).

Damaged Soil Systems

Soils are far more damaged and damageable than we realize, but the problem is often hidden. The cumulative effects on forest systems and other environments, of acid rain, nitrogen deposition, global warming, ozone thinning, unnecessary grading, and stormwater changes have left a legacy of severely altered soil conditions and totally modified soil food webs. The consequences and remedies are still largely unknown.

Many of these changes are so pervasive that we take them for granted. Take earthworms, many nonnative, which now are abundant throughout the urban forest system. In fact, they are not part of the historic community of living creatures in native forests and are typically associated with more disturbed landscapes. Earthworms in general increase soil fertility by initiating the breakdown of organic matter, aerating and mixing the upper soil, and creating a microenvironment that stimulates the bacteria that convert ammonium to nitrate. High earthworm populations also foster nitrification by supplying the oxygen necessary to convert ammonium to nitrates. They take a system already disturbed by added nitrogen and push it farther from normal by consuming the litter layer five times as rapidly as fungi and converting excess food into nitrate. The same kind of self-reinforcing cycle can be seen when aquatic systems fill with algae (Nixon 1995). Each shift in the soil character will in turn ripple through the entire system. Unfortunately, in many woodlands that look mature because they have larger trees, there is a lag in the succession of the soil, which may still be dominated by earthworms and bacteria and impoverished in terms of types of fungi, invertebrates, and other, more efficient paths for nutrient cycling.

Building Soil Systems

The objective in restoration is to restore the nutrient cycling and energy flow of the historic soil system. First, work to protect existing soil resources and then explore techniques to increase the overall biomass of the soil and to foster the diversity of native soil flora and fauna.

Recommendations

• *Identify, protect, and monitor areas of native soil that are relatively undisturbed.*
Most areas contain places where there is less-disturbed soil that can serve as rough models of local soil conditions. Studying the more natural soils at the same time remediation is being documented in a disturbed landscape will provide a standard for measuring the success of different approaches. The natural sites also serve as propagation sources for locally adapted microorganisms.

• *Reduce local sources of soil contamination, including added nitrogen.*
Evaluate local air pollution impacts, especially that of automobile exhaust. Removing roads wherever possible is of paramount importance, especially in more natural areas. What is convenient, even to the restorer, such as easy access, may be lethal to the most jeopardized species. Educate the community about regional air pollution impacts. Many other management practices, such as pesticide use, also affect the realm of the soil. The most popular herbicide, for example, glyphosate, which is often used to control exotics, enhances conditions for bacteria but makes a poor substrate for the development of forest fungi.

• *Recognize that the user is inseparable from the solution.*
No treatment of soil will make it impervious to compaction, erosion, and other such disturbances. Confine all use in forests and other natural landscape fragments to designated trails to minimize degradation from feet, hooves, and wheels. Prohibition alone never is enough. Users will stay on trails to the extent that trails create the elements of satisfaction that keep them there and provide access to desired destinations. The gradual building of the litter layer and the absence of bare soil off the trail are hallmarks of success.

• *Minimize "working the soil."*

Despite a lot of knowledge about the damage done to living systems by constant perturbation, there is still a tendency to overwork soil. Beyond the familiar structural damage, such as that caused by working a heavy soil while it is wet or the erosion that accompanies any soil disturbance, the soil's level of microorganisms is also severely affected. For example, plowing and any mechanical disturbance to the soil will tend to foster the rapid growth of bacteria, which in turn generate exopolysaccharides, which cause the soil to slump in rain. Other substances make soil hard to wet, or hydrophobic. Cultivating soil is almost always deleterious to natural areas and constantly resets the time clock back to disturbance rather than allowing more complex, stable, and diverse soil systems to develop. We need to try new techniques, such as planting new seedlings in logs or stumps, to avoid soil disturbance while enhancing survival. Another technique is vertical staking, wooden twigs driven vertically into the soil. Vertical staking serves to aerate and loosen the soil without damaging the roots of existing vegetation, and it avoids the need to completely turn the soil. In addition, it favors the development of fungi instead of bacteria because it incorporates wood into the soil.

• *Reevaluate the usefulness of current methods of stockpiling topsoil.*

Harris, Birch, and Short (1993) describe the progressive impacts of stockpiling, which is a frequently used method to retain a site's topsoil during construction. The first phase is an instantaneous kill of many of the living creatures in the soil that occurs with the initial removal and stockpiling. During the next few months there is a flush of bacterial growth as well as fungi but only in the upper soil on the outside of the pile, the new "topsoil." During the next half year or so the soil stratifies in layers. The primary distinctions reflect the amount of oxygen in the soil because of its depth in the pile or level of saturation with water. The developing layers consist of both near-surface aerobic and deeper anaerobic zones as well as a shifting transition area between them. When the soils are restripped and replaced elsewhere, there is another instantaneous kill of most living organisms followed by a flush of bacterial growth.

• *Experiment with alternative strategies that better preserve native soil food webs when moving soil is necessary.*

Experiment with methods that keep soil horizons intact, such as moving blocks of soil. Practitioners are using and modifying equipment like old sod forks and front-end loaders as well as developing new equipment, such as John Monro's soil mat lifter, described in Part III, for this purpose.

• *Reevaluate the addition of organic matter to enrich disturbed soils.*

The continuous rain of airborne nutrients onto soils in the form of acid rain and nitrogen deposition from air pollution raises serious concerns about many traditional management practices with regard to the use of organic matter as a soil additive and our almost automatic addition of nutrients to disturbed soils. Researchers have shown repeatedly that fertilizer benefits weed species. Creating less-hospitable conditions in the conventional sense can actually enhance the performance of native species. Using elemental sulfur on test plots, Jean Marie Hartman at Rutgers University and her co-workers (1992) lowered the pH and reduced nutrient availability in a mixed meadow to foster native species over exotics. Many invasives, both native and exotic, are nitrophiles and do poorly under such conditions.

• *Reevaluate the use of mulch and soil amendments that are harvested from landscape communities other than those native to the site.*
Because to a great extent soil organisms are what they eat, bringing in organic material from other sources will not necessarily foster the growth of the same soil organisms as are in the desired native community. In an artificial soil such as made land or a highly contaminated soil, it's not the addition of organic matter but what kind we use that will impact the nature of plant succession on the site. The more indigenous the existing landscape, the more important it is to minimize the use of dissimilar materials.

• *Reevaluate the conventional management of brush, dead wood, and leaves.*
Even where no additional fertilizer is added, it is important to modify our management of dead wood and vegetative debris to more closely mimic natural conditions. This sounds obvious, but how often is organic matter collected from a site, taken to another location to be composted, and then used at still another location when it is "well rotted"? Under more natural forest conditions, however, the major contribution of organic matter is not well-rotted compost but rather wood, twigs, and leaves that slowly break down in the place where they fall. Adding wood and raw, rather than composted, leaves more closely mimics the natural scenario.

• *Develop new ways of observing and monitoring soil health.*
Unfortunately, standard soil tests are of limited assistance to the restorationist. For example, nitrogen levels are poorly evaluated when they are measured only as concentrations at any one time rather than as total flux over time. Conventional tests also ignore the biotic component altogether. A number of researchers are working on new methods. One, Jim Harris, of the University of East London in England, who has been monitoring soil changes associated with restoration, has developed a set of techniques for measuring the size, composition, and activity of a soil's microbial community. These measurements can be used for comparison with a less-disturbed target community to assess the level of recovery of the soil system. He and other researchers have developed methods that, at least in England, have increased fungal populations with significant beneficial impacts to soil development and nutrient cycling.

• *Build populations of soil fungi.*

As noted earlier, heavy nitrogen enrichment from air pollution and increased compaction, erosion, and sedimentation have tended to favor the growth of bacteria over fungi and invertebrates. Thoughtful management promoting the development of fungi, through appropriate treatment of the soil, soil surface, and litter layer, can help restore indigenous food webs in forest soils.

Management to Foster Fungi and Other Forest Organisms

Because only fungi can break down lignin, the woody component of plant matter, allowing dead wood and woody debris to remain on the ground layer is a major component of the effort to rebuild soil fungi. Raw woodchips and small limbs on the soil surface provide an ideal matrix for the rapid development of a dense fungal network in the soil that, unlike bacterial decomposers, also provides surface stabilization. The webby, sticky quality of the mycelia of fungi serve to knit the surface particles and litter to reduce erosion and conserve moisture that is vital to the life of forest soil. While a deep layer of woodchips can create a growth-suppressing mulch that later floods the area with nutrients, a very thin layer of woodchips stimulates the development of more complex soil biota while limiting the overall rate of the addition of nutrients. Wood's slow rate of decomposition is also important where rates of decomposition have accelerated dramatically. Because lignin has a very low decomposition rate, it is a more durable ground cover that promotes the development of a stable litter layer.

Occasionally it may be necessary to inoculate the soil or vegetation with mycorrhizal fungi, although in most cases local sources of inoculum are likely to be available from wind and animal dispersal. Where soils are high in nutrients it may be more important to manage nutrients and foster fungi than directly inoculate, especially if inoculation is not required to establish plant species. Small amounts of soil from analogous sites nearby or woodchips colonized by local mycorrhizae may be used to inoculate sites where natural processes have not been effectual, where there is a substrate limitation, such as thin soil over bedrock, or where plant-specific requirements do not occur.

Jim Harris recommends using thin blankets of fresh woodchips from 1/2 to 1 inch thick, which create ideal surface conditions for the development of fungi. Within weeks, a network of fungi colonizes the surface so densely that the woodchip layer can actually be shaken loose from the soil by hand and moved elsewhere to inoculate an area nearby with local fungi. This method, local harvesting and dispersal of indigenous fungi, should become an important part of soil management programs and is preferable to using a mass-produced commercial inoculum for restoration purposes.

We can also manage blowdowns better than by simply removing fallen trees, as is the current convention. Instead, we can minimize the hazard of a falling tree to area walkers while mimicking more natural processes of decomposition that encourage the growth of fungi and invertebrates in the soil by partially up-ending the stump. The upended root mass reveals a near-perfect seedbed for native species and maintains enough of the tree's still living roots to maximize the extent to which its nutrients are passed directly to neighboring trees.

Commercially produced mycorrhizae have been very successful in reforesting drastically disturbed lands, such as mine spoils, all over the globe. Sites in Kentucky, for instance, where soils were extremely acid, with pH values as low as 2.8, have produced pulpwood for harvest in just fifteen years from inoculated seedlings (Cordell, Marx, and Caldwell 1991). When considering such products, however, evaluate their potential impact on native subspecies of mycorrhizae. Like commercial plant propagation, this approach risks the hastening of extinctions of local varieties. We still need to develop appropriate procedures and protocols for disseminating fungi and other soil organisms as much as we do for larger plants and animals. Such techniques are well developed in the western states but have only recently been applied in the East.

Fire also acts as a stimulus to many wood fungi and invertebrates and reduces bacteria, which in turn fosters the growth of fungi. In a study of changes in beetle populations following fire in boreal coniferous forests in Finland, scientists found a sudden appearance of a diverse group of beetles that feed on wood fungi (Muono and Rutanen 1994), which in turn implies an even more rapid response by fungi. These wood-fungi-feeding forest beetles are fire specialists and represent an important evolutionary adaptation at an ecosystem level to recurrent fires of the past and are a side-benefit of restoring natural patterns of fire to the forest.

Native soil conditions and biotic communities and processes need to be the models for our interventions in restoring native habitats. The remaining remnants of native soil are, therefore, bioreserves for the richness that once characterized our soil heritage. The approach should be to restore, rather than replace, soils. Soil made in place is favored over the imported topsoil. Instead of reintroducing missing components with inputs from outside the environment, we should instead focus on fostering the restoration of remnant and indigenous communities of soil biota, which furthers the general goal of "restoring-in-place" to the extent feasible. By doing so, we also minimize the casual dispersal of local subspecies of soil microorganisms and exotic soil organisms. In the worst-case scenarios, such as areas where soil is completely depleted, some materials from outside will be needed, but even in these situations the soil-building resources inherent to the site should be used to the maximum extent possible.

Soil Protection and
Restoration During Construction

Construction for utility lines, roads, and rights-of-way are frequent intrusions into parks and natural areas. Typical pipeline construction, utilizing a 75-foot-to-100-foot wide construction zone and equipment that sometimes weighs more than 20 tons, causes serious soil compaction, damaging the roots of trees adjacent to the right-of-way and destroying the soil as a living medium and reservoir of forest reproduction. In Morristown, New Jersey, when the Federal Energy Regulatory Commission mandated that the Algonquin Gas Transmission Company route a new 36-inch natural gas pipeline through a mature oak-beech forest in the Loantaka Brook Reservation, the Morris County Park Commission negotiated for an opportunity to develop alternative strategies.

In protecting this site, the commission's primary goal was to preserve the forest as an intact ecosystem. Protection measures focused on maintaining a continuous tree canopy overhead to minimize forest fragmentation and on preserving soil biota and structure. Working closely, Andropogon, the contractor Napp-Greco, and Algonquin, the pipeline engineers, devised and implemented the following strategies:

• *Realign the route.*
Realigning the proposed route to follow existing park trails and other existing seams of disturbance reduced forest clearance and the extent of soil disturbance.

• *Minimize disturbance to the adjacent habitat.*
Construction did not begin until the construction zone was fenced along the entire route through the park. Qualified arborists felled the trees, which were lowered with ropes into the construction zone, to prevent damage to the adjacent forest. Wherever possible, they left the tree stumps in the soil (Figure 17.1).

• *Reduce the width of the construction zone from 75 to 35 feet.*
Working with heavy equipment on top of the linear pile of soil that was excavated from the trench eliminated that portion of the corridor needed for separate soil stockpiling. This method also reduced soil compaction and reduced most damage by serving as a cushion for heavy equipment. Welding the pipe in the trench instead of alongside it also limited the area of disturbance (Figure 17.2).

Figure 17.1. The pipeline right-of-way was fenced and cleared before any work to limit damage to adjacent woodlands took place.

Figure 17.2. Welding the pipe in the trench rather than alongside it further reduced soil damage.

• *Retain the soil structure intact to the extent feasible.*

The restoration goal was to reestablish the rich forest ground layer, which required that the soil structure and its reservoir of life remain intact. Regulatory agencies in charge of soil protection mandate only minimal restoration, concentrating primarily on separating topsoil from subsoil and stabilizing the bare ground with turf grasses after the soil is replaced.

Andropogon's alternative restoration proposal was to harvest the upper soil layers in blocks, before digging the trench, to preserve the existing stratification of the soil and to ensure the continuity of soil microorganisms, woody rootstocks, bulbs, corms, and seeds. Napp-Greco, the contractor, developed a specially adapted front-end loader blade to remove the soil blocks from the forest floor over the pipeline trench (Figure 17.3). These were stored alongside the open trench for a few days and then replaced over the backfilled trench. Upon completion of each day's run of pipe, workers narrowed the fencing to just the path width to protect the newly transplanted blocks of soil. The spaces between the reinstalled mats were mulched with leaves from the local Shade Tree Commission (Figure 17.4).

After construction, the impact on the forest floor was limited. Canopy cover was largely intact and a thick regrowth of ferns, wildflowers, shrubs, and small trees resprouted on the forest floor (Figure 17.5). The contractors used the same technique along the disturbed stream corridors, except that a rich

Figure 17.3. The modified bucket of this front-end loader lifted large blocks of soil intact, which were stockpiled beside the trench for a day or two before they were replaced.

Figure 17.4. After each section was completed, a narrower fenced area allowed for access but protected the replaced woodland sods.

Figure 17.5. One season later, the ground layer rebounded. New trees were also planted.

Figure 17.6. Wetland sods, rescued from a construction site and transported on skids, replaced the degraded vegetation along a disturbed stream corridor with a more historic plant community.

Figure 17.7. Lush growth a season later has resisted exotic invasion well.

herbaceous ground layer was harvested and rescued from a site where a parking lot was under construction. These soil blocks were stockpiled on wooden skids and then staked in place along the steep stream banks by diagonally split lengths of two-by-fours (Figure 17.6). Regrowth was dense and lush and has resisted invasion by weedy species (Figure 17.7).

CHAPTER 18

Plants for Restoration

The restoration of native plants in disturbed forest ecosystems challenges all our knowledge and ingenuity. Locally and regionally we are losing species at accelerating rates. We can no longer assume that plant communities will recover on their own, nor that we will be able to transplant wildlings effectively. In many past rescue efforts, most transplants typically survived only a few years before disappearing from their new locations. Present commercial propagation includes only a small percentage of native species in any given area in the East.

Because restoration of native plant communities is still a developing field of knowledge, every restoration project is really a living laboratory. Every project presents learning experiences in how to propagate local ranges of species, knowledge necessary if we hope to return these plants to, and sustain them in, the wild. It also presents opportunities to develop landscape designs and management practices that are bioregional in focus and more sustainable over time.

Where you cannot restore conditions for effective natural reproduction of indigenous species, you will need to become an active agent of recruitment, assisting the reproduction and growth of plants in new areas. Protected landscapes and other biological reserves may ultimately be vital as sources of seeds and other propagules that can potentially be introduced into new ranges as well as for the creatures that depend on and support vegetation. Germ plasm and seed banks are an important component, but it is even more vital to sustain plants (and animals) on the ground, by fostering their recruitment, reintroducing them when needed, and controlling invasives.

Our knowledge of native plants is riddled with large gaps and pervasive misperceptions. For example, one widely held misperception is that native species are less suited to many modern landscape conditions than exotics. This is simply not true. Natives are, by definition, very well adapted to the local climate and terrain. The real problem is that severe disturbance reduces the landscape potential to a very few species whether they are invasive exotics or aggressive natives. While the full breadth of native communities is not sustainable in the urban environment, many native species thrive there and can be used to create rich and compelling landscapes if appropriately managed. It is also important to note that while exotic food sources are used by wildlife, they do not appear necessary to their survival. Native plant communities, on the other hand, are vital to indigenous animals.

No ready-to-use, comprehensive national inventory of ecological resources is currently available, although organizations such as the National Institute for the Environment, the National Biological Survey Act, and the National Ecological Inventory have been initiated to gather this information. The Nature Conservancy, in a public-private partnership with state fish and wildlife agencies and scientists, has been documenting species to develop the Natural Heritage Network, an ongoing record of the status of potentially endangered plants and animals. A Natural Diversity Index is being compiled for each state. This effort released a *1997 Species Report Card: The State of U.S. Plants and Animals.* To obtain information about indigenous local species, contact your state chapter of The Nature Conservancy and other environmental agencies and organizations, as well as botany departments at universities and native plant groups, and keep the lines of communication open with them. As your project progresses, your discoveries and learning will provide new information that you can share with researchers and organizations, adding to the "terrestrial ark" that we need to save the richness of our forests.

Working with the System

Enormous problems face any plant restoration effort, whether it is at a governmental or individual level. Soil conservation, fish, wildlife, and forest services at the federal and state levels are already involved in plant propagation, distribution, and planting programs and determine much of what is planted through recommendations incorporated into the regulatory process. For example, the lists of species currently recommended by the Natural Resources Conservation Service for erosion and sedimentation control serve to define what type of seed is widely available in the nursery trade. Unfortunately, these lists also tend to stifle the use of native plants, in part because plant specifications developed for state departments of transportation are typically used as the standard conventions for the vast bulk of construction projects. Therefore, these species dominate the inventories of nurseries and discourage the introduction of natives. In New Jersey and Pennsylvania, for instance, only one of the thirty or so plants recommended is native. When confronted with this fact, the agencies' response is often that these are the species that the nurseries have. Such circular arguments guarantee that native species will not be introduced into the trade and thus our landscapes.

One important problem is that some of the plants most commonly used as quick stabilizers are not only invasive, but they also retard the return of native plant communities. Although tall fescues—including the popular variety "K-31," are very invasive, they are among the most widely recommended, used, and researched grasses. Included in almost every specification for lawns and road-sides, they are difficult to eradicate from the landscapes they invade and have strong allelopathic effects on other species, especially trees and shrubs. Fescue toxicity is estimated to cost livestock producers up to $200 million annually (Burchick 1993). Ironically, at present, some native milkweeds are listed as agricultural weeds despite the fact that they are vital native host plants to the monarch butterfly.

Anyone wishing to use native species often must spend considerable time and effort justifying any deviation from accepted conventions, and obviously it would be difficult to totally revise the lists overnight, but their emphasis could be shifted gradually to phase in native plants. Even a small experimental plot can demonstrate that native plants are generally more practical, easier to maintain, and more stable for use in landscaping than introduced species.

One of the most important initiatives at this time may be the Federal Native Plant Conservation Committee, which was established in 1994 in a memorandum of understanding. The memorandum was signed by seven governmental agencies—the U.S. Forest Service, Soil Conservation Service, Agricultural Research Service, Bureau of Land Management, Fish and Wildlife Service, National Park Service, and National Biological Survey—as well as five non-governmental organizations—The Nature Conservancy, Center for Plant Conservation, National Association of Conservation Districts, Soil and Water Conservation Society, and Society for Ecological Restoration. Their eventual recommendations could be instrumental in addressing the growing crisis of biological pollution. On April 26, 1994, President Bill Clinton also signed a memorandum, motivated by the *Report of the National Performance* led by Vice President Al Gore, that directs the federal government and federally funded projects to use "regionally native plants for landscaping" and "construction practices that minimize adverse effects on natural habitat." The government recognizes the cost-effectiveness of such an approach and directs the Department of Agriculture "to conduct research on the suitability, propagation, and use of native plants for landscaping." Demonstration projects will encourage reduced use of fertilizers, pesticides, and irrigation and increased energy efficiency.

The possibilities for existing institutions and agencies to play a positive role in the preservation of native plant communities are almost limitless and require only a shift in program and product emphasis. The state of Illinois has demonstrated real leadership in this area. As in most other states, the Illinois state-run nursery program once grew a wide array of fast-growing, easy-to-propagate, invasive exotics. In fact, after twenty years of producing autumn olive, the nursery was producing 20 percent of the total state output, or more than 1 million seedlings a year by 1982. But all that changed in 1983, when the Seedling Needs Committee, representing the divisions of Wildlife Resources, Forestry, Public Lands, Planning, and Natural Heritage, and area nurserymen, decided that producing exotics was not necessary if suitable native substitute species could be found. The committee shifted its emphasis to native species.

Today, the state nursery is now propagating fifty-two species of native woody plants as well as prairie grasses and wildflowers that are started from seed collected in state parks and conservation areas by biologists, foresters, volunteers, and maintenance personnel. Several dozen species are made available each year and no exotic species are subsidized. Public education has been an important part of this effort and has included very accessible materials such as a one-page flyer on controlling the invasive garlic mustard. The Illinois Department of Natural Resources also shares seed with the state's Department of Transportation. This department has advised commercial growers of the growing demand for regional native seed because of their new planting policy, which calls for planting native communities beyond the mowline along state roads.

Other states are recognizing the benefits of going native. The Virginia Department of Conservation and Recreation in cooperation with the Virginia Native Plant Society has held conferences on invasive plants and developed informative flyers such as the "Invasive Plant Species of Virginia." The Wisconsin Departments of Natural Resources and Transportation are preparing to grow native seed on three prison farm sites. The Minnesota legislature defeated—but may reconsider—an innovative program to develop native seed as an alternative agricultural crop and also allows for repayment of start-up loans with native seed supplied to state agencies. The Natural Resources Group of the New York City Department of Parks and Recreation has developed a native species planting guide for New York City and vicinity that is organized by plant community types suited to different environmental conditions and is intended, in part, to stimulate a native-based nursery industry.

Wild-Collected Plants

In addition to all the other challenges to their survival, popular native species face the serious problem of being overcollected in the wild. Unfortunately, at this time it is difficult for the consumer to avoid wild-collected plants because of the lack of regulation at the state or federal level. In many cases, a collected plant simply has to be held in a nursery for a single season to be labeled as "nursery-grown." Look instead for a "nursery-propagated" label.

In an attempt to prevent the overcollection of wild plants, nine organizations formed a coalition in 1991 to petition the Federal Trade Commission (FTC) to amend its *Guides to the Nursery Industry* to eliminate confusion and possible deception about the original source of plant material; make the guide-lines conform with standards applied to international trade laws and the Convention of International Trade in Endangered Species of Wild Fauna and Flora (CITES); and eliminate the unfair competitive advantage enjoyed by nurseries that sell relatively cheap wild-collected plants instead of more expensive propagated ones.

At this writing, the FTC had yet to respond to the petition, which was signed by the Natural Resources Defense Council, California Native Plant Society, Environmental Defense Fund, Garden Club of America, Mt. Cuba Center for the Study of Piedmont Flora, National Audubon Society, Native Plant Society of Oregon, New England Wildflower Society, TRAF-FIC (USA), Native Gardens (Greenback, Tennessee), Niche Gardens Nursery (Chapel Hill, North Carolina), Montrose Nursery (Hillsboro, North Carolina), and the American Association of Nurserymen. In the meantime, the Mail-Order Association of Nurserymen adopted a labeling code of ethics, but it is voluntary and not legally enforceable.

Some progress is also being made at the local level. Rhode Island, for example, has a "Christmas Greens" Law, which protects evergreens commonly used as winter decoration. The Virginia Native Plant Society, in conjunction with local officials, has developed a comprehensive program directed toward the protection of ground pines.

Extraordinary Efforts, Positive Results

The challenges of restoration will require all the skills of the horticulturist and the botanist. Once-common species close to the edge of extinction, such as the American chestnut, deserve special efforts, not unlike those directed toward rare and endangered animals. There are encouraging signs that progress is possible. The American Chestnut Foundation is among several groups endeavoring to breed disease-resistant strains suitable for reintroduction into natural forests. Another fruitful line of investigation is the use of the so-called hypovirulent strains that are nonlethal strains of the blight organism that do not injure the trees but prevent more virulent strains from infecting them. And most recently, the firm of Hoffman LaRoche has announced some progress in developing disease-resistant American chestnuts via genetic engineering. The restoration of this once-predominant canopy tree of much of the eastern forest with its phenomenal utility to wildlife may happen in the foreseeable future.

The survival of a few species is undermined by our lack of knowledge about their reproductive processes, which in turn restricts our ability to reintroduce them to a site. For many plants the form of the juvenile is strikingly different from the adult and may have different pH, nutrient, and moisture requirements. In many cases the juvenile forms have never even been seen or at least have not been recognized. The seedling may not only look different, but it also often germinates under different conditions than it requires to grow as an adult.

To cite one example, until Jean Marie Hartman, an ecologist and landscape architect at Rutgers University, began to study swamp pink, the juveniles had almost never been seen in the wild. More important, in trying to propagate this plant, she and her students realized that seedling mortality for transplants was nearly 100 percent until they reached a certain size, at which time they could be planted successfully in the environments favored by either juveniles or adults (personal communication). This kind of information is vital for conservative species, especially before field plantings are undertaken with valuable plants.

The real work of restoration is to learn from mistakes while keeping the goal in sight. This all takes time, but remember that time is relative. Many of us, horticulturists and foresters included, are often quick to abandon a native plant under stress and substitute an exotic species. For example, the current vogue to replace plantings of flowering dogwood with Korean dogwood, the host of the anthracnose disease that infects native dogwood populations, only serves to further jeopardize the native species. There is a similar rush to find hemlock species from Asia that are adapted to the hemlock wooly adelgid as substitutes for the native hemlock.

The Crosby Arboretum in Picayune, Mississippi, which is dedicated to the plants of the Pearl River Basin, exemplifies the concept of a bioregional arboretum. At the very outset the arboretum's board looked beyond the boundaries of their site to seek to preserve the biodiversity of the region by acquiring crucial natural areas and joining in shared conservation agreements to protect others. At the same time, the arboretum is restoring natural habitats on the Pinecote Visitor's Center site on land that was previously both logged and farmed. Using innovative technologies such as prescribed burning (Figures 18.1–18.5), the arboretum is pioneering the management of southeastern coastal plain landscapes in a gardenesque setting. This applied research as well as the growing collection of indigenous plants that is fully documented and monitored will be invaluable in the effort to restore local ecosystems and to develop a regionally appropriate landscape vocabulary. The ultimate goal is to re-create conditions where natural recruitment processes are reestablished.

The Role of Nurseries

For many people, nurseries are a primary source of information about plants and play an important role in local landscape selections. Some nurseries are already making an effort to behave more responsibly when it comes to growing and promoting plants. An article about purple loosestrife in the April 1994 issue of *Network News,* the newsletter of the Eastern Native Plant Alliance,

Figure 18.1. The managers of the Crosby Arboretum research and demonstrate prescribed burning to restore successional habitats suitable for natural areas and home landscapes alike. (Photo by Ed L. Blake Jr./The Crosby Arboretum)

Figure 18.2. The meadow around the Pinecote Visitor's Center just after prescribed burning at the Crosby Arboretum. (Photo by Ed L. Blake Jr./The Crosby Arboretum)

Figure 18.3. Within weeks of the fire a rich meadow again glows in the afternoon sun. (Photo by Ed L. Blake Jr./The Crosby Arboretum)

Figure 18.4. This pitcher plant bog, known locally as a buttercup meadow, is sustained by periodic burning; it would otherwise succeed to forest. (Photo by Ed L. Blake Jr./The Crosby Arboretum)

Figure 18.5. Longleaf pine, once the predominant community, requires fire to stimulate growth of its seedlings. (Photo by Ed L. Blake Jr./The Crosby Arboretum)

illustrates one East Coast nursery's growing awareness of its role in sustaining natural systems:

> White Flower Farm [Litchfield, CT] will not offer purple loosestrife after this spring, the (1994) catalogue announces, because it has become a major pest along waterways throughout America, "crowding out native plants that don't share its vigor." To the nursery's credit, the statement adds, "Because marshy areas in the Northeast have been filled with Lythrum for as long as we can remember, we were slower than we should have been in taking the problem to heart."
>
> Changes over the last several years in the catalogue's description of purple loosestrife provided a glimpse of the way this nursery's decision evolved. Throughout the late 1980s "good for naturalizing" was given as one of loosestrife's merits, though in 1989 the catalogue noted that seedlings had never been found near stock plants.
>
> For several years the description said, "Yes, there are wetlands in the Northeast where Lythrum is a common weed and some will stick their nose in the air at it for this reason." In the Fall 1990 catalogue that statement was replaced by, "Yes, there are wetlands in the Northeast where Lythrum salicaria is a common weed and it should not be encouraged by further planting." By 1992 the suggestion of defensiveness was gone, and the listing emphasized that the plants offered were not L. salicaria but sterile hybrids of L. virgatum, "no threat to wetlands."
>
> The [1993] catalogue, noting bans on the sale of Lythrums in some states, expressed the opinion that existing vast populations would "almost certainly overwhelm" local control efforts and that the circumstances "were not occasion for

excluding Lythrums from all gardens." It did warn, however, that the "self-sterile hybrids" offered "can and will interbreed with local populations if not deadheaded." Meanwhile the number of states to which varieties offered in the catalogue could not be shipped had grown from one, Minnesota, in 1989 to six in 1993 and eight in 1994, four in the Midwest, one in the Southeast, and three on the west coast (2–3).

The times are changing rapidly, as this response written by Julie Manson, Environmental Council on behalf of Smith and Hawken to a letter of complaint by the authors illustrates:

> Thank you for your letter of concern regarding the Porcelain Berry.
> I referred your question to our plant buyer who replied that there are areas where this plant is not invasive. Nevertheless, due to the concerns you mention we will not offer Porcelain Berry in the future, either through our stores or catalog.
> Here in California we wage our own battle against invasive species such as Pampas Grass. I am reading with interest the information you sent on this topic, and will keep it for future reference.

Both consumers and nurseries play important roles in what plants get planted. In time, the growing interest in native species may drive the phase-out of problem exotic species as much as regulatory restriction.

Changing Markets

The market for native species differs from the current market in important ways. One of the most difficult problems for the landscape restorationist is obtaining locally indigenous species and subspecies for use. Unfortunately, the economics of plant production has resulted in large centralized growing facilities that serve very extensive regional markets. "Superior" native varieties, like nonnatives, are propagated vegetatively and distributed widely. These are often significantly different from local native varieties and, in places, may jeopardize indigenous species. There is no seed production nursery for native grasses and wildflowers in the East, so even when we try to plant a native meadow in New York or New Jersey, we cannot use local subspecies unless we have collected local seed. The differences are often strikingly apparent. To cite just one example, Indian grass grown from midwestern seed dwarfs its eastern counterpart. These plants may persist and/or hybridize with native subspecies and become a problem for restoration.

There are several reasons for these nursery trends that fit poorly with the needs of indigenous species. First, there is pressure to develop varieties that meet the industry's level of predictability. Second, growers are increasingly relying on production methods that produce a salable crop the quickest—that, unlike the slower methods of reproduction, such as seed, reduce the genetic diversity of the population and increase the risk of creating a supercompetitor.

Large-scale propagation is in some cases the answer to curbing wild collection of native plants. Modern propagation efforts have already reduced the collecting of wild populations of carnivorous plants used indoors in terraria or as decorations. Many of the newer technologies, such as tissue culture, have greater application in this market where the full range of natural diversity is not required. Nonetheless, it is advisable to maintain adequate genetic diversity in cultivated populations as well. Tissue culture also offers the opportunity to expand the numbers of threatened local subspecies to better ensure their survival.

The responsibility to think ecologically is as great when one is trading in native species as it is with nonnatives. The possibility of creating mutant strains that may threaten local subspecies diversity is very real because of the kinds of practices that characterize the plant production industry today. A more ecological approach would be to foster smaller-scale, localized propagation and production in a larger context that is monitored and managed. To do so, propagation standards for both industry and institutions are needed to protect local levels of subspecies diversity and ensure responsible propagule collection, informative labeling, and appropriate distribution.

There is, fortunately, a vigorous and growing trend toward the bioregional nursery that is focused on a specific regional landscape. These nurseries often show great creativity in their marketing programs, informing the consumer about unfamiliar native plants and the role the home gardener can play in sustaining natural systems. The Pinelands Nursery in New Jersey, for instance, holds an annual daylong conference that informs local landscape architects, contractors, agency personnel, and interested individuals about native plants and techniques for their establishment.

Not only are educated consumers and industry personnel embracing the concept of biodiversity, but they also are beginning to grasp its economic benefits. Simply having well-managed habitats is becoming a valuable asset. The Native American community on Walpole Island in Michigan, which has maintained the high quality of its indigenous prairie by burning, is now collecting and supplying native seed in a program developed by the School of Natural Resources and Environment at the University of Michigan in Lansing.

But even a well-developed regional nursery system may not be adequate to meet the needs of those restorationists working on the most critical habitats. The most fruitful course of action is often to propagate locally indigenous plant material using seeds and propagules from the site itself and from the immediate locale under similar habitat conditions. Skilled collectors and propagators are needed to meet this demand, and your project may be just the place to do it.

For the innovative contractor there also is a growing market for a wide array of landscape management services related to restoration that provide less-costly and environmentally better alternatives to conventional maintenance. The same corporations that once sought lawn-mowing services for vast areas of turf are now seeking those with skills in woodland management and meadow installation. Sustainable landscape management and sustainable horticulture are in. Pesticides, peat, inorganic fertilizers, and mowers are out.

Landholders and regulatory agencies are increasingly looking toward the reestablishment of stable vegetated riparian corridors to address non-point-source pollution control and flood management that also provide multiuse greenways. New products related to these services include a wide array of soil stabilizers made entirely of organic materials, such as the fiberschine, a coconut fiber roll that is used to anchor wetland plants along shorelines. In addition to the growing market for specialized plant materials and products for soil bio-engineering and other innovative techniques, there is a demand for qualified installers as well as contractors with specialized skills.

The Home Gardener

By being a thoughtful and informed consumer the home gardener plays an important role in restoration. What we do in the home landscape mirrors our actions in the community landscape and is a reflection of how we perceive our relationship to the larger natural world. It will take a great many fine and innovative gardeners to wean us away from lawn and provide attractive alternatives to current landscape conventions. Concurrently, there is a great need for public education to link the public's preferences to a broader understanding of the environmental consequences and benefits of the choices they make.

When we are aware of what is happening in the landscapes around us, the possibilities are endless. The Florida atala butterfly, presumed to be extinct in 1965 when the last known colony was destroyed, probably owes its existence to nursery plants and homeowners. In the 1980s, when a small number of larvae of the subspecies were found, the repopulation success story was supported by a widespread community effort to plant its host species, a cycad called coontie.

There is something quite preposterous about our devotion to the landscape styles of the seventeenth, eighteenth, and nineteenth centuries at a time when we must address the future of our landscape. Beneath our lawnmowers and asphalt is an invisible forest that represents the richest and most complex landscape type that can be attained given our climate. After centuries of beating back the forest, we now find ourselves sadly winning this age-old battle. It will take equal effort to bring the forest and all of its richness back.

CHAPTER 19

Living with Wildlife

Wildlife is dependent upon *whole* ecosystems. The success of wildlife in the future will depend in large measure on how we succeed in learning the lessons of ecosystem management and promote the restoration of whole ecosystems.

At present there are two major approaches to managing wildlife. The first is to modify the environment; the second is to remove or add individuals and/or species. Both approaches are necessary in restoration, but any technique that requires trapping and handling wildlife, including monitoring, goes beyond the scope of this book and should not be undertaken without the involvement of wildlife specialists. As you discover the issues at work in your region, you'll find that the need to coordinate with agencies and scientists is actually an opportunity for better integrating the project into the larger community.

The approaches recommended in this book are primarily geared to enhancing the habitat value of the landscape for many kinds of wildlife as well as protecting the landscape from excess damage where populations of a few species, such as deer, are unusually high. The objective is to increase the opportunities to support wildlife in the larger environment by changing the patterns of development and providing for more secure, separate, and continuous environments for wildlife to reduce both contact and conflicts with human beings. Good-quality landscapes, however, do not always support diverse or historic wildlife. Once eradicated, many species of wildlife do not necessarily return with the plants that once supported them.

Fostering Continuity and Reducing Barriers

Conserving biodiversity depends upon positive interconnections between landscapes, so that the characteristics of landscapes of different types do not present barriers to living things. The effective size of a barrier increases with the distinctions between landscape types. In other words, a tall-grass corridor through a forest presents a greater barrier to the movement of wildlife than a shrubland of the same width, but far less a barrier than a swath of asphalt for the species that inhabit forests. Landscapes with the following characteristics diminish the differences between native and managed landscapes to benefit wildlife:

- **Native plant communities.** Indigenous wildlife are dependent on the native plants with which they have coevolved. Even though a few native animals use some exotic species heavily, a wider array of indigenous creatures will benefit from restoration of whole communities of native plants.

- **Layered landscape structure.** The natural landscape is typically layered, from the ground to the canopy, rather than reduced to a single layer or two such as turf and shade trees. A forested landscape includes several layers of young trees, shrubs, vines, and herbaceous species, not just specimen plants in mulch or turf. In a grassland the difference is between relatively uniform turf and a richly layered meadow of grasses and wildflowers.

- **Living ground layer.** In a natural environment, the ground surface is a vital realm, the cradle of the environment's biodiversity, quite unlike asphalt, turf, or thick mulch. The life in the soil contributes greatly to the presence of larger organisms as well as fosters natural plant recruitment and dispersal of species.

- **Natural cycles of disturbance.** Every stage of forest succession from initial disturbance to old growth is vital to sustaining wildlife biodiversity. Natural cycles of disturbance provide important opportunities for species variability.

These characteristics reduce differences between intentional and natural landscapes, to create a more continuous matrix of forest vegetation at various stages of succession throughout the landscapes that we use and manage, thereby providing habitat for ranges of species. Barriers in the landscape, however, are also important. The continuous, relatively predator-free deep forest, for example, serves as a barrier between canopy gaps where migratory birds can nest with some safety. The same birds that are attracted to the circular edges of gaps separated by forest expanses are also attracted to the continuous edges of the fragmented landscape, which in contrast are high in predators. The patchwork of small holes in the forest fabric is a very different place from the slim corridors and islands of forest that characterize today's landscape. The natural landscape is both continuous and patchy, reflecting a delicate balance between gaps and corridors.

The continuous corridors of the built environment are largely barriers for wildlife, except for those species that thrive in storm drains and in lawns. One look at a modern highway reveals how little thought goes into accommodating the needs of wildlife. To reduce costs, road construction usually evens out the terrain by cutting and filling so that the road is at the same grade and elevation as the adjacent land. Animals are forced to cross the highway because there is no other route available to them. Waterways are in culverts and pipes, except for larger streams, where animals may have access under bridges. The elevated stretches are often over other highways rather than streams. The final severance comes with sound barriers and concrete medians. The mown grass along the roadside may actually attract deer and other browsers to the hazards of traffic. Road deicing salts may even lure in animals from some distance. The carnage is especially visible when a new road opens but diminishes as wildlife populations in general succumb to all associated highway impacts. The amount of land area affected is staggering. Even by 1969, according to Worldwatch, there was already one mile of road for every square mile in the United States.

Imagine how we could diminish the impacts of a highway or other barrier if we simply create appropriate crossings suitable for most wildlife. The success of modifications to Interstate 75 through Big Cypress National Preserve in Florida demonstrates what kinds of changes can occur when reducing the impacts of fragmentation is a design goal. Continuing deaths of the endangered Florida panther killed by vehicles spurred the retrofit of 40 miles of highway after warning signs along the roadside were ineffective. The plan includes both extensive physical modifications and ongoing monitoring of wildlife populations, habitat, and the effectiveness of the management.

To allow for the movement of panthers and other wildlife, the entire stretch of highway was fenced with 10-to-12-foot-tall chain link, and wildlife corridor underpasses were installed about every mile. The specific locations were determined by radio-telemetry studies of bobcats, study of the terrain, and panther surveys as well as roadkills. The underpasses were designed to be wide enough (100 feet) to allow for unobstructed viewing of the landscape on the other side. Light from open medians between each two-lane bridge of traffic further entices other wildlife and diminishes the perceived distance and hazard. Gary Fink of the Florida Department of Transportation noted that the "overall effectiveness of the corridors is evidenced by the fact that since their completion there have been virtually no panther, bobcat, deer, or bear fatalities along the project roadway." (Yunt 1995.)

Similar projects are being implemented across the globe. Finland is now spending $26 million on special underpasses to keep elk from being hit by cars, accidents that now also claim about six human lives each year. Here in the United States we build countless miles of sound walls every year, especially in the most densely populated regions. We could expand the concept and design of these barriers to expand opportunities for safer crossing. By requiring that streams be bridged rather than piped and culverted, we could funnel wildlife to the available crossings, in the process sustaining both valuable habitat and well-used natural corridors.

Barriers are equally important where they help avoid conflicts. The resurgence of beaver populations throughout much of the Northeast has led to much heated debate in many urbanized areas, not unlike the situation with white-tailed deer, except that beavers alter the landscape in different ways, such as flooding lowlands and felling trees. But many of the problems can be avoided with the use of devices that prevent the animals from building a dam across a pipe opening and/or lower water levels if dams are backing up. Beaver bafflers and beaver dam restrictors, which are placed inside of pipes and culverts, are made of concrete wire with openings that allow for passage of fish and wildlife but that are too small for a beaver to maneuver a branch through. There are even methods that encourage beavers to build in more suitable locations.

Enhancing the Landscape for Selected Species

Landscape management and planting efforts in restoration projects often target selected species to help focus the community's attention. The public's awareness of more complex habitat issues often follows their concern for signature species such as butterflies, wolves, wild turkeys, or spring peepers.

The effort of the New Hampshire chapter of The Nature Conservancy to restore habitat and populations of several endangered butterflies serves as an excellent example of a comprehensive management program for specific wildlife. The Concord Barrens is home to over thirty species of endangered plants, moths, and butterflies, including duskywing skippers and Karner's blue and frosted elfin butterflies. Although the landscape is entirely fragmented by suburban development, restorers are seeking to manage a complex of grassy openings within the fabric of the larger landscape to continue to support these species. Even under natural conditions, the successional habitats these species rely on occur in patches anyway, shaped by the spotty patches of fire that once characterized this landscape.

The management program developed by The Nature Conservancy is a good model that includes vegetation management, plant propagation, and captive rearing as well as community education. It is based on the best scientific information available and is expected to take decades to implement. The mosaic of sites, consisting of over 400 acres, is managed by a combination of clearing, mowing, and burning. The core of this project is 271 acres at the Concord Airport, a cooperative partner in the restoration effort. The Nature Conservancy has made a commitment to propagating local genotypes, including 10,000 plants of the threatened lupine. Other species being propagated include spreading dogbane, New Jersey tea, common milkweed, and meadowsweet. A captive breeding program for the Karner's blue butterfly as a flagship species has been invaluable in introducing this restoration work to the public (Van Luven 1995). Already, visitors, residents, and workers alike experience some of the state's rarest and most beautiful butterflies along specially managed trails.

Europe has experienced even more dramatic declines of butterflies than North America. The Netherlands is typical, having lost fifteen species in this century alone. In the fifty years between 1934 and 1984, a period in which the British grasslands declined by 97 percent, half of the fifty-nine resident butterfly species either went extinct or declined precipitously.

Because many butterfly species inhabit early successional environments, their habitats are especially suited for replication in urbanized landscapes. Where the landscape has been managed for an extended period, however, changes in the way the landscape is managed may not be advisable. Stopping grazing in a protected area led to the demise of the British race of the large blue butterfly, which was dependent on a turf-dependent ant. To be effective at sustaining butterflies and other species, we will need to better understand and implement appropriate management of seminatural areas as well as natural sites.

Repeated restoration efforts have led to more successful models that strongly rely on extensive study and documentation of each species and its particular habitat requirements (Pullin 1996). In one successful effort in England, the population of heath fritillaries grew from a few individuals to over 2,000 after four years. The management included a system of broad swaths cut through the woods to foster the movement of the adult butterflies, which require successional habitat in the midst of a maturing landscape. The Joint Committee for the Conservation of British Insects published guidelines for insect reestablishments in 1996, and the prominence and popularity of butterflies is being used in Britain to spearhead restoration in general.

Similar programs are needed in each region for key species. Susan Bratton and Albert Meier (in press) note that some species, of both plants and animals, seem especially prone to disappear with disturbance and suggest they should be a special focus of restoration efforts. Salamanders, for example, are a diverse and geographically distinct population of creatures that can serve as both indicators of health as well as standards for evaluating restoration success.

The citizen's role in monitoring and protecting biodiversity is fundamental to any program implementation. The Friends of Acadia in Maine raised funds to monitor peregrine falcon populations and succeeded in closing a cliff trail during the nesting and hatching season. When disturbance was reduced, so too were nest abandonment and nest failure. The single pair at the beginning of the project produced seventeen chicks in the first few years and were later joined by a second pair.

The longest-running species protection program has continued in Wales for nearly a century. Every active nest of the red kite is under twenty-four-hour surveillance. After struggling to sustain as few as just three pairs for seventy years, the Royal Society for the Protection of Birds has seen the number of breeding pairs rise to over 100. The goal of the society is for this raptor to become as numerous as it once was when kites snatched food from school-children on London streets. John Gower, a program coordinator, noted: "The kite is the national bird of Wales and is symbolic of something in Wales. Like the history of the Welsh language, it has retreated into the hills, consolidated and spread out again (Woodward 1993)."

New Habitat Opportunities

There are unexpected opportunities for wildlife that we can recognize and capitalize upon.

It's hard to believe today, but buffalo (or bison) persisted in Pennsylvania's forests until the mid-eighteenth century. Today, the Wildlife Recovery Council, based in Pennsylvania, is advocating the restoration of herds of elk and bison in eastern states such as in the Allegheny National Forest. As Director Scott Thiele said in 1993 in the *Philadelphia Inquirer:* "It's time to bring the buffalo and the elk home. How can we expect Africa to save its elephants or India its rhinos or Mexico its jaguars if we neglect to restore our own prominent missing animals, including bison, once native from New York to Florida, and elk which used to range from New York to Georgia and Alabama (Associated Press 1993)." In the Netherlands, researchers are evaluating the introduction of primitive domesticated cattle and horses to replace the ecological function of the now-extinct aurochs (wild cattle) and wild horse, two once-important herbivores. They hope that other still-surviving species dependent on the same successional habitats, such as cattle, can replicate the conditions that sustained the local grasslands historically.

Other opportunities are just waiting to be discovered. The endangered Indiana bat was recently found, not in an area with caves, as in Kentucky or Indiana, where it is usually sighted, but in the shaft of an old iron mine in northern New Jersey, where a few dozen were holed up amid tens of thousands of little brown bats. Hal Bryan and John MacGregor (1988), bat specialists, have prepared an unpublished paper for the Alabama Department of Industrial Relations describing how to assess old mine shafts and natural caves for bat potential. They recommend that potential sites can be checked for bat suitability and that entrances to potentially important populations be gated to prevent people from disturbing the bats without inhibiting the bats' movement.

We can miss no opportunity to restore a more continuous and contiguous pattern to open space and forestlands. Large sites are required for most wildlife reintroductions, as well as abundant food plants. We need more, not less, habitat consolidation, and with that we should be able to expand options for predator reintroductions. Indeed, there are few reasons for wilderness area management better than those supporting the reestablishment of large predators such as wolves.

The federal government is considering the reintroduction of wolves into the 5.4-million-acre Adirondack Park, the largest remaining wild area in the East. Sixty percent of the land is privately owned and has a population of 130,000 people. Over 40,000 locals signed petitions supporting the wolves. Proponents of the reintroductions often cite the tourist appeal of a chance to hear a wolf pack howl, and they hope the wolves will help with a runaway beaver problem. There also have been sightings of wolves in Maine, where it may be returning naturally.

Farther south in the Highlands of Hyde County, North Carolina, the native red wolf, which was nearly extinct, has been reintroduced despite the old fears about attacks on livestock and competition for game species. Although there is no evidence that the wolves have ever taken any livestock at all, the state legislature has passed a law that would allow the killing of wolves that are perceived to be attacking farm animals or pets, in direct conflict with the Endangered Species Act. The reality is that owners are compensated for perceived livestock losses to minimize the controversy even when the case is erroneous or when it is proved that local dogs, not wolves, were the killers. Similarly, the Defenders of Wildlife compensates ranchers near Yellowstone.

Nevertheless, a large number of local people welcome this remarkable opportunity. Between public and private lands the forty or so wolves have over 440,000 acres of land to range across. The red wolf eats largely rabbits, raccoons, and fawns, all of which are plentiful in this landscape. Ironically, at the same time, the North Carolina Wildlife Resources Commission has undertaken research to determine whether "mammalian predator control" would increase the supply of rabbits for hunting.

Even farther south, on Cumberland Island, Georgia, a designated National Seashore, bobcats were successfully reintroduced in 1988 and 1989 (Diefenbach et al. 1993). The management plan for this site, developed by the National Park Service in 1983, addressed the possibility of reintroducing extirpated species as part of a mandate to "maintain the abundance, diversity, and ecological integrity of native plants and animals in national parks" (16 United States Code §§ 1, 2–4). At least in the short term, the animals and their offspring seem to be thriving and reproducing. Longer term, the reintroduction will provide a rare opportunity to study an isolated population over time and to evaluate different management strategies and test models of population dynamics and genetics.

In the course of this project, researchers noted the importance of the source of the reintroduced animals. Animals from analogous habitats and adapted to local conditions appear to have been most successful while the captive-born or raised fared less well. Visitors have been very interested in the reintroduction, which is reflected in the Park Service's interpretive programming.

In the meantime, a smaller predator, the coyote, has been expanding its range from the West. The first free-ranging coyotes appeared in Pennsylvania in 1910, as well as hybrids that are a distinctly larger, eastern variety, a mix of western coyotes, wolves, and domestic dogs. While capable of bringing down a sick or injured animal, the coyote does not appear to exert a population-control effect on deer.

Some wildlife reintroductions may require extraordinary interventions to restore those functions that have been interrupted. In fall 1993, Bill Lishman led a brood of eighteen hand-reared Canada geese on their first migratory journey behind an ultralight aircraft that had been their surrogate parent for weeks beforehand. Indeed, the first sound these chicks heard upon hatching was a tape recording of an ultralight. In 1995, a clutch of hand-reared sandhill cranes took off after an ultralight to start their first migration. Such research may prove invaluable in retracing safer migratory journeys, especially for the rarest of birds, such as the whooping crane.

Conflicts with Wildlife

Where we protect and sustain natural landscapes for ourselves we will also foster other creatures, not all of them desirable from a human point of view. The continuous cleanup of the waters of the Delaware River, for example, has led to greatly increased populations of gnats, which were, until recently, very reduced by water pollution. Only one of the seventeen species of gnats or black flies in New Jersey swarms peskily around our heads, the *Simulium jenningsi,* but it is bothersome enough to have everyone clamoring for control. Many communities have started using the bacterium Bt *(Bacillus thuringiensis),* which is touted as safe and selective but which actually kills the larvae of many species. When applied to river corridors, it can reduce gnats by 90 percent as well as the aquatic larvae of an entire order of insects, the Diptera. In turn, it affects dragonflies and other predators and important pollinators such as crane flies, as well as many fish, bats, and insect-eating birds, including swallows, swifts, flycatchers, and hummingbirds, which rely on this annual bug bonanza.

A highly controversial issue in urban areas, hunting by feral cats and dogs as well as domestic pets, has profound impacts on wildlife, especially birds, small mammals, reptiles, and amphibians. Domestic cats kill at least 26 million wild birds annually. The problem is that many pet owners don't realize that keeping their animals indoors is safer for them, and many probably believe that their pets' predation is minimal or even desirable. No doubt this is an issue that deeply affects restoration. If a wildlife restoration project is to succeed, feral populations of cats and dogs cannot be maintained. Your local humane society may be able to assist in community efforts to cope with the problem.

Bird feeders can bring disease to wild birds by unnaturally concentrating them and exposing them to high feces accumulations and moldy seed. In general, any effort to feed wildlife should at least do no harm. Before undertaking any feeding, it is important to understand and monitor the effort. Feeding may, in times of crises, be of assistance to some birds; however, long-term solutions should stress the availability of naturally occurring food sources.

Bird feeders, it should be mentioned, attract Norway rats, which also prey on birds. Uncontrolled food sources are the major factor supporting high rat populations. It is no surprise that rats are the most common mammal in Central Park. In an effort to control rats, many park areas are routinely treated with pesticides that have not been authorized for outdoor use. There are clearly impacts to wildlife and, in every likelihood, to humans as well. Controlling food sources is probably the most effective weapon against rats.

Encounters with wildlife are inevitable and often considered problematic. Bear populations in Pennsylvania tripled in fifteen years to about 8,500 animals in 1995. New Jersey's population has grown in twenty-five years from about 40 to between 400 and 600. In 1993, a bear in New Jersey killed one lamb. In the first four months of 1996 alone, bears were blamed for the loss of 29 rabbits, 24 chickens, 2 goats, 1 lamb, and 1 poodle. Most of the conflicts are caused by the easy availability of food in developed areas. Of course, to minimize the risks both to humans and the bears, the animals need large-enough habitat and corridors for traversing the larger landscape. Perhaps, too, we will learn to accustom ourselves to the occasional encounter and not to be so anxious. The shifting emphasis toward restoration introduces us to a new way of relating to other species that sees our fates as shared.

The White-Tailed Deer Dilemma

No aspect of ecosystem restoration will present greater challenges than confronting our relationship to wildlife. The current controversy over white-tailed deer exemplifies the problems of wildlife in the urban/suburban landscape. As deer gather at our doorsteps, we are becoming all too aware of the imbalances we have created. The known association of white-tailed deer with the tick carrying Lyme and other diseases has added urgency to what was already a crisis from the point of view of preserving natural habitats.

A recent conference in New Jersey, convened by the American Association of Arboreta and Botanic Gardens, was devoted to the management of white-tailed deer, as part of a comprehensive effort initiated several years ago by the Morris County Park Commission. Its goal was to bring together all those concerned about the current plight of the deer and the landscapes they occupy. The conference attracted participants from all over the eastern United States and Canada representing a broad array of interests, from wildlife biologists and landscape architects to animal rights activists and the agencies that issue hunting licenses, and, more than anyone else, those who manage landscapes affected by deer. Only a few years ago, a similar group spent much of their time arguing about whether or not there was a problem, but now all concede the need to work together to find solutions to what is clearly a serious problem. At least ninety-eight plant and animal species that are threatened or endangered in this country are negatively affected by deer (Miller, Bratton, and Hadigan 1992).

The deer problem is not the only factor, but it is a major one, affecting the very serious fate facing many other species of plants and animals for whom the remaining habitat is not adequate to sustain them into the future. The ephemeral wildflowers do not have time for us to wait for an easy answer to the deer problem. The deer are a keystone species—that is, both their behavior and numbers exert a significant influence on the biodiversity of the ecosystem. At densities exceeding ten animals per square mile, for example, songbird populations decline. We must deal with white-tailed deer soon.

A look at the fluctuating numbers of white-tailed deer in the Northeast during the last century gives us a broader perspective on the current situation. Before European settlement, when forest cover was dense, population levels were relatively low. The management of the landscape by native peoples appears to have favored reproductive rates of deer by increasing available browse while mortality was maintained by hunting pressure. The population was relatively static. The hunting free-for-all that occurred after European settlement and the habitat loss that came with the initial wave of deforestation decimated the deer and their predators alike. By the turn of the century, deer were so rare in the Northeast that newspapers reported local sightings and park managers began to reintroduce them. At the same time that horticulturists were planting 50,000 Norway maples in Wissahickon Park in Philadelphia, wildlife enthusiasts reintroduced white-tailed deer. This scenario was repeated throughout the region, so that today the deer are the most controversial animal in the forest.

From these small beginnings deer populations slowly recovered with the regrowth of the young forests in the first half of this century. The landscape continued to change in character in favor of deer, and the deer responded. At first, a combination of factors, from road kills to hunting, kept populations relatively steady or slowly increasing. But in the last twenty-five years, when more farmland and forest was converted to development than in the previous three centuries, the population growth rate experienced sharp increases. The deer reached what is called a "threshold population"; numbers increasing rapidly at the same time that food increases in their environment, not unlike human populations.

Deer populations do not necessarily "need" to be managed to stay within sustainable levels. Even though major predators are absent, dogs, vehicle collisions, and other hazards are sometimes enough to keep the numbers from growing rapidly if the base population is relatively small. The cessation of hunting with residential sprawl in many places did not alone account for increased numbers of deer. Scientists have noted that many parks had, until recently, populations that were not fluctuating or exploding, even in areas without hunting to help control populations. In many rural areas where forest cover is more expansive and food sources less abundant, the reproductive rate of females is lower than in more suburban areas, where food sources for deer are more ample and more fawns are produced in younger does. But a small change in the balance of birth and mortality, such as the control of poaching or the conversion of a large forested tract to succulent turf, horticultural plantings, and lots of new edge, can trigger a very rapid growth rate that then spirals out of control, even where legalized hunting persists. This scenario has been repeated in many places in the last decade, especially in large metropolitan regions.

Deer have the capacity to regulate their own population to some extent, but mostly in favor of increasing numbers rapidly rather than reducing them. When food is scarce, the individual animal may be maintained, but its growth and reproduction will be limited. When populations suddenly drop, such as after hunting season, deer are capable of reproducing very rapidly by giving birth at an earlier age and through twinning and tripletting.

Hunting does not necessarily affect overall numbers. In the Midwest, for example, doe hunting was added as early as the 1950s in response to growing numbers of deer and associated impacts. Nonetheless, the population has grown from less than 400 in 1925 to 700,000 in 1991 in Missouri. The Pennsylvania deer harvest as managed by its game commission went from an average of 2,000 animals a year in 1930 to over 1,600,000 by 1994. The practice of hunting primarily or only bucks still leaves enough males to impregnate all the does and so is ineffective as population control. Lifespans are short, with few animals living beyond a few years. The resulting population is much the same as that produced in the modern livestock industry—primarily juvenile with a high birth rate and a high death rate. It is as if deer were being raised by agribusiness under current game commission's policies.

Overpopulations of deer resulting in overbrowsed landscapes are actually detrimental to the animals themselves as well. The evidence is decreased body weight, reduced winter survival, and increased disease and parasitism. Overabundance of deer is also associated with high incidence of automobile collisions. And not only is the predominance of the most desirable and nutritious food reduced by overbrowsing, but undesirable plant species, such as white snakeroot, which is toxic to deer, also become abundant. As the habitat degrades further, the deer are forced to be less selective. Plants once shunned by deer, such as shrub honeysuckles, may become the main course later on. The most favored species, such as hemlocks and many hardwoods, disappear from the forest along with songbirds and seed-eating birds.

But it is equally important to remember that deer herds cannot be blamed for all the failure of native plant recruitment. Many woodland ephemerals are also affected by the loss of pollinators and seed distributors or by herbicides. The reduction in the incidence of ground fire has minimized the occurrence of reproductive conditions suitable for many species. Maturation of many forest stands has also changed species composition. The deer, like exotics, are symptoms of the larger problems, not just a problem in and of themselves.

An unfortunately common but understandable homeowner/landscaper reaction has been to plant those species that are generally perceived as unpalatable to deer. Some of the species least favored by deer include invasive exotics such as barberry and autumn and Russian olives. Even when unpalatable natives are used, such as spicebush, this planting strategy only serves to favor those species that are already benefiting from the loss of their competitors. The preferable strategy would be to plant and protect those species most favored by deer, to ensure their adequate recruitment.

There is a great deal of discussion about how many deer there should be and what "target" population management should be directed toward achieving. The number of deer that a given parcel can support in good physical condition over an extended period of time is generally termed the "biological carrying capacity" (BCC). That number represents an average of the range of historic population fluctuations. The biological carrying capacity is the starting point that determines the measure of deviation and the level of change. In some suburban landscapes with high deer populations, the number would have to be reduced by 80 percent at once or by taking very high proportions of does each year for several years to reestablish the biological carrying capacity. While not a measure that is likely to be undertaken in many places, it underscores just how substantial the required shift would be.

Some people urge that at least further deterioration of the landscape be arrested, an interim objective that requires less-dramatic reductions in herd size. But even this goal will require difficult decisions. Remember, too, that the long-term survival of dormant propagules of many native species is not guaranteed indefinitely. Sustaining whatever is feasible of our remnant biotic heritage will require that we restore what can be restored, not only that we stop destroying further. It is not known what the biological carrying capacity of the retrofitted and restored eastern forest landscape might be like or just how many deer it could sustain for extended periods of time. At densities of fewer than about ten deer per square mile, understory vegetation is dense and varied. Researchers estimate, however, that at least six to eight years of low deer populations are needed to reestablish vegetation enough to even assess what species still persist in past grazed and browsed woodlands.

The actual number of deer in a given area more often than not is at what is called the "cultural carrying capacity"—that is, the number of deer that the local human community has tolerated. This density is too high to prevent habitat degradation and may, as recently in so many places, lead suddenly to a population explosion that greatly exceeds what the public finds acceptable. Increasing numbers of traffic accidents and the high incidence of Lyme disease (even though mice, chipmunks, and other mammals are also important Lyme tick hosts) typically provoke public reaction. The toll on the highways is substantial—120 deaths and 8,000 serious injuries of passengers and drivers annually. In Pennsylvania, over 30,000 deer are killed on the roads each year, and one in four vehicles involved was towed.

Many methods have been tried to control deer and their impacts, all with variable success:

• *Trapping and transferring to other locations*

This technique may sound good but usually does not work. Even assuming that there are places where deer are wanted, they are frequently seriously injured or traumatized by transport, and many die soon after because they are unable to adapt to the new range. The methods required for trap and transfer are better suited to a population of rare animals you are trying to reestablish rather than a large population you are trying to relocate. Animal translocation is an infant science, and we have much to learn about when it is feasible.

• *Fencing and repellents*

These alternatives are relatively costly and only applicable to specialized site situations. In general, repellents work best when changed frequently and where densities are relatively low. Where no other measures are in place, fencing, or exclosures (Figure 19.1), may be the only recourse to protect valuable landscape features. It is important, however, to remember that fencing only intensifies the problem somewhere else. Temporary protection can also be used to shield small new plantings, which are especially vulnerable to browsing. The most important role of exclosures is probably to protect rare and remnant indigenous populations. You can also use selective exclosures to minimize human exposure to Lyme-carrying ticks. As many as 83 percent fewer host-seeking nymphs and 90 percent fewer host-seeking larvae (Daniels, Fish, and Schwartz 1993; Stafford 1993) are possible under this method.

Figure 19.1. An exclosure in the Allegheny National Forest reveals the potential richness of the landscape that is now being eaten by an overabundance of deer. (Photo by Ann F. Rhoads/Morris Arboretum)

• *Selective killing and culling*

Applied to specific age classes and sexes of deer, selective killing and culling can reduce a deer herd substantially and rapidly. But this approach has been resisted by hunters and game commissions alike in favor of limited doe hunting and other such policies that have not proved effective as management tools. If hunting is to be used effectively to manage deer populations, the trophy mentality must go. Hunting policies that encourage a selective cull by age and sex serve as an effective interim control.

Another consideration is that cultural attitudes toward hunting vary widely and influence how deer are, or are not, managed. Support for hunting as a management tool remains typically high as long as the open space is privately owned. But as urbanization increases and open space is increasingly publicly held or set in an urban context, hunting declines in popularity and is more difficult to implement as a management tool.

• *Fertility-control agents*

Immunocontraceptives offer some of the most "natural" sounding solutions to the problems of animal management. Despite limited information on long-term effects in deer, results of using PZP *(porcine zona pellucida)* have been promising in small, contained herds. PZP, the sperm-attachment hormone from a pig, stimulates females of other species to produce antibodies that then block fertilization. It is delivered by barbless dart gun annually. On a captive herd in Ohio, for example, PZP was 100 percent successful. Current research centers on problems related to the effective treatment of larger, free-ranging herds. To date it has demonstrated no known physiological or behavioral side effects, and its effects are reversible in deer. The Food and Drug Administration, which has repeatedly withheld necessary approvals, is the primary obstacle to further testing in the wild.

• *Reintroducing large predators*

The most natural and desirable solution to some of the deer problem is to reintroduce natural predators and the continuous forest necessary to sustain them. Many dismiss this solution too readily on the grounds that the eastern forest landscape is too urbanized. Nevertheless, we are still eliminating predators from areas where they still persist as if they did not have critical natural value. For example, the Sterling Forest on the New York–New Jersey border, a 17,000-acre parcel that still supports bobcat and bear, is the subject of a proposed development that would irreparably fragment one of the last large blocks of forest in the region. It is going to take an act of Congress to save this irreplaceable landscape, even though it still supports large predators. Wherever these animals of the old forest persist, they should be protected, and areas nearby should be the subject of recovery programs that consolidate and expand the potential wild habitat areas to the extent feasible to ensure that that population lives under as natural a set of conditions as possible.

Even without predation, landscape character can influence the numbers of deer supported. Consider allowing large areas to mature until they become poor deer habitat in order to protect more vulnerable species. Because deer do not range widely over the landscape, it is estimated that mature forest areas as large as 75 square miles might have half the density of deer at their core, thereby protecting other species more effectively. A doe may have a home range as small as 24 acres; a male, about 4 square miles.

For most landscapes, the need for some response to the dramatic increase in deer populations is immediate and the solutions are not ideal. We propose the following program as an interim response to this problem.

Overall Guidelines for a Deer Management Program

The following program rests on three important interrelated goals: community education, habitat protection, and population management.

Community Education

Between the widely divergent opinions people in a community may have about their deer problem and the frequent lack of scientific information informing the discussion process, reaching consensus on a management plan is extremely difficult. You therefore will probably need to begin your effort with a community education program that includes as wide a group as possible, from schools and municipal officials to area scientists and regulators. Managing agencies have as much to learn as residents.

The program of the Morris County Parks Commission in New Jersey provides an excellent model. Their program includes a conference series with associated educational videotapes, a wildlife biologist on staff who was recommended by area animal rights groups, monitoring of deer populations, and the use of fencing (exclosures) in field tests to evaluate the impacts of deer, and many other integrated efforts. The Morris County Park Commission has taken the lead and financed this countywide initiative with excellent results. The problem is far from solved, but they have reached a point of greater community consensus and support.

Because there is so much disagreement about the numbers of deer and how they impact the landscape, surveying the residents as well as involving them in monitoring is often worthwhile. The borough of Fox Chapel in Pennsylvania has adopted a wildlife management plan that includes residents' surveys as well as citizen-based monitoring of the population dynamics of local deer with regard to density, age structure, sex ratio, reproductive rate, and movement. This kind of information can be coordinated with monitoring of vegetation so that the effects of management can be recorded as well.

Habitat Protection

Until deer populations are controlled at a regional level, habitat protection must rely on local exclosures in areas ranging from the size of an entire park to multiple, small sections throughout a site.

• *Large-scale exclosures.* Before deciding to fence an entire site, you must evaluate the potential impacts to adjacent areas where grazing pressures will be increased and patterns of movement will be altered. Simply shifting damages to surrounding habitats is inappropriate; at the same time, protecting the most valuable and/or vulnerable landscapes is vitally important. And it is important not to disrupt the movement of other animals, including those whose activities play an important role in native plant recruitment.

On some larger properties a combination of fencing and terrain may create an actual enclosure that contains a herd of deer. Such situations are considered ideal for using birth-control programs. Consider applying to wildlife management agencies for status as an experimental site where the herd is effectively captive.

• *Multiple small-scale exclosures.* Two types of small-scale exclosures are recommended for use in most situations: large-gate and small-gate exclosures. Large-gate exclosures, made of large-gauge box wire and range fencing, for instance, are intended to keep deer and other large animals out while allowing the free passage of smaller creatures. Small-gate exclosures, such as chain link, allow access to the smaller herbivores, such as mice, while excluding rabbits, woodchucks, skunks, and other larger creatures as well as deer. Exclosures that exclude all animals, such as some plastic mesh, promote less diversity than those offering access to selected animals.

We recommend that you use exclosures that are temporary as well as relatively permanent, depending on the species being protected and the intensity of browsing and grazing pressure. For example, you might seek to provide long-term protection to a rare herbaceous plant and use an interim exclosure elsewhere only until the tree saplings are large enough to withstand some browsing.

Population Management

No effective landscape restoration is possible where population levels of white-tailed deer greatly exceed the landscape's biological carrying capacity (the level required to sustain indigenous communities and biodiversity). To be unequivocal, it is unlikely you can exclude population management as a major consideration in your program. Hunting is still the most economical and effective method of control in practice. The Natural Lands Trust (NLT) in Pennsylvania has developed a model deer management program for the preserves it owns.

Because the preserves are in populated areas and well used by the public, safety is a primary concern. Hunters must pass mandatory proficiency tests, and are confined to bow and arrow use near residences. They carry permits, wear identifying NLT armbands, mark their locations on preserve maps, and sign in and out. The number of hunters is restricted and all hold a county antlerless permit to harvest does. The program is monitored with field trials on every property to determine effectiveness and actual deer populations and impacts. Despite early controversy, the program is appreciated by hunters and locals alike, and you can once again see bucks with large racks as well as wildflowers on these NLT-managed lands.

We also recommend birth control because hunting is not feasible in all locations. Give serious consideration to this option and begin to thread your way through the red tape. Expect to be stonewalled by the agencies, in part because contraception and population control are highly politicized subjects and this spills over into the realm of wildlife management.

These highly charged conflicts make coordination with game management agencies very difficult. Every individual and agency represents differing premises and goals, which is one of the primary reasons you need an extended and continuous educational and scientific focus. All parties have a long way to go to resolve these issues. You and your project will have to be advocates for bringing a focus on ecosystem restoration to wildlife management. For assistance in this endeavor, contact the Humane Society of the United States, which has excellent information available on most management strategies as well as a newsletter devoted to wildlife management, "Wildlife Footprints," which is dedicated to helping groups become more effective in protecting wildlife and their ecosystems.

Perhaps the greatest barriers still are our attitudes toward wildlife and the deep divisions between people and wild animals in modern culture. Our ancestors who lived more closely with wildlife knew them well. We, too, must come to learn the ways of wild creatures if we are to sustain biodiversity.

CHAPTER 20

The North Woods of Central Park

The ongoing restoration of the North Woods of Central Park illustrates a comprehensive approach to the challenge of establishing a more positive relationship between people and the landscape. After decades of neglect, misuse, and overuse that had left the park in shambles, the North Woods is a model for other urban wildlands.

The renewal of Central Park began in 1980 with the creation of the nonprofit Central Park Conservancy. In 1985 the conservancy published *Rebuilding Central Park: A Management and Restoration Plan.* Since then, the conservancy has raised millions of dollars and coordinated with the Department of Parks and Recreation to renovate the park.

Despite successes elsewhere in the park, by 1984 all efforts to restore its more natural areas had come to a halt, stymied by controversy and disagreement over what constituted restoration. The initial efforts began as restoration of the historic vistas that were part of Frederick Law Olmsted's design, but those who valued the site as a seminatural area feared that design-oriented management would seriously affect wildlife. When some large trees were cleared to reopen one historic vista, opponents brought the proposed work to a halt.

There were what seemed to be irreconcilable differences among the various users and caretakers, as if the disparate individuals and agencies involved could never share a similar vision of the landscape or cooperate to address management issues. Those responsible for long-term care of the site questioned the maintainability of a natural area in Central Park. Some, for example, favored the use of nonnative plants recommended by fish and wildlife agencies to attract wildlife while others saw these plants as invasive pests. Security problems also loomed large. Nor were there any successful models of a habitat restoration in an urban forest. The degradation continued unabated.

In 1988, the conservancy began working with Andropogon Associates to break the deadlock and develop a restoration and management plan for the woodlands of Central Park, including the 90-acre forest at the northern end of the park that is now known as the North Woods. Andropogon's objective was to integrate the restoration of the landscape into the daily workings of the agencies responsible for its care and to involve the public in the project. The following account describes many of the strategies that have been integral to the growing success of this project and the steps that were taken in the process.

A Participatory Process

Because so many different people and agencies have effects on the landscape, the restoration process must provide for long-term continuity and be participatory and broadly representative if it is to be effective. The work process therefore began with a set of interviews to identify the key players and their concerns and perceptions about the landscape. The initial summary report described areas of agreement and disagreement and led to the establishment of the Woodlands Advisory Board, individuals representing the Sierra Club and Audubon Society, the Parks Council, the Institute of Ecosystem Studies, the U.S. Forest Service, and the New York City Department of Parks and Recreation as well as conservancy staff and local citizens.

From the outset the board determined that all decisions would be made by consensus. If the group could not reach agreement, no action would be taken at all. There was a lot of incentive to reach consensus because if the board could not come together to address urgent issues such as security, then others, who might not value the woodlands so highly as a natural area, would eventually do something.

Since its inception, the board has met monthly. Despite some initial fears that a consensus-driven, participatory process would be too cumbersome, the board has made timely decisions and has been able to take immediate action on several fronts at once. At the same time, the board has provided a solid foundation for pursuing long-term design development that is informed by monitoring.

While many innovative city projects are brought to a sudden halt when administrations change or when the one individual who spearheaded the effort moves on, this project has already survived severe personnel and funding cuts to the parks department. The process has led to a remarkable degree of integration of the project into the workings of the parks department and the conservancy. Within a few years of the board's creation, every member was well informed about this landscape, and multiple layers of in-house expertise had been developed in plant identification, exotics management, and path construction. A large constituency, both public and private, now participates in and supports the restoration of the North Woods.

Developmental Process

By proceeding one step at a time and then reassessing our strategies, we can be more responsive to the special and unique qualities of each place.

• *The restoration plan evolves as it grows from the knowledge gained in the process of managing the site.*
Because there was so little agreement about how to implement a restoration plan, the advisory board agreed to take a developmental approach in which each step depended upon the success of the previous stage. The board started with field trials and established short-term goals and criteria for success.

• *Each phase of restoration opens new opportunities and reveals previously hidden aspects of the site, both positive and negative.*
The advisory board made no attempt to craft a comprehensive, grand master plan. At the outset no one had enough information or knew the capacity of this landscape to recover, if natives would reproduce when exotics were controlled, or if inappropriate user activities could be modified at all. The board's willingness to agree to remove exotics depended on whether previous removals met the goals of no increase in erosion and sedimentation or trampling. The board did not agree to a natives-only policy until it could demonstrate that an adequate array of native species would survive and persist on the site.

Prioritizing Local Damage Control

• *Your first priority is to reduce the levels of ongoing damage and stress to the site.*
The board knew it had to take some action immediately, before all data and planning were complete, or the deterioration of the landscape would continue. It targeted four problems that required ongoing efforts to manage:

1. **Off-trail biking, trampling, and vehicles.** There can be no forest recovery, whether natural or managed, with constant disturbance to the soil surface. If restoration is the goal, then virtually all the millions of visitors to Central Park need to stay on paths in the woodlands. This issue cannot be addressed without involving the entire community. Those who use the landscape must take part in order for an effective solution to be developed. The issues are multifaceted and range from needed infrastructure improvements to public relations and policy enforcement.

2. **Poor stormwater management.** Over the years, poor maintenance of the path system and drainage infrastructure as well as extensive areas of denuded soil led to spiraling levels of damage. The original Olmsted plan included a drainage system for adjacent streets that separated urban stormwater from the park's water features. Over time, however, the outlets clogged without maintenance and stormwater overflowed road margins, severely damaging the park's landscapes. The path drainage features designed by Olmsted also collapsed without maintenance. The proliferation of unofficial desire-line trails created by off-trail users, especially mountain-bike trails that ran straight down the slopes, further exacerbated runoff and erosion. Restoration of the infrastructure was a prerequisite to restoring the landscape.

3. **Proliferation of exotic plants.** Norway and sycamore maples as well as Japanese knotweed were overrunning the woodlands and were among the few species reproducing successfully. Although the board initially had little confidence that native communities could be sustained in the park landscape, even with management, it agreed to suspend planting invasive exotics in the North Woods and to limit their expansion by management, which included monitored removals.

4. **Lack of maintenance and security.** Although Central Park is relatively safe compared to the city at large, every incident occurring there is front-page news. It is as important for people to *feel* safe as it is to *be* safe. The need for adequate visibility provided incentive to remove the exotic maples and knotweed where they formed virtual tunnels over the pathways, creating a very uninviting aspect with limited views. The board was very cautious, however, because the clearing of understory undertaken in the 1960s and '70s to increase visibility for security reasons had lasting negative effects on native understory and ground-layer vegetation. Trash and debris also undermined perceived security in the woods, so maintenance was given high priority.

Building Local Expertise

• *Restoration is a local effort that requires new skills and expertise in the local community.*

There is pressure on every major park system to cope with inadequate budgets and, often, a shrinking workforce, at the same time that park use in general is increasing and outdoor recreation is expanding exponentially as an industry. Hiring freezes and cutbacks take an annual toll, heavy equipment is often used instead of laborers, and the overall quality of care is diminished. In the face of these challenges, the Central Park Conservancy instituted many innovative solutions to deal with the bureaucratic and financial hurdles that characterize the management of public land. One of the most important was its commitment to building its own staff to perform the restoration work. The shift in emphasis toward restoration requires a great deal of specialized knowledge and skills by those who are responsible for the landscape but in turn offers the chance to work smarter, not harder.

A crucial step toward the actual implementation of the program was the hiring of a woodlands manager who began to document site conditions, coordinated volunteer activities, and served as an advocate and a caretaker whose first concern was the landscape. This role is crucial to the success of a restoration and may be filled by an individual or a team, but it cannot be assigned to those for whom the landscape is only a peripheral concern. Even though time and budgets are limited, the woodland manager's priorities are very different from those of others in the park. Today, it not hard to understand why this landscape is improving when you walk through the forest with a person who knows every tree seedling personally, who will remark about which volunteered naturally or which were planted and when, and where the seed came from. The woodland manager of the North Woods, Dennis Burton, has recently written a book titled *The Nature Walks of Central Park* (1997) in which he describes the sometimes-surprising wildness of the park.

From the outset, it was clear that volunteers were not enough to complete a forest restoration. The park needed increased funding for maintenance and, more important, increased staffing. The conservancy pioneered the use of in-house crews in addition to outside contractors on capital projects as a way of building their organization. The restoration of the Glenspan and Huddlestone arches that bridge the stream that runs through the North Woods was the first capital project in the woodlands (Figure 20.1). The conservancy bid on and won the contract to perform the landscape management and planting, which permitted the hiring of three individuals for the eighteen-month contract period. The goal was to use that time not only to initiate extensive on-site management, but also to demonstrate how important it is to have workers in that part of the park. This approach provided flexibility and facilitated a more experimental approach to the work than would have been possible in a conventional contracting procedure. The conservancy worked closely with the Department of Parks and Recreation throughout the project. The team modified the design in the field based upon what was discovered daily in the course of the work, but each decision was guided by clear overall policy and guidelines.

The woodlands manager reviewed all the work and supervised a planting crew that today continues as the woodlands management crew for ongoing maintenance of the landscape. The woodlands crew provides a comprehensive array of physical services, from litter removal and limited stonework, to seed and plant collection, new planting, watering, and exotics control. Crew members also provide impromptu tours for visitors and participate in informal educational and public relations programs. The presence of the crew in the

Figure 20.1. The restoration of portions of the stream called the Loch and the Glenspan and Huddlestone arches at either end of this creek provided opportunities for restoration of the surrounding woodlands. (Photo by Sara Cedar Miller/Central Park Conservancy)

area also greatly enhances real and perceived security, and is reflected in the funding received from the Use and Security Task Force of the Department of Parks and Recreation to maintain the crew. The project built a high degree of in-house expertise, perhaps the greatest asset of all.

Providing Appropriate Access

• *Without appropriate infrastructure for access, almost any use of a site can be damaging.*

Walkers and other users tend to stay on trails that are inviting and go to places of interest. If a design seeks only to restore past character and does not take current uses and environmental conditions into account, it will fail. While there was a general goal to retain Olmsted's original path design in the North Woods, the board agreed to several modifications, including a few "adventure" trails in areas where extensive trampling had occurred. Andropogon conducted a workshop on trail construction at the site with the conservancy staff designers, the construction manager, and the construction crew. The narrow trails they designed are as natural in appearance as possible, constructed of large stepping stones, and are also expected to have lower maintenance costs over their useful life than dirt or woodchip trails. The new trail (Figure 20.2) near the Huddlestone Arch replaces over five separate outlaw trails that once coursed down-

Figure 20.2. In-house staff in Central Park completed a new stone trail to replace several unofficial trails. The trail retains and recharges runoff in several infiltration pits along its length. (Photo by James Yap)

slope to the stream that runs through the valley but have now been regraded and replanted. Designed to convey stormwater downslope without gullying, the trail provides access to the water's edge. It gives the visitor a sense of leaving the main trail to explore while staying on a reinforced ground surface. New bridges and large boulders bring the walker close to the water while protecting the soil surface (Figure 20.3).

Creating Positive Uses

• *The park user must play a key role in restoration if we expect to change how the public uses its parks.*

Volunteer labor is a crucial component of the restoration process, not because it saves money, but because it builds positive uses and educates (Figure 20.4). In Central Park, volunteers control exotics, stabilize slopes, and gather and plant seeds of native species from elsewhere in the park; they are also a growing and well-informed constituency that advocates for the natural values of the landscape. In Central Park, what began as a restoration project now has grown to become a major focus of visitors services in the North Woods, providing them with the opportunity to learn about and participate in the woodlands restoration.

Figure 20.3. Replacing bridges and adjacent stonework protects the site while providing access for visitors. (Photo by Dennis Burton/Central Park Conservancy)

Figure 20.4. A volunteer crew under the direction of Dennis Burton, the woodlands manager of the North Woods in Central Park, spent many hours stabilizing eroded soil and removing trash and exotic invasives. (Photo by Sara Cedar Miller/Central Park Conservancy)

At the park's interpretive facility on 110th Street, volunteers keep a wildlife log as part of the growing bio-journal of the site. Volunteers note what species they've sighted and where, any special conditions, their names, and the date. Two members of the advisory board developed a checklist of the birds of Central Park and a field card for the growing numbers of younger, local bird-watchers in the park. Nikon, in response to a plea in the North Woods newsletter, "Woodswatch," generously loans binoculars to children for this effort.

At the present time, over sixty-six schools in the city are involved in North Woods programs. Students are planting trees, shrubs, and wildflowers and measuring the growth and survival of both planted and volunteer vegetation; they also are digitizing and keeping monitoring records for the Biodiversity Project, a program to monitor any deliberate gaps created by removing exotic maples and assess the management of gaps aimed at promoting natural communities.

The conservancy's participatory approach is especially important in its effort to prevent off-trail biking. In the past, careless riders cut numerous new trails into the slopes, which then become conduits for stormwater. The public education program emphasizes the role that cyclists can play in restoring the landscape. Today, young cyclists are signing pledges not to go off-trail or ride recklessly.

The Ramble Project

The North Woods is not the only woodland in Central Park. The Ramble, another small woodland area in Central Park, was one of Olmsted's favorite designs. A jewel-like landscape, the Ramble is a highly dramatic terrain that centers around a constructed rivulet called the Gill. Hills steepened with added fill surround this artificial stream, but by 1995 much of their surface was barren. Mountain bikers, who love the steep slopes, were responsible for high levels of damage, but many other off-trail users, ranging from casual daytime strollers to evening cruisers, were also responsible. Although the landscape had been deteriorating rapidly, it was still known as one of the hottest birding sites in the Northeast. It is strategically important to birds crossing the megalopolis and one of the best places for a stopover.

The board had agreed years before that a comprehensive approach was needed in the Ramble and decided to couple a program of public education and enforcement with a highly visible restoration project in the heart of the Ramble. This approach required substantial groundwork before any physical work projects were begun. A series of meetings with a range of groups from police to bikers publicized the ongoing restoration program and the effort to keep bikers and walkers from going off-trail in the natural areas of the park. New signage in the Ramble alerted the visitor to its status as a protected New York City ecosystem and spelled out the rules for appropriate use of the area. The next step was to restore a highly visible site as a demonstration area. The advisory board selected one of the most damaged sites in the Ramble, where the exposed roots of a few cherry and locust trees still clung to barren soil crisscrossed with ruts and trails. In 1996 the ground crew started by installing fencing that ran right across a popular shortcut and were prepared to repair it daily if necessary. Signs were placed on-site to explain the Ramble Project to the visitor.

Since then the staff has been evaluating two techniques for dealing with compacted soils on the Ramble site. The first entails reworking the entire compacted surface and protecting it with "erosion blanket," affectionately called "trauma blanket" by park workers. Jute mesh is staked in place over a thin woodchip layer covering the bare soil, and undecomposed leaves are put on top of the jute. The second method, "vertical staking," consists of branches driven into the ground like wooden stakes every 6 or 8 inches. Installation requires minimal surface disturbance. The branches convey water and moisture downward into the root zone and loosen the surface as they decompose.

As it turned out, the fence never needed repairing. The ground layer has rebounded beyond all expectations—which demonstrates just how much of the problem is due to disturbance and how resilient a landscape can be if stress is reduced. Sometimes an exclosure works as well with people as it does with animals.

A Monitored and Documented Process

• *Documenting and evaluating the consequences of management are necessary parts of management.*
In the North Woods project, the board decided that all actions at the site should enhance its biological health and diversity and that ongoing monitoring was required to assess impacts on its ecological systems. To this end, the Natural Resources Group, a division of the New York City Department of Parks and Recreation, completed a baseline vegetation survey of the North Woods. A map delineating all major vegetation types is updated periodically during the restoration process, and detailed field notes record site conditions and species present at the site.

The conservancy and parks department staff and volunteers have begun numerous other monitoring projects and field trials as management questions arise. They also maintain a management log and photographic records. The documentation process daily grows more integral to the ongoing care of this landscape, although there is still a long way to go before the North Woods will be considered an adequately monitored site.

Although there was no initial agreement on all implementation strategies, the board agreed to several initial principles to guide its management directions in the interim, one of which was to monitor all actions. Among the first issues to arise was the debate over native versus exotic plants. The group held many divergent opinions but agreed upon an interim strategy that new planting, at least for the time being, should be indigenous to the area and that some limited exotics removal should be undertaken and evaluated. The results of the monitoring have so far been encouraging and suggest that native species can be sustained if exotics management is undertaken. Since all landscapes in the park require some maintenance, this approach appears to be successful.

The cautious removal of exotics was the first step in the process. Norway maple, sycamore maple, and Japanese knotweed were targeted as the three most aggressive and prevalent exotics. The initial goal was to prevent the incursion of exotic invasive plants by removing exotics from the few relatively healthy remaining areas where native plants were still present and reproducing. The next objective was to remove exotics from those areas where they were just becoming established. Lastly, in degraded areas where no natives were reproducing, the goal was to simply contain the invasives and prevent their further expansion.

The Natural Resources Group established field trials, including control plots, to assess conditions resulting from the removal of exotic saplings under 4 inches in caliper. Deborah Lev, the biologist who set up these plots, remarked on the difficulty of finding any two sites in the North Woods that were sufficiently comparable to set up valid experimental plots. Like other urban woodlands, Central Park is both quite simplified yet highly variable because of its past uses.

Soon after exotics removal began and when the competition from invasives was lessened, the native species returned. Each year brings new surprises. In the first year there was a wavelet of tulip poplar seedlings, and by the second and third years, young red oaks started to appear. One year conditions favored sassafras. Red maple, white ash, sycamore, and sweetgum are also appearing on their own. The next generations of native trees are now present, not because canopy was removed to stimulate reproduction, but because the understory was opened up by the removal of the exotic saplings. Of course, the exotic maples have also found opportunities in the newly opened understory after their larger siblings were removed, and they still are the most numerous understory plants although their numbers are reduced. Management is still necessary to sustain the newly appeared native saplings, and the long-term success of the natives remains uncertain, but progress can be measured in the growing number of natives.

One of the most surprising species that is reproducing well in the North Woods is mockernut hickory, a plant rapidly disappearing from many suburban and rural woodlands. Its appearance here is probably due to the heavy and fertile seedfall of the mature specimen hickories elsewhere in the park. A more detailed and quantified study of hickory recruitment is under way and includes mortality assessments on oak and hickory seedlings.

Over time, support for the natives-only policy has grown steadily as the habitat quality has improved and the aesthetic character has been enhanced. Early fears that native communities could not reproduce the drama of the Olmsted landscape have largely evaporated. People now recognize that the North Woods is the only place in Central Park dedicated solely to native plant and animal communities.

When the board decided to tackle the large exotic maples whose seeds provide an endless and immediate source of propagules, the woodlands manager developed a detailed inventory that was intended to help prioritize removals. Because more than one-quarter of the North Woods' canopy trees over 6 inches in trunk diameter are Norway or sycamore maple, the board needed to plan removals so that woodland conditions were sustained. The inventory included information on the size and condition of each tree and noted trees in the understory that would grow rapidly if overstory were removed.

Then one day, literally out of the blue, came the inevitable change in plans. A severe windstorm coursed through the park, bringing down over 150 trees, including two of the largest oaks in the North Woods. While the board was developing the Biodiversity Project, to monitor gaps resulting from removal of the maples, its task shifted to monitoring real gaps. The board elected to replant trees in the Red Oak Gap, as one of the study sites is named, to see if that would jump-start succession; in the Black Oak Gap, they decided to simply remove exotic maples and knotweed as they appeared. Now, each summer, youth interns measure the growth of all understory species, including both planted and volunteer vegetation. In addition, the woodlands manager and staff keep detailed records of all management to permit ongoing assessment and revision of the techniques and strategies. The information from these studies, in turn, informs the ongoing evolution of the restoration and management plan.

Enhancing Recruitment

• *Center your planting strategy around enhancing natural recruitment rather than replacing it.*

If we want to sustain native communities in their habitats, we must apply our propagation skills to restoration. If we make a commitment to this goal, it is possible to maintain the gene pool of each genetic neighborhood.

The changing priorities in Central Park illustrate planting strategies centered on enhancing recruitment. At the outset, the object was to plant something native that would survive and to remove seedling- and sapling-size exotic maples. As native species appeared, priorities shifted. Very early on tulip poplar was especially widespread, so the Woodlands Advisory Board decided not to plant it anymore and to concentrate on species that were not reproducing. Each season is somewhat different. Tulip poplar seemed to establish in a wet year, but a summer of drought saw numerous sassafras stems appear in the understory. Oak and hickory are so special in the park that every seedling is an event.

The same shift in priorities occurred with the ground layer. The project members first used native woodland aster, divided and plugged as well as seeded, because it worked. Now that the ground layer is better established, they are focusing on other species. As native species appear on their own or establish successfully, they are deleted from a proposed planting list consisting of native species known to have been on the site in an 1857 survey of the landscape before the park was built. While many of the tree species found in 1857 are still present in the area, most plants of the ground and shrub layers are absent and no natural means of transport to the site are available. The policy is to reintroduce plants several times, knowing that different seasonal conditions will favor different plants.

The board is increasingly aware of the importance of sustaining local subspecies and the need to find plant sources in the genetic neighborhood of the park. Some native species that are currently unavailable in the trade are being propagated in place in the park and also grown under contract with local propagators. Although initial baseline monitoring showed that there was little environmental rationale to past plantings, all new plantings reflect native communities and natural patterns. As the potential of this landscape is revealed, those responsible for its care raise their expectations.

Because the managers of Central Park are trying to look at whole systems, it is not surprising that they have turned their attention to the soil. They first tackled the problem of extensive areas of bare and compacted soil. The solution included a restored path system that focused on meeting the needs of the visitor while stabilizing and protecting the soil. But even where they had successfully reestablished a continuous litter layer, the ground was still very different from that of a more natural forest and far from a habitat for salamanders and other creatures of the woodland floor. The webby mycelia of fungi binding leafy mats that are so conspicuous in the litter of a natural forest were noticeably absent. Mushroom fruiting bodies were also uncommon, but earthworms were as thick as in a cornfield. The standard, routine soil tests gave no information about the living components of the soil.

The Central Park Conservancy and the New York Department of Parks and Recreation initiated several new tests intended to examine the soil more closely. Margaret Carreiro of Fordham University and her graduate students (Figure 20.5) are completing the first study ever of mycorrhizae under urban conditions. It should be noted that this is literally a groundbreaking study, for not only is it the first of its kind in the eastern temperate forest, but the effort required purchasing increasingly powerful augers to deal with the tremendous levels of compaction in the soils. One of her students, Emer Macguire, looked at the ectomycorrhizae, which are common fungi associated with forest trees. The initial results suggest that fungal populations are reduced but not absent,

Figure 20.5. Researchers in Central Park take soil cores to study the impacts of disturbance and urban conditions on the soil fungi associated with oak trees in both lawns and woodlands. (Photo by Sara Cedar Miller/Central Park Conservancy)

although they occur at greater depths than would be typical of a less-disturbed soil with less surface compaction. Carreiro et al. will later compare their frequency with that on oaks in less-disturbed wooded areas that serve as the site restoration models on the outskirts of the city.

Another student, James Butler, who is working with Steward Pickett, an ecologist from the Institute for Ecosystem Studies, is evaluating the mycorrhizal potential of differing soils, including those of Central Park.

Early on the Woodlands Advisory Board set the goal of becoming a research center for urban forestry. Today, important research is being performed in the North Woods.

There is no end of interesting ways to look at restoration in the context of this complex site. To the historian there is a wealth of uninterpreted information in this park, from the Revolutionary-era soldiers' encampments and the ruins of the famous Mount Saint Vincent's Hospital to the remnants of a Native American village. For the landscape architect the design by Olmsted and Vaux is still timely and yet consistent with the current focus on restoring the natural ecosystems of the park.

The ultimate success of this project is dependent upon changing how the park is used and perceived. Nothing conveys this message so well as coming upon the woodland crew and volunteers at work. Much has been accomplished in only a few years, and it appears probable that, with effective management, the North Woods can sustain a remarkable amount of wildness in the heart of the city and the grandeur of Olmsted's original vision.

PART III

Management Manual

This manual does not describe a single, step-by-step program that you can apply to any site, for two reasons. The first is that every situation is unique and requires that we customize all implementation to each site's specific conditions. The second is that the information will soon be out of date. Techniques and approaches are rapidly changing and being refined as more and more people undertake restoration efforts.

Part III seeks to strike a balance between the uniqueness of every site and the need for some basic approaches and methods. The object of the following chapters is not to give a rigid formula for restoration but rather to prepare each team of restorers to ask the right questions. Although new approaches and products are interesting to learn about and will prove useful to your efforts, what is more important is the knowledge that you and the rest of the team gain about the landscape and the effects your actions have. At every step of the way, seek as much professional, scientific, and amateur help as possible. And as this book recommends repeatedly, implement your plan in the context of monitoring, using field trials to evaluate and refine the specific solutions for each site and documenting your actions.

Remember, you can't do just one thing when it comes to restoration. While your goal may be narrow, such as to control one particular species, many reactions follow any action. Take a small thoughtful action and then test and monitor your hypothesis in the field before implementing it at a larger scale. We may not be able to predict the future, but trying to do so, and then observing and assessing the actual results, is important.

This manual does not recommend brand names or manufacturers for materials that are generally in the market. Suppliers are indicated for those products that are less widely available, but because local markets are always changing, you will need to stay aware of what is available. Modified versions of these products as well as new products turn up frequently and need to be assessed. At the same time, you should urge local suppliers and contractors to consider products and techniques that are not yet in use in your region.

CHAPTER 21

Monitoring and Management

All landscape management is purposeful and reflects specific goals, such as the restoration of an historic plant community. Nevertheless, given the current state of knowledge about ecological restoration and the deteriorating condition of many landscapes, there is no clearly defined path to achieve ecological objectives. We are embarking on a journey that must be informed by science and continually monitored in order to be effective.

At the 1996 annual conference of the Society for Ecological Restoration at Rutgers University, a group of scientists gathered to provide peer review for several ongoing forest restoration projects in the region. The first question was about monitoring, and the topic did not change all afternoon. All those who attended were involved in on-the-ground restoration projects and already facing the difficulties of documenting a site adequately. They pressed the scientists repeatedly for answers about what is most important to monitor and how to do it. At the end of several hours the following key recommendations stood out:

- *Monitor and record all restoration projects.*
The best way to monitor and measure progress and learn through experience is to closely observe and record what is happening on the ground. Without the ability to assess the consequences of management, we cannot justify any action we take. In addition, monitoring provides new knowledge that we can pass on to others.

• *Consider keeping a complete photographic record with an accompanying descriptive narrative of site observations as a first step in site documentation.*
You can do this with minimal tools, such as a camera and a USGS (U.S. Geological Survey) "quad sheet," a topographic survey at a scale of 1 inch to 2,000 feet (1:24,000). On some sites it may be easier for you to record site information on an aerial photograph. Both quad sheets and aerial photos are generally available from public agencies, such as your local county planning commission or state agency for environmental protection, as well as private suppliers. Taken regularly, the images record a broad picture of landscape succession, including unintentional but valuable observations. Your photographic record and written narrative will form the basis of your site database.

• *Aim high despite current limitations on time, funding, and experience.*
Your ultimate success will depend on how well you know the site, so give the issue of monitoring adequate creative attention. You will be grateful later that you invested the energy in keeping records despite how cumbersome it seems at the outset. If time or budgets are small or even if you are an individual working alone, it is important to remember that if you monitor nothing else, at least monitor whatever interventions you make and what you observe to follow them. Lack of funding is not an excuse for failing to monitor, although it requires setting priorities concerning what things are most important to monitor and how to do it.

• *Work closely with local scientists.*
Monitoring, more than any other aspect of restoration, requires effective communication with the scientific community. Involve local scientists with expertise in water systems, plants, animals, soils, geomorphology, and other areas in the restoration team and the project's advisory group. You will not be able to strategize effectively without accurate scientific information, so you must obtain the expertise you need.

• *Develop a system for recording site information that is permanent and convenient.*
A data recording system's ease of use often determines whether or not it is used as intended. If sites are difficult to locate on a map or on the ground, it discourages documentation. Consider installing permanent metal markers, as well as visible physical features, to identify your major plot sites. Many parks use a grid system to identify locations, but specific sites may be difficult to locate in the field in forested conditions. In such a case, you might consider the use of a "Global Positioning System" (GPS), which connects to a satellite network in order to pinpoint specific locations in the field. With a GPS you can digitize your field position almost automatically, eliminating many difficulties of lo-

cating your data accurately on a site map, especially on larger sites, and allowing you to record observations at the designated locations.

Three strategies are key to developing an effective monitoring program: a *restoration team* that is responsible for monitoring; a *site database and management log* that documents conditions, actions, and their consequences; and a *landscape restoration model,* which serves as a reference ecosystem and a standard for evaluating success of the project. In addition, you will need to institute procedures for the following:

- Damage control and periodic site reviews
- Updating earlier information
- Recording conditions on the ground and testing soils
- Recording and evaluating vegetation
- Documenting wildlife
- Monitoring succession

The Restoration Team

An early step in developing a monitoring program is to develop and review your restoration objectives and proposed management. Ideally, this should be done with a broadly representative advisory group that meets regularly. Everyone who interacts with the site should be encouraged to participate in this continuous process. The advantage of this approach is its broad base: considering all concerns will enable the team to review as many potential strategies and outcomes as possible.

In defining your objectives, be as explicit as possible: identify the expected results of management and refine them as your project proceeds, so that your assumptions can be compared later with the actual changes that occur in the landscape and vegetation. Learning when our expectations are incorrect helps to ensure that we learn as much as possible from both mistakes and successes.

Volunteers can implement many of the actions described in this book if they start small, have adequate supervision, and monitor with care. But you will probably want assistance in designing a monitoring program, and you may want to implement actions that require specialized expertise. Your success in large measure will depend upon the breadth of experience and creativity of the people you are able to bring to bear on the problems at hand. Much specialized knowledge is almost certain to be necessary, including engineering, horticulture, botany, zoology, and soils science. When the situation warrants, you should obtain professional advice.

But finding good advisors can be hard, especially in a new field like ecological restoration, and especially for a grassroots organization, a grammar school, or an underfunded agency. You really can't expect one individual to know all you need; you'll need to assemble a team. If you are just starting out, the most effective way to find the right expertise is probably to go to conferences with a restoration theme and meet people face to face, listen to their approaches and findings, and ask questions about shared problems. More than likely, either they or someone they know will have the background you need.

Try to seek out local scientists in area institutions who have an interest in related fields and who might become involved in your restoration. Local schools are ideal for monitoring efforts because they can serve as public repositories of information that in turn can contribute greatly to both the academic curriculum and the community knowledge base. Sometimes all it takes is a mini-grant to make it possible for a class to pay the limited expenses necessary to initiate a special project on your site.

In the Northeast you can get a very broad education by attending the region's many conferences and workshops. Consider a regional meeting of a national or international organization as well as local conferences that may be geared to your specific region and landscape type. Contact your local nature center, your state's chapter of The Nature Conservancy, your state's native plant society, or the Society for Ecological Restoration, and you will soon be well informed about conference and workshop opportunities. If you go to a few sessions, you will likely end up on restoration-related mailing lists, which will help you find scientists and other individuals who are already focused on restoration.

As you become more involved in your restoration project, you will also likely become aware of monitoring projects already under way in your community. The state of New Jersey is using 180 specially trained volunteer census takers to complete the New Jersey Herptile Atlas documenting seventy-two species of frogs, turtles, lizards, salamanders, and snakes. The Nature Conservancy is monitoring butterfly populations through the Butterfly Monitoring Network. To learn about existing programs, call your local chapter of the Nature Conservancy and other local environmental organizations.

The following discussion offers guidelines for a monitoring program that those who are managing the landscape develop incrementally. It is intended to give an overview and to help you coordinate with local specialists, schools, agencies, and institutions.

The Site Database and Management Log

The site database and management log are a continuous inventory of the environmental factors of the site and a record of all site interventions. They may include, but are not limited to, the following:

• *Systematic photographic log with accompanying narrative descriptions.*
A photographic record is important not only because it captures the site's char-

acter, but also because it records many things that are not otherwise intentionally documented. A picture is literally worth a thousand words, especially if it is actually a series of photographs of the same locations taken at regular intervals over time. The accompanying narrative augments the photographic record and provides for additional anecdotal comment. There are many ways to keep these records, from digital photography to snapshots pasted in a looseleaf binder. One of the simplest methods is to make a videotape log with verbal narrative at the time of the periodic review.

• *Research*

In most instances, direction on how best to document a site may come from the past. You will therefore need to do some research, comparing what persists on the site today with historic data such as plant lists, photographs, and trees identified in early surveys and used as markers, as well as deeds, herbarium specimens, and studies. Your local historical society may be a good place to start. Even a small advertisement in a local newspaper asking for historic landscape information may yield fruitful results.

Look for information at varying scales. Area agencies and organizations have a surprising array of information for the restoration-minded. If you are in Brooklyn, for example, you might start with the *Native Species Planting Guide for New York City and Vicinity*, prepared by the Natural Resources Group of the New York City Department of Parks and Recreation (Luttenberg, Lev, and Feller 1993), which describes local plant communities. At the next scale you might want to look at *Ecological Communities of New York State*, published by the New York State Department of Environmental Conservation, Natural Heritage Program (Reschke 1990). At a far larger scale, the *Landscape Restoration Handbook* (Harker et al. 1993), produced in collaboration with the New York Audubon Society, describes the landscape communities of the entire United States.

• *Baseline monitoring*

A baseline study records current conditions of the landscape for comparison with both past and future site conditions as well as with other sites. It is literally the basis for ongoing evaluation of change in the landscape, from plant and animal communities, soils and microclimate, to fire patterns. The inventory should include a recent topographic map depicting all natural and made features and information on the geology and soils, surface water and groundwater, and all plants and animals, such as mammals, reptiles, amphibians, birds, soil microbial communities, and invertebrates. While the usual goal is to complete baseline monitoring for many factors, both biotic and abiotic, and then repeat it regularly, the reality is that a monitoring program is often not instituted all at once but incrementally because of time and financial constraints.

- *Periodic site reviews*

Both short- and long-term landscape management objectives and the methods used will need to be evaluated and reassessed for each site on a regular basis. Some of these reviews need only be very brief and note only what the general conditions are; any active intervention may have to be postponed until other, higher-priority, areas are dealt with. The central point of these reviews is to ensure that regular strategic planning occurs, not just a quick reaction only when damage has progressed. This process also fosters regular reevaluation of the management approach.

- *Ongoing site monitoring*

Ongoing comparative analysis of information in the baseline and historical records is desirable. The process of expanding and updating information about the site needs to be continuous over time.

- *Ongoing record of regular maintenance activities and amounts of time spent*

Include records of trash pickup, weeding and watering, fence repair, tree maintenance, and whatever other activities occur on the site. Simplified forms help limit the time taken away from hands-on site work.

- *Ongoing records of special-purpose management activities to restore and maintain cultural landscape features such as vistas and gardens, including a description of existing conditions, recommended action and rationale, and actions taken; key this information to your site maps*

Many natural areas are part of historic sites, and so their management must be integrated. Preserving both is not mutually exclusive: Restoring historic native communities enhances the authenticity and context of other historic and archaeological restoration efforts.

- *Ongoing assessment of all management*

Follow-up is needed to evaluate actions. In addition to recording what occurred according to your observations, include your recommendations for further action and monitoring.

- *Site journal*

This continuing record of the observations of the restoration team should note daily and seasonal variability, such as blooming schedules, as well as ephemeral, occasional, or unusual events such as a population explosion of a small mammal and migrations. Often the importance of what is recorded is not known until later.

Making a habit of documentation combined with inquisitive observation is basic to restoration. All management is an effort to direct succession, so try to see the landscape changing before your eyes. What did the landscape look like ten years ago? What will it look like next year? Ten years from now? Fifty or a hundred years from now? How might the landscape be altered with management? Ask yourself as many questions as possible about the region in general and the relationship of your site to the larger landscape. Ideally, this exercise should be a dialogue engaging a variety of different people who are familiar with the site, a process to inform the monitoring program. Test your observations with other people and other sites. Ask a lot of questions of people, over and over.

Landscape Restoration Models

Where a site has been altered from its historical condition, it is important to identify other landscapes that can serve as suitable and feasible restoration models. Research into historic site conditions on these reference landscapes derived from literature, site analysis, or local residents and users usually serves as a primary source of information for a restoration effort. Historic plant surveys, for example, allow you to perform an invaluable comparison with current conditions.

You may also be able to locate in your region similar but more natural examples of the habitat you are restoring to serve as valuable models for your restoration effort. Look for natural landscapes within your region where ecosystems are like those on your site, places that share similar geology and soils and that occupy similar terrain but are less disturbed than your own site. These landscape restoration models are a source of information not only on soil, vegetation, and wildlife, but also on the larger natural processes that support local biodiversity. A site restoration model also can be used to determine criteria for evaluating the success of management. Monitor the landscapes of the model sites for the same factors you monitor at the restoration site; the records will serve as a source for comparative analysis and a measure of success. Opportunities to compare the more natural area with the restoration site are limitless and range from temperature shifts of the forest floor to invertebrate populations, from appearances of resident and migratory birds to leaf predation, from recruitment patterns to nutrient cycling. It is also very useful to look at similar but more disturbed sites to see likely future problems.

Landscape restoration is not about creating a completely new landscape that is an idealized version of nature, any more than a child can be viewed as a lump of clay to be molded into whatever form we choose. Rather, the object is to optimize the potential for "self-design" of the site, the site's ability to recover and develop by means of its own inherent processes, to allow it to become the highest-quality natural landscape that can be achieved given its condition and the context in which it is set.

For any area there is a wide range of landscape types that might occur over time, as well as many that would not be at all likely. A large, relatively undisturbed site will present a very different range of options than a narrow corridor in a developed landscape fabric would, no matter how similar their original conditions. The wider the deviation of the desired outcome from the intrinsic conditions of the site, the greater the chance for failure. For example, a created wetland in an upland landscape is far less likely to be supported by an appropriate hydrologic regime, despite creative engineering, than a wetland reestablished in its former setting.

For the North Woods restoration in Central Park, the natural landscape that defines the nature of the project is the band of oak forest that begins on the granite highlands at the north end of the park and extends into upstate New York and New England. The site in Central Park is one end of a gradient of wildness that ranges from this island of forest in an urban area to less heavily disturbed wildlands hundreds of miles from the city. Central Park is part of this landscape, and its central management themes recur in more natural areas throughout the region because the factors that threaten the larger ecosystem are magnified in the city but also pervade the entire matrix.

Those working in Central Park are studying regional natural areas that share conditions similar to those found in the park to gain perspective on what can be accomplished and what species should be considered. One site, Saxon Woods, off the Hutchison Parkway, is subjected to less air pollution, is somewhat older, and has been less disturbed by trampling and vehicles. It is a good model for what we can aspire to if we reduce the level of disturbance in Central Park. In addition to studying this and other sites, the managers are reconstructing the historic record in an effort to ensure that propagules of species most representative of the ecosystem are available on their site. The objective is to provide opportunities for self-design of this site by ensuring the presence of all the species that would be there if the site had not been isolated from the larger systems.

In Prospect Park in Brooklyn, New York, managers are using local soils to determine the reference ecosystem or model that would help define what plant species are most suited to the site. Plants communities are very closely linked to the subtle differences in soil types in a given area. Even where the more specialized vegetation of the past has been largely eradicated, the remnant soils can help determine appropriate new plantings.

The northern end of Prospect Park occupies the Harbor Hill Moraine, the terminal moraine of the Wisconsin Glacier about 18,000 years ago. The flatter southern portion of the park sits on the edge of the outwash plain of the glacier. Despite significant modification to the soils by grading, the soils in these parts of the park still express their past composition: the hilly northern sections still consist of a boulder-strewn stony loam, while the coastal plain soils of the southern areas are largely remnant sandy and silty loams. To restore vegetation, the managers have developed two separate lists of recommended plant species that reflect the differences between these historic landscape types.

Some sites are so disturbed that a completely different habitat than the previous condition may be more appropriate. For example, the Freshkills Landfill on Staten Island, New York, rises hundreds of feet above what once was tidal wetlands that are unlikely to be restored, at least in the near future. It would have been difficult, to say the least, to establish largely native grasslands and young woodlands that would in time succeed to forest. A well-managed native meadow may be a better interim step on a journey to mature forest than skipping all early-successional steps and reintroducing mature-forest species in the site's highly modified soils.

The best approach is to learn everything you can about the site's former role in the landscape. Beware of the microcosm approach in setting your restoration goals—that is, avoid trying to miniaturize the larger landscape and to represent the entire ecosystem on a single site. Each and every site has an inherent role based on its place in the terrain and its relationship to the larger ecological context. A mini-region might be a reasonable goal for an arboretum exhibit, but most sites need to again fill their historic roles, as a floodplain forest or tidal meadow, for instance. A stream valley is more important as a piece of the larger hydrologic system than as a place to create all of the habitats of a region. The goal is to reweave whole cloth, by stitching the patches together.

A restoration project may begin as a specific site, such as a park, but the focus and context always need to include the larger ecosystem. The most appropriate management area may not necessarily conform to property lines or political boundaries. Therefore, try to keep records both of your project area as well as the larger, once-whole system. Today, what is left of a forested stream valley may be part of a patchwork of properties in varying levels of naturalness, each offering different opportunities and constraints, or it may be confined to a narrow strip of park in an otherwise totally urbanized setting. Think of the initial site as your foot in the door to the restoration of an ecosystem. As the authors of *Restoring Diversity: Strategies for Reintroduction of Endangered Plants* (Falk, Millar, and Olwell 1996) note in their Introduction, "[W]e suggest that the restorationist's view must be simultaneously on the big picture and the small, on the grand design and the fine details."

Damage Control and Periodic Site Reviews

One of the first tasks of management should be to arrest damage so that no more ground is lost. The restoration team must determine priorities. Just as one purpose of long-term monitoring is to notice impacts and trends that might otherwise go unobserved in the short term, a purpose of a periodic review is to notice those more sudden changes and new impacts that require timely action. One of the first actions taken by the restoration team in Prospect Park was a detailed map of soil disturbance (Figure 21.1).

To be effective, the site review team should consist of people who have been a part of the regular review process to ensure continuity as well as a few fresh faces added occasionally to bring new ideas and perceptions to the project.

In conducting periodic site reviews, paying attention to first impressions is essential. What has changed since the last inspection? Have site conditions improved or worsened? If so, why? How do conditions on the site today compare with those recorded in the last set of photographs and comments?

While damage control is the first priority, you should also notice areas of richness that might not have been observed previously and are inadequately protected. Even where the site appears undisturbed, evaluating the existing level of protection and establishing an adequate review schedule are still important.

Often the restoration team must establish priorities to do the most good with often-limited resources. One way to do this at the outset of a project or where the level of impact seems overwhelming is to categorize the sites as stable, declining, or degraded. This step will help ensure that the most valuable habitats are given the best protection and that further damage is effectively contained.

The specific conditions for these criteria will vary with each landscape, but often, as a first step, you can briefly map just the most serious impacts, such as the presence of exotics, erosion, bare soil, or deer browse, to provide a quick overview of the site's status. If you are especially concerned about exotics, for example, stable habitats might include those areas that are largely native. Declining areas would have native species present but exotics reproducing more vigorously, and degraded areas might be largely exotic. These terms are of course relative. In the North Woods Project in Central Park, where almost no stable, largely native patches existed, a more important distinction was between those areas in the early versus late stages of decline.

In stable areas, typically the most important goal is to prevent to the extent possible any future damage by means of management. In some cases, however, management in declining sites usually centers on fostering remaining native species and reducing ongoing stresses. A degraded site may have no native elements to rescue or sustain. There the goal might initially be to ensure no further expansion of the invasive plant without actually attempting to eradicate it. A priority of management would be to prevent that area from serving as a source of infestation to surrounding landscapes.

The condition of the ground is another useful measure of current stresses, which may include drainage problems as well as use. Look for overt impacts such as dumping or trampling.

The next step is to develop a protection program for those areas that are declining. The site team must identify crucial actions to take to curb damage and to initiate repair on the site. You also must try to identify the source of the

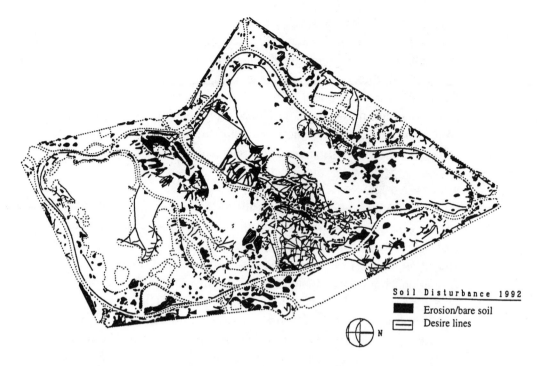

Figure 21.1. The staff at Prospect Park undertook detailed mapping of existing ground conditions to document the extent of erosion and trampling damage for a comprehensive restoration plan. Note that the steep slopes may be readily discerned.

damage as well as the full extent of the area of impact, both of which may extend beyond the project's boundaries. Stormwater damage, for instance, usually crosses many boundaries and jurisdictions. An invasive exotic might be a more localized problem or of regional significance.

Some Considerations in Monitoring

Instead of describing the many monitoring options and techniques, this section focuses on the objectives of monitoring and the kinds of questions that may arise. In the course of monitoring you will be documenting many specific factors in the landscape, but the primary goal is to better understand the place as a system embedded within a larger system.

The Watershed

All issues related to water in your landscape will also relate to the larger watershed and sub-watersheds of your site. You will be able to obtain a significant amount of information from federal, state, and local agencies for your area, as well as your local watershed association, to develop a detailed survey of the watersheds in your landscape.

In the Wissahickon Watershed Stewardship Program in Philadelphia, volunteer local stewards are designated for each sub-watershed of Wissahickon Park. The volunteers are preparing detailed maps of the watershed identifying drainage conditions throughout the landscape as well as areas of severe erosion, debris and dumping, and a variety of other factors related to water management. In addition, they have developed a photographic manual including examples of every condition they are documenting as well as different levels of severity to ensure consistent and accurate mapping.

The Ground

Review available soil data where available, beginning with your county's soil survey, which you can obtain from the Natural Resources Conservation Service, formerly called the Soil Conservation Service. In urban areas labeled "made land," your prospects for good soil are poor and you will have no real information about specific site conditions. In metropolitan areas, where the landscape is typically a fragmented patchwork of relatively undisturbed to highly degraded areas, vegetation may give you important clues to the relative naturalness of the soil. For instance, remnant native plant communities are likely to be associated with remnant patches of less-disturbed native soil profiles, where there has been minimal grading, filling, plowing, liming, or fertilization. Compare your present soils and associated vegetation to native soils in the region's less-disturbed areas, evaluating how they are different or similar to historical soil conditions.

In cities like Philadelphia, where strict fire codes meant stone and masonry construction, there is often enough lime from residual mortar in soils to elevate the pH above historical levels. Rather than moderately acid, like the native soils in most of the surrounding area, the soils in many places in Philadelphia's small parks are circumneutral—that is, neither acid nor alkaline—good for lawn but not for native forest communities, which are associated with acid soils. Consider requesting complete soil mapping for your site if all of it is presently labeled as made land. Information on soils and parent material will be one of the best indicators of appropriate planting strategies.

Soil monitoring today typically analyzes physical parameters such as nitrogen, phosphorus, and potassium, moisture, heavy metals, chlorination, and other hydrocarbons. For standard agricultural soil tests contact your state land-grant college of agriculture or your county extension service of the Natural Resource Conservation Service, U.S. Department of Agriculture.

The soil's living components such as mycorrhizae and other fungi and soil flora, gastropods and invertebrates, arthropods, and nematodes and soil macrofauna are equally important in restoration as the soil's physical parameters, but less commonly documented because the techniques for measuring them are still new. Look for researchers specializing in soil food webs in your region. One company, Soil Foodweb, Inc., for example, provides assays of soil and compost microbiota including active and total fungi and bacteria, as well as protozoa and nematodes. An interpretation of the results is available. For information, contact Soil Foodweb Inc., 980 N.W. Circle Boulevard, Corvallis, OR 97330, 541-752-5066.

The status of the soil microbial community is a good indicator of the status of the larger landscape system. Research has demonstrated that measurements of soil microbial activity can be used to evaluate the levels of both ecosystem damage and recovery. Three characteristics are crucial to this assessment: the amount of biomass, the rate of turnover of organic materials, and the level of biodiversity in the system. All three factors are very low in a highly disturbed site (Harris and Hill 1995; Bentham et al. 1992).

Jim Harris and Tom Hill (1995) measure both biotic and abiotic factors in assessing the status of soil microbes. Using the levels of enzyme activity of living creatures, they assess total biomass and total microbial activity. They then determine the proportion of fungal biomass. These measurements, when combined with total nitrogen, soil organic carbon, the ratio of carbon to nitrogen, and the ratio of microbial biomass carbon to organic carbon, give scientists a clear picture of the status of soil communities and system dynamics. While these sophisticated tests are not yet widely available to restorationists, it is important to be aware of what kind of information you may ultimately want to obtain. For those who wish to know more about microbial succession as well as restoration in general, see *Land Restoration and Reclamation: Principles and Practice* by James Harris, Paul Birch, and John Palmer (1996), published by Addison Wesley Longman Ltd., Longman House, Burnt Hill, Harlow, Essex CM20 2JE, England.

Sometimes you will not have all the relevant data you need or want in order to make decisions. Even when specific expertise is not available to you, consider an interim solution like mapping all fungal fruiting bodies, such as mushrooms and puffballs, that appear in the landscape by making a photographic record. At some later time, it will provide valuable data for analysis. Map all rotten logs and woody debris that might determine areas where additional management to favor soil microorganisms would be useful. Jim Harris pointed out another method at a conference: you can take soil cores now and dry and store them for analysis when funds become available.

Vegetation

One of the most useful activities anyone can engage in is to make a *vegetation map* of the site. This usually is a very revelatory process that familiarizes the participants with the site's conditions. The most important task is determining what vegetation types to map. Unfortunately, no federal agency has comprehensively mapped plant community types in the United States as the Soil Conservation Service did in its nationwide soil mapping program. The information available on vegetation is variable and in many instances may be little more than the extent of woodland that is plotted on USGS quad sheets and wetlands maps from the National Wetlands Inventory.

The most useful vegetation mapping is probably of plant associations, or plant communities. A "plant association" consists of species that are frequently found together under similar environmental conditions as well as in roughly the same population proportions. They are usually described briefly by the names of the most common species in the order of their predominance, such as an oak-hickory forest or a beech-maple forest. Because vegetation is a reflection of many site factors, mapping plant association types can also reveal important information about soil types, drainage conditions, and potential animal habitat.

This type of mapping is usually done by delineating areas of apparently similar vegetation on aerial photographs and then verifying the "signature" of that type in the field. Different vegetation communities have subtly different textures and characters on an aerial photograph that are recognizable to a trained interpreter. Different kinds of photography, such as true and false color, different scales, and the season or hour of the overflight permit different levels of refinement. Some vegetation types are more readily discerned than other kinds. For a large site, "Global Positioning Systems" are invaluable, but a small site with many identifiable site features can be mapped entirely on the ground. Wherever plant-community-type mapping has been done in the past it offers an exceptional opportunity for comparison with new mapping to see what changes have occurred.

In less-disturbed forests, consider mapping the ground-layer vegetation, especially the herbaceous layer, which is the most vulnerable, so you can identify and protect areas of biotic richness as quickly as possible and get an overview of the general level of variability within the landscape. Sue Bratton and Albert Meier (in press), who research forest succession, propose a procedure for determining which native ground-layer species are likely to need additional propagation and reintroduction based on current levels of recruitment success. They suggest that the ground-layer plants that merit special attention include mesic species, species dependent on gaps for recruitment, plants dependent on another species that has been extirpated, and species that have narrow microhabitat preferences, require deep organic soils, disperse or grow slowly, or have only small and scattered populations. You will need to sample in both spring and fall to adequately survey ephemeral plants.

In more degraded landscapes, there may be little correlation between existing plants and where they would occur naturally. In Central Park, for example, the legacy of past plantings bore no relation to where a plant would occur naturally even when native species were planted. The baseline vegetation mapping documented current conditions but did not reveal much about the landscape's intrinsic suitability for native species because there was little rationale to the pattern of vegetation. The goal of mapping was simply to record what was there for comparison in the future. The park and conservancy staff developed two types of maps: one recording the major canopy vegetation, the other recording the ground layer, including both herbaceous species and woody seedlings, as well as levels of soil disturbance. All canopy trees over 6 inches in diameter were located on a grid map of the site to document those trees over time. The trees also serve as reference points for recording locations of planting and other management activities in the management log and site database.

In addition to vegetation mapping, the most common tools used in baseline monitoring are "transects" and "plots." Site transects are especially useful for assessing and documenting vegetation at a large scale when a broad overview is needed. Sampling plots usually serve to give a more detailed picture at a smaller scale. A transect is an imaginary line marked out on the landscape, along which the occurrence of plants or any other factor can be counted, such as all woody stems that are on the line or every plant that occurs within a set distance from the line at certain points. Often very detailed sampling of a small area, called a "quadrat," is taken at either random or regular intervals along the transect. The metric system is usually used for scientific sampling, so a quadrat might be 1 square meter. Where the research focus is narrow, the transects might be determined randomly, but in the highly variable and often-patchy quality of the real landscape today, setting the transects intentionally is often advisable to ensure that the transects intersect all habitats and different site conditions. Research and documentation do not necessarily have the same goals or exactly the same requirements, so also consider taking photos along the set of transects or plots you develop and keep a continuous photographic log of these cross sections of your landscape.

Sampling plots are designated smaller areas within the larger landscape in which you count frequencies of plants or take other measurements to indicate conditions in the larger system. Like transects, sampling plots are not necessarily located randomly in an evaluation of a restoration effort. Quadrats are usually located randomly within a plot.

You can also use plots and transects to assess specific stress factors. You can assess deer browse or the amount of defoliation by gypsy moths that season, for example, in different parts of the landscape or from season to season. Or you can evaluate the relative density of understory stems that are crossed by branches or leaves.

You can vary the scale of both plots and transects to fit the scale of the site. In Louisville, Kentucky, for example, a local scientist is assessing the effectiveness of ongoing management in several parks designed by Olmsted by using plots at multiple scales sited intentionally throughout the parks. The plots range in scale from large sites (one-twentieth of an acre) for counting all woody stems to smaller sites (one-fiftieth of an acre) for counting small shrubs and vines, to even smaller sites (6 feet by 6 feet) for counting herbaceous plants and calculating the percentage of the ground that they cover. Together, those scales will portray both the larger character and smaller details of the parks' landscapes.

An exclosure, as described in Chapter 24, "Planting," is a fenced-off site. Used as a specialized sampling plot, it helps determine the impact of a factor by excluding it from a portion of the landscape. The fenced-off restoration sites in Central Park, for example, reveal how much will grow simply by excluding human foot traffic; they are not unlike deer exclosures in the countryside. An exclosure is an excellent tool for revealing new aspects of your site's plants and processes that were obscured by continuing damage.

Floyd Swink and Gerould Wilhelm (1994), botanists at the Morton Arboretum, developed a method for assessing the quality of plant communities that is based on what they call an "index of conservatism." The index of conservation measures what proportion of the species present in the landscape are conservative species, rather than generalists and opportunists. The process is rather simple in concept and begins with gathering the best area botanists and ecologists to evaluate each individual plant species based on their knowledge of its requirements and associates. They then assign a value, called a "coefficient of conservatism" to each species, based on the extent to which that plant is adapted to specialized habitats. A weedy plant found everywhere has a very low value while a highly specialized plant has a very high value. A combined value, called the "floristic quality assessment" or "index" is then assigned to a given landscape site based on the values of all the native species present. No re-created landscape scored higher than 35 in their study. A score of 45 or higher almost certainly indicates a site with natural reserve area potential. A landscape with an even higher value merits the greatest protection.

Determining the relative conservatism of species is vital in order to prioritize restoration efforts. The approach can also be used as the standard of a restoration's success and so can shape your entire restoration program. It has been used, for example, to restore tallgrass prairies, and the state of Michigan has recently completed a floristic quality assessment (Herman et al. 1997).

An interesting aspect of the method is that it is based on the quality of remaining communities without regard to the presence or absence of exotics. This way of valuing a landscape is rooted in the recognition that once a specialized landscape is disturbed, its biotic integrity is obliterated forever, as if it were an extinct species.

Wildlife

Baseline surveys of existing wildlife often begin with birds because of the extraordinary numbers of skilled amateur birdwatchers in many communities and the degree of organization, leadership, and documentation provided by the Audubon Society, including its annual Christmas bird counts and breeding bird surveys. Surveys of mammals, reptiles, and amphibians, however, usually require the assistance of specialists as well as permits from state and federal wildlife agencies in many cases because the animals are often trapped for census taking. You may find it helpful to contact local organizations to learn about ongoing wildlife studies that may apply to your own project.

Many techniques that do not require the handling of live animals, such as counting footprints or roadkill surveys, also can be extremely useful sources of data. You can regularly review what tracks are made on a freshly raked patch of ground. Sometimes a scent lure such as urine or a visual lure, such as a suspended, shiny disk, is helpful. Specific approaches vary with species. Frog counting is often done by ear because most calls are easily distinguishable. You can often get a good deer census in an area with an aerial photograph or overflight after a fresh snow, when they stand out clearly. Another useful approach is to survey the landscape for habitat elements that benefit selected wildlife, such as dead trees, called snags, and fallen logs, vernal pools, and other seasonal wetlands.

The succession of animals, even very small ones, proceeds in concert with plant succession; therefore invertebrate population surveys provide special opportunities for monitoring the success of restoration efforts. Microbial invertebrate communities in soils give a better reading on the successional status of a forest restoration than the more obvious features of gross vegetation, since they are representative of a tremendous diversity of species that play an extraordinary array of different roles critical to ecosystem function, including detritivores, predators, and prey. Success in reestablishing rich insect communities, for example, suggests that many other natural functions have been restored along with the more visible elements of vegetation. Such methods, however, require real expertise, not always available to a land manager. If expert help is not available, pay attention to the most obvious elements, which are the plants and the habitat's characteristics.

It is also important to recognize that animals, like plants, also have different levels of conservatism. Ron Panzer of Northeastern Illinois University noted that about 25 percent of local insect species are missing or absent from degraded landscapes (Panzer 1995). He calls these species "remnant-reliant"— that is, their survival depends on existing relict populations. Remnant-reliant insects often depend on conservative plants, unlike remnant-independent species, which are less exacting in their requirements.

Kathy Williams (1993) of San Diego State University devised a program to compare arthropods from a restored riparian woodland to populations in adjacent natural wetlands. The restored site was intended as habitat for the endangered least Bell's vireo, but no pairs had returned after the first three years. Nevertheless, her monitoring of arthropods showed rapid establishment of most of the common insects, although their population ratios were different from those in the natural area that served as the model. The re-created landscape had far fewer larger insects compared to the natural one. When the planted landscape matures, the trees are taller, and the landscape structure is modified, however, the numbers of large insects will increase and food sources may then be adequate to support the vireos.

Arthropod monitoring also revealed information that will be useful for future restorations in the area intended to benefit least Bell's vireos, such as what species supplied the most abundant food sources. In Europe, monitoring the status of land snails has been a useful indicator of the restoration of natural associates, especially in limestone landscapes. Different species favor woodland conditions, that is, greater than 50 percent closed canopy, and others favor earlier successional stages (Magnin and Tatoni 1995).

As noted earlier, Susan Bratton and Albert Meier (in press) suggest that in the eastern forest we should focus on salamanders as indicators of the conservation of native species and as standards for evaluating the success of wildlife preservation and restoration efforts. There are dozens of species of salamanders and reptiles in the eastern forest and even greater diversity at the subspecies level. Despite the occurrence of up to six species in a single square yard under undisturbed conditions, their return to a cleared landscape may easily take a century or more, making them excellent indicators of both forest quality as well as recovery. Despite the scientific controversy over how quantifiable the use of species as indicators is, each landscape has some species of special concern that help to reveal the status of the larger system. In Pennsylvania, for example, the regal fritillary, a once-common butterfly, has suddenly all but disappeared for reasons that are not fully understood, raising serious concerns about the future of many other species.

Although there is a tendency to overstate the conclusions that can be drawn from the monitoring of indicator species, selective documentation of specific elements of the system as a gauge of the larger system is very useful. Consider focusing on species or a group of species whose success may relate to another species or group of species, such as ants, which are seed dispersers of many woodland ephemerals.

Monitoring Altered Succession

The restoration of an oak-hickory forest being undertaken at the Brecksville Nature Center demonstrates the way a management strategy based upon altered succession is defined, monitored, and evaluated. The site in question lies about thirteen and a half miles south of the city of Cleveland and is part of the Metroparks system. Until the 1920s the land was in agriculture and now supports an even-aged oak woodland, a canopy so dense that only sugar maple and beech reproduce beneath it. The historic landscape supported maple and beech on the side slopes, but the hilltops were oak-hickory, the forest that the park managers want to restore and sustain. The regrowth of the oaks after agriculture produced an even-aged condition that was not typical of these forests historically. This condition, in turn, led to three significant changes in the population dynamics of the habitat: a decline of understory and ground-layer species, a decline in overall tree growth, and the invasion of beech and maple. The management plan that the Brecksville Nature Center developed included five major strategies:

> 1) monitoring of species presence and their vigor, 2) prescribed burning with a ground fire during the unleaved condition, 3) cutting of Beech and Sugar Maple that are too large to be damaged by the burning, 4) cutting of some oaks to open up the canopy and 5) planting of seed from local ecotypes (Smith 1990, 108).

In addition, the managers determined four goals for evaluating the success of the project:

> 1) seventy-five per cent of the species listed will set seed at least once out of three years and produce offspring at least once in every five years, 2) Beech and Sugar Maple will produce seedlings less than once in ten years, and will be less than 15 per cent of the total basal area, 3) White Oak and Shagbark Hickory will produce seedlings more than once every fifteen years, and 4) non-native species will be less than ten per cent of the total above ground stems (ibid.).

These goals define clear objectives whether or not they are achievable with the current management. They set specific parameters for assessing and—more important—*modifying* the strategies based on what occurs. The standard of measure is the landscape and what happens on the ground.

The managers in Central Park are evaluating how best to manage successional change along the Loch, the stream in the North Woods. The old willows that once lined parts of the channel have died and fallen over and apart in place. At first the landscape looked somewhat trashed, but quite soon it attracted a wide diversity of birds. For a while the gaps were rich in new plant species as well—that is, until the young ash trees took hold and started to grow apace. Now the questions are not about removing exotics but whether or not to remove natives, albeit very common species, to sustain these special places. The answers will be found only by setting clear objectives, monitoring outcomes, and then refining strategies. Ultimately, all of our efforts will matter to the extent that the landscape recovers. Monitoring is the crucial tool that brings the details of the real landscape into focus.

Ground Stabilization and Soil Building

Restoration usually begins on the ground. Restabilizing the ground surface is a prerequisite to revitalizing the soil, which is essential to reestablishing native plant and animal communities. Existing problems tend to worsen without soil restoration. Once eroded, a path continues to channel and concentrate runoff, increasing its erosive force. A compacted area attracts more off-trail traffic that further inhibits germination and root growth. Although the damage to the ground in many fragmented landscapes often seems overwhelming and difficult to prioritize, it is essential to initiate repair of the landscape as quickly as possible.

The issues involved in soil restoration are complex and range from reducing negative impacts by users and managers to reintroducing invertebrates and fungi. All are complexly interrelated. Your first priority should be to focus on protection and restoration in those areas that support the most diverse native communities. Therefore, you will want to begin your restoration efforts in the areas that will best serve to protect the most intact natural communities that remain. Second, you should always avoid or minimize any new impact.

You will likely encounter many problems related to soil restoration on your site. This chapter describes strategies and techniques for stabilizing and building soil:

- Providing appropriate access
- Minimizing disturbance
- Retaining stormwater
- Removing fill and debris

- Stabilizing slopes and gullies
- Replacing and amending soils
- Repairing compacted soils
- Stabilizing the surface
- Managing dead wood and brush

Providing Appropriate Access

Restoration and management require appropriate access to the landscape so that visitors and managers may engage in positive uses of the site. The object is not to wall off natural areas, leaving nature to take its course, but to intervene intentionally to manage resources. Access is necessary; without good paths, people trample at will and do far more damage than they would otherwise. But determining what level and kind of access are appropriate is a difficult task. We generally overestimate our need to access a landscape, especially with vehicles, and underestimate negative impacts. Many natural areas are far too easily accessed and vandalized.

Guidelines for Access

• *Evaluate the nature of current access, its impacts, and the real need for access.*
Evaluate both legal and illegal access as well as the source areas. Correcting the problems may require working with local community groups to create new opportunities for positive uses and improve local site surveillance.

• *Do not accept damage as a consequence of access; rather, accommodate access in ways that do not degrade the resource.*
Focus on establishing *positive* access, such as multiuse trails. Determine the most minimal access requirements and evaluate how to meet them with the least impact.

• *Where unauthorized vehicular access occurs, effective barriers as well as signage and periodic enforcement may be necessary.*
Solutions range from fences and chains to berms and rock piles. The availability of materials should determine the nature of the barrier.

Guidelines for Nonvehicular Trails
A path is an attempt to balance impacts: it offers protection to the surrounding ground surface as well as a means to confine pedestrian or vehicular use, but it is also a source of disturbance. Open any new trail with great care. It will have long-term impacts on the landscape and entails a commitment to maintenance. Most paths in urban wildlands are undermaintained and become

sources of stormwater damage, trash, and exotics. Often deterioration proceeds so far that outlaw trails are indistinguishable from once-paved paths.

A multiple-scale trail system is often the most practical approach to dealing with differing site conditions and providing for visitor needs. The system might include a main trail accessible to vehicles for maintenance and security purposes as well as to wheelchairs and other assisted access. The secondary trail system might be for pedestrians only, or it might also include bike loops and horse trails. Consider a single-file adventure trail to permit limited access to special places that walkers consistently favor, such as an overlook or pond edge where a wider trail would be inappropriate. The use of large stepping stones or wood rounds on adventure trails minimizes construction of hard surfaces (Figure 22.1). Boardwalks and bridges are often the most successful access in lowlands.

Figure 22.1. Well-set stepping stones create an adventure trail that gives some of the feeling of going off the trail but without the impacts.

Wherever new pathways are proposed, first lay them out on the ground so they may be evaluated and modified as needed. The design of the trail is a crucial component in controlling erosion. Well-sited trails designed with the proper surface and drainage can minimize environmental impacts and are simply more maintainable. Signage is also important to inform the visitor about access. On heavily used trails, consider a low rail to contain walkers and protect adjacent vegetation (Figure 22.2).

• *Monitor and repair all trails on a periodic basis.*
The major purpose of trail maintenance is to control erosion. All access results in impacts. The only way to ensure against unacceptable levels of damage is to monitor and then take action when the first sign of damage is noted. Evaluate the condition of the infrastructure, identify potentially hazardous conditions, and recommend appropriate actions. Evaluate the condition of trees and branches overhanging major public walkways to help minimize the risk to visitors from falling limbs and trees.

• *The path surface must be adequate to carry the level of traffic it receives.*
Over time, the condition of the trail itself will illustrate the need for a more durable surface or better care. Periodic graveling may be needed where use is moderate. Where use is heaviest, a paved surface is typically required.

Bare soil is not an acceptable path surface in a heavily used park because it wears poorly and cannot be distinguished from an outlaw trail. When bare soil is visible anywhere, people feel free to trample far more casually than when paths are well maintained and erosion well controlled. Therefore, you will need to eradicate and revegetate all bare-earth, desire-line trails and restabilize all areas of bare soil. Boardwalk is an excellent solution in any fragile habitat, not only wetlands. It conveys a strong message to the visitor about the fragility of the landscape. A boardwalk effectively contains pedestrian movement and can often provide barrier-free access in uneven terrain with less site modification than is required for a shallow-grade path. Many simple boardwalk designs can be implemented by volunteers.

Several excellent trail maintenance handbooks are available. For copies, contact the publishers listed below:

Trail Building and Maintenance, 2d edition, by Robert D. Proudman and Reuben Rajala (1981). Appalachian Mountain Club, 5 Joy Street, Boston, MA 02108, 617-523-0636.

Trail Design, Construction, and Maintenance, by William Birchard, Jr,. and Robert D. Proudman (1981). The Appalachian Trail Stewardship Series. Appalachian Trail Conference, P.O. Box 236, Harpers Ferry, WV 25425, 304-535-6331.

Appalachian Trail Fieldbook—A Self-Help Guide for Trail Maintenance, by William Birchard, Jr., and Robert D. Proudman (1982). The Appalachian Trail

Figure 22.2. In the New York Botanical Garden a low wooden rail substantially reduces off-trail damage along the main path. (Photo by Marianne Cramer/Central Park Conservancy)

Stewardship Series. Appalachian Trail Conference, P.O. Box 236, Harpers Ferry, WV 25425, 304-535-6331.

Trails Manual—Equestrian Trails, by Charles Vogel (1982). Equestrian Trails, Inc., 13741 Foothill Boulevard, Sylmar, CA 91342, 818-362-6819.

These manuals are geared to larger, wilder areas than many of those found in more urban areas, but the techniques are still applicable, especially where paths are largely unpaved. Also very worthwhile is a series of English trail manuals designed for volunteer conservation corps. The numerous topics covered illustrate the very sophisticated level of maintenance sustaining a public and private trail network in Britain that is centuries old in places. Compiled by Alan Brooks, *The Practical Conservation Handbooks* were revised in 1982 by Elizabeth Agate. Titles include *Waterways and Wetlands* and *Footpaths.* The handbooks are published by the British Trust for Conservation Volunteers, 36 St. Mary's Street, Wallingford, Oxfordshire OX10 0EU, telephone: 0491-39766.

For a thorough guide to developing and maintaining multiuse trails, consider *Trails for the 21st Century: Planning and Design,* edited by Karen-Lee Ryan et al., published by Island Press, 1993. This manual was developed by the Rails to Trails Conservancy in cooperation with the National Park Service.

Temporary Access

Before starting any site work, think carefully about its possible impacts. Well-meaning weeders trample soil no differently from walkers wandering off the trail. Select your route carefully, keeping to permanent trails to the extent possible.

When working off-trail, consider using a bucket-brigade line to convey materials, including plants, to minimize trampling (Figure 22.3). Instead of everyone walking back and forth, the carefully placed individuals in the line pass materials back and forth from person to person. Route access through areas of disturbance or walk over exotics while avoiding those areas where native reproduction is still occurring.

Any access, even by small-wheeled vehicles, is potentially damaging. Use planks on the ground, for example, to keep from making wheelbarrow ruts. Several workers can carry fairly large loads, such as rocks in nylon cargo slings. For heavier loads, try using rollers. Study the site before starting work, and be sure everyone is informed about each step of the procedure.

Minimizing Disturbance

Finding a construction site in the Northeast that is not surrounded by dusty roads and muddy creeks is next to impossible. The use of protective site fencing to preserve natural areas is even rarer. Parks and wildlands often end up as dumping grounds and staging areas for adjacent construction projects. Even restoration work entails at least some level of additional disturbance. Consider advocating and/or instituting the following steps to minimize disturbance:

• *Install fencing wherever potentially damaging activities, such as site work and construction, are carried out.*

Fencing and enforcement are required to effectively contain large equipment. On an active construction site, inspect and repair fencing three times daily: once prior to starting work, at midday, and upon closing the site after work each day. Enforcement is important, it must start on day one, and must be reinforced repeatedly.

There are several major kinds of site fencing, each of varying strengths, from chain-link, for the most extreme situations, on sites where no access at all is permissible, to plastic site or wooden "snow" fencing (Figure 22.4) in the least problematic situations. Range and box wire fencing (Figure 22.5) are adequate for intermediate levels of protection. Be careful not to damage existing vegetation in the course of installing or removing fencing. For most types of fence, the shorter the distance between posts the sturdier the fence. But while strong fencing seems desirable, it is also expensive and may involve more damage during installation. A complementary program of education and enforcement may be more effective than increasing the strength of fencing.

Figure 22.3. A bucket brigade minimizes damage by workers on-site.

Figure 22.4. Site fencing keeps visitors on the renovated path while the landscape beside it grows after restabilization and planting.

Figure 22.5. Range fencing in Central Park protects small channel islands from disturbance while vegetation becomes established.

- *Where fencing is used, either permanently or temporarily, extend protection to as large an area containing root systems as possible.*

Begin the effort to protect roots by trying to determine where they may be. Assessing the exact pattern in which roots grow is not easy; people usually underestimate their extent. Roots are notoriously opportunistic and go where the gettin' is good. The patterns of forest plant roots are complex, ranging over extensive ground in adaptive patterns rather than predictable ones. For an open-grown tree the rule of thumb for estimating the area of important roots is to start with the area of the outermost limit of the branches, extended by half again. Remember, though, that variability is high. Root systems in the forest often overlap. Study the visible indicators of the root zone, such as raised soil over roots that are close to the surface.

Remember, too, that fencing often alters walking patterns and may even lead to the creation of new desire-line trails. Walkers, for example, often follow the outside edge of a new fence.

- *Evaluate the potential for damage and develop a strategy to minimize root damage where disturbance is inevitable.*

Damage exceeding one-third of the root system is likely to lead to the death of even the most hardy specimen. Even when you decide to dig manually rather than using earth-moving equipment, you still may not be able to reduce root damage sufficiently and will need to explore other options. Where the damage will be too severe to save the tree, consider leaving the tree to die in place if it presents no likely hazard. Where a larger tree presents too great a liability, try cutting the top to remove the wide canopy only, leaving the trunk as a snag to serve as a den tree and insect habitat. Elsewhere it may be advisable to consider cutting down a tree that will be excessively damaged and encourage new shoots, called "coppice growth," from whatever roots remain. In a few cases, horticultural root pruning undertaken at least a year or even several years prior to possible damage may be used to try to save special vegetation. In general, however, root pruning is very expensive and entails damage to adjacent plants and therefore should be avoided except in the most exceptional situations. Tunneling beneath roots is an option in some cases.

- *Hand-dig wherever root impacts are probable.*

Hand-digging is costly but not as expensive as attempting to replace the value of what might be lost. When excavating manually, attempt to save all roots 2 inches or greater in diameter. Wherever a root is cut or accidentally damaged, be sure to trim any ragged ends to leave a clean cut. Do not paint or treat the wounds in any other way. Be aware, however, that even hand-digging will do damage, especially to smaller plants, shallow-rooted species, and plants growing in thin soils.

• *Keep any roots exposed by excavation damp or they will die.*
Backfill as soon as possible; use wet straw or burlap as an interim cover if there
is any delay. Remove any excavated trash from the spoil, and make sure no trash
is disposed over the area of root excavation. Compact the backfill a few inches
at a time to the original firmness, except at the surface if the soil had been overly
compacted. Water the backfill to eliminate air pockets and moisten roots.

• *Where root damage has occurred, evaluate the need for additional irrigation
during drought conditions for at least a two-year recovery period.*
Plants do not recover immediately from stress; it takes time for roots to regrow.
Monitor the soil's moisture to ensure that you water only when necessary to
sustain valuable plants through a drought immediately after root loss. Do not
fertilize a damaged plant because the resulting new growth typically creates an
increased demand for water that the compromised roots are not able to meet. If
the plant was not pruned beforehand, pruning the shoots to compensate for
root loss may be advisable, although it should be done with caution. This once
popular approach has recently been called into question by experiments at
Cornell University demonstrating that pruning to make up for root loss does
not benefit a newly transplanted tree. Instead, a new transplant seems to benefit
from any and all leaves because of the net energy gained from its greater leaf
area. Similar studies have not yet been carried out under forest conditions.

• *Where trenching or excavation must be undertaken, take special measures to
minimize damage.*
You may wish to refer to a simple handbook on trenching, *Trenching And Tun-
neling Near Trees: A Field Pocket Guide For Qualified Utility Workers.* [Tree City
USA Bulletin], edited by James R. Fazio. National Arbor Day Foundation, 100
Arbor Avenue, Nebraska City, NE 68410.

• *Silt fencing and other runoff controls are needed wherever soil is exposed to
increased erosion.*
Erosion and sedimentation in natural areas caused by poor runoff control in
nearby construction areas is commonplace. Be sure to work closely with your
local soil conservation service agents to ensure better enforcement. Maintaining adequate erosion and sedimentation control on your own site is equally
important, even for very small scale projects that are not subject to regulatory
review.

Silt fencing may consist of hay bales staked in place or fabric barriers staked
along the land's contours to retard runoff and trap sediment. The most effective
silt barriers start below grade in a shallow trench, but their installation means
damage to soil structure that would not be acceptable in any relatively undisturbed area. Simple straw bales staked into the ground to secure them are less

effective but usually entail much less damage. You can somewhat compensate for the lack of effectiveness by decreasing the spacing between the lines of bales. Always try to minimize the erosion potential overall by minimizing the amount of land area exposed at any one time. Likewise, do not clear the entire work area at once if it is not necessary.

Retaining Stormwater

You can make a variety of small changes to terrain to modify drainage characteristics of the landscape. Some techniques are only suitable where existing plants can be replaced with vegetation that is equally valuable. Many effective methods entail some soil disturbance. Soil bioengineering techniques such as brush layering and live staking, described later in this chapter, are exceptionally useful where flows are heavy and the establishment of vegetation would otherwise be impossible without added protection. These techniques are less suitable for a wooded stream corridor, where only limited soil disturbance can be tolerated. The most difficult areas to restabilize are woodlands because even minimal grading damages vegetation, and reestablishing plants is difficult because of shade or the specialized requirements of many forest species.

Wherever there are open areas such as lawn or even pavement, there are likely to be opportunities for collecting and storing excessive surface runoff until it can naturally infiltrate the ground. The most obvious choices for water storage impoundments are places where there is already occasional standing water and where a slight increase in the water level could improve retention and infiltration. This is especially true for lawn areas, where even a gentle slope can result in rapid rates of runoff.

Small-scale impoundments can retain rainwater long enough to greatly improve infiltration rates locally. They are advisable for lawn areas near woodlands, streams, and wherever rapid runoff presents a problem. A small berm, a dike to contain and direct runoff, even only 4 to 6 inches high, placed on the low side of a patch of lawn or open field can change drainage characteristics significantly (Figure 22.6). While no single such impoundment will make a great difference, many small sites placed throughout the landscape can decrease runoff at a regional scale.

Where the slope is long, low, and largely open, a sequence of berms and trenches can be effective in slowing stormwater movement downslope to effect better infiltration. The number, spacing, and sizing of the trenches and berms will vary with the site. The idea is to create many small opportunities for storage and infiltration. Remember that digging trenches and making berms are forms of disturbance that are not suitable where less-intrusive methods would work as well or where native soils or vegetation are undisturbed. On most sites you can build a berm with the soil from the trench and do not need to transport additional soil to the site.

Figure 22.6. This flat depression in the turf is also a retention area that fills with water for a few hours after a rain to create a round, ephemeral reflecting pool.

A series of berms and trenches as well as bands of tall grass to filter runoff can diminish damage due to runoff from such areas as athletic fields and other parklands to adjacent seminatural areas. Allow the berms and trenches to grow tall grasses and wildflowers if the site must otherwise be kept open for some purpose; otherwise the goal should be to manage for a gradual return to forest.

Under woodland conditions where the construction of berms and trenches on a slope may be too damaging, you can minimize the downslope movement of water by establishing a shallow berm along and outside the upslope edge of the forest in adjacent less-vulnerable landscapes. Depending on soil conditions, the berm can sometimes be augmented by a shallow, gravel-filled infiltration trench that runs parallel to the berm; otherwise the trench may silt in rapidly. A filter fabric lining in the trench between the soil and gravel will help maintain the berm's permeability longer. Filter fabrics allow water but not small particles to pass through and therefore keep sediment from filling the gravel bed. Another important factor to consider is any potential impact to the stability of the slope (Figure 22.7). Remember that the berm must be at right angles to the slope or it will convey water in the downslope direction.

Although only a small volume of runoff can be held and recharged in this manner in any one place, if these measures are repeated throughout a watershed or continuously along a slope, you can achieve significant restoration of historic drainage patterns.

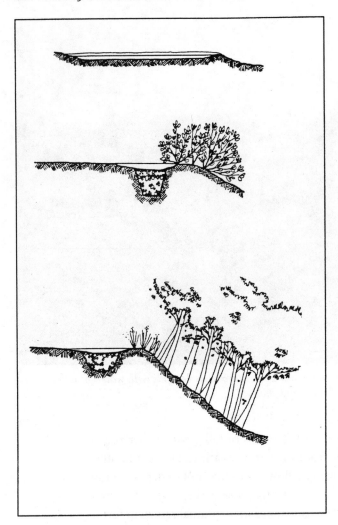

Figure 22.7. Typical sections for shallow bermed impoundments.

Shallow excavations are sometimes used to augment retention in upland areas. Shallow depressions maintained in turf grass tolerate periods of brief inundation. Low spots planted as a native wet meadow of ferns, sedges, rushes, or lowland trees and shrubs tolerate longer periods of standing water (Figure 22.8). As before, it is important that the runoff does not carry high sediment loads or these areas will silt in too rapidly. If bare soil areas are appropriately stabilized, however, sedimentation should not be a problem.

The sources of water for upland wetlands and shallow pools include floodwaters from adjacent streams and rainwater. If the outflow of a pool is between 1 and 2 feet higher than the new wetland's lowest elevation, it will retain some water seasonally. If at all possible, construct a simple water-control structure to enable you to manipulate water levels, especially in the early years when vegetation is becoming established. All slopes should be shallow enough to be stable even without vegetative cover. If you vary the bottom elevations, you can avoid monocultures such as cattail.

Figure 22.8. Volunteers in Wissahickon Park in Philadelphia combined exotics removal with stormwater management by diverting small amounts of runoff to the depressions left behind where knotweed roots were grubbed and by planting native species, such as witch hazel and skunk cabbage.

Natural regeneration is likely to be best in new wetlands that are adjacent to existing wetlands. The suitability of plant species depends on the depth and duration of standing water. Some tree species, such as persimmon and buttonbush, will tolerate flooding for more than one year, while swamp white oak and basswood will tolerate less than a month during the growing season. Other species, primarily herbaceous, occur in areas that are permanently underwater. Blue iris and lizardtail, for example, are found in shallow water about a half foot or so in depth, while sweet flag and pickerelweed are more likely to be found between 6 and 18 inches in depth. Water lilies are still deeper. There are many excellent references on wetlands planting available. Restoring a more natural, that is, seasonal hydrologic regime to a wetland area can be very beneficial for germinating seeds that often require a relatively dry period.

Choose wetland plants with care. In some instances grasses, such as fescue, planted as temporary cover in basins persist to become future pests. Switchgrass is an excellent native substitute for exotic redtop and annual rye for stabilization purposes and will soon die out wherever there is standing water, permitting other species to become established.

When revegetating wetlands, consider using both seeds and plugs. Plugs, ready-to-plant seedlings, are more reliable and develop faster but are more costly. Seeds are prone to rotting in wetlands and may require reseeding.

In less-disturbed forests where erosion and sedimentation are not serious, consider creating a seasonal wetland to help restore historic hydrologic conditions. "Ephemeral pools" are seasonal wetlands. Because many are too small to be protected under wetlands legislation, ephemeral pools are being filled or converted to open water habitat. For some species, like salamanders, pools like these are essential because they dry up part of the year and so cannot support fish that prey on salamander larvae. Ephemeral pools require very low levels of sediment and minimal fluctuations in water level in order to sustain rich amphibian populations, so they are not necessarily good options for controlling excess runoff. Ephemeral pools also require proximity to forest because salamanders and many frogs cannot cross far over open fields, much less roads.

Many ephemeral pools are the result of seasonal high groundwater or perched water tables created by spring rains and snowmelt over soils with low permeability. These are called "vernal pools." Groundwater-dependent pools generally have the least fluctuation and may therefore be quite shallow. Where pools must rely on surface water only they will also require refugia. "Refugia" are simply deeper areas in the pool that serve as a safe haven during periods of fluctuating water, which is likely when you have only surface water to support the pool. Where there is flowing water at least seasonally, it is also important to ensure that there are also areas of still water with vegetation where species can attach their eggs.

Improvements to in-stream habitats can also help to increase retention and reduce runoff as well as manage sediment. The general goal is to re-create the varied environments of a natural channel when they have been modified by excess runoff or channelization. Boulders placed in a stream channel provide cover for small creatures and vary the current, which in turn varies the habitat. Similarly, you can place channel-lining rocks to re-create natural patterns of riffles and pools, which typically recur approximately every six times the width of the stream. You can also excavate small pools in the channel bottom or create shallow plunge pools using low checkdams to restore more normal flow patterns.

The best journal on watershed management is *Watershed Protection Techniques: A Periodic Journal on Urban Watershed Restoration and Protection Tools,* published by the Center for Watershed Protection, 8737 Colesville Road, Suite 300, Silver Spring, MD 20910, 301-589-1890. They also publish an excellent book entitled *Site Planning for Urban Stream Restoration.*

The best journal on establishing wetlands is *Wetland Journal,* published by Environmental Concern Inc., P.O. Box P, St. Michaels, MD 21663, 410-745-9620.

For an excellent manual on how to protect streams, see *Living Waters: How to Save Your Local Stream,* by Owen D. Owens, published in 1993 by Rutgers University Press, 109 Church Street, New Brunswick, NJ 08901, 732-445-7762.

For stream restoration techniques, see *Better Trout Habitat: A Guide to Stream Restoration and Management,* by Christopher J. Hunter (Island Press, 1991), and *Stream Analysis and Fish Habitat Design: A Field Manual,* by Robert W. Gastonbury and Marc N. Gastonbury (1993), published by Newbury Hydraulics, Box 1173, Gibsons, British Columbia, Canada V0N 1V0, 604-886-4625.

An excellent manual on restoring a variety of habitats, including lowlands and streamsides, is *Restoring Natural Habitats: A Manual for Habitat Restoration in the Greater Toronto Bioregion,* prepared by Hough Woodland Naylor Dance, Ltd., and Gore and Storie, Ltd. (1995) for the Waterfront Regeneration Trust, 207 Queen's Quay West, Suite 580, Toronto, Canada M5J 1A7, 416-314-9490.

Toronto, Canada, has an excellent school program geared to amphibian conservation and restoration that is described in *For the Love of Frogs Build a Pond,* published by the Metro Toronto Zoo (1992). Contact the Adopt a Pond Program, P.O. Box 280, Westhill, Ontario, Canada M1E 4R5.

Removing Fill and Debris

Reestablishing the historic terrain is an important aspect of restoring natural patterns. Therefore, the first step before initiating any ground stabilization is to evaluate the site to determine where and the extent to which grading has occurred and whether the natural terrain can be restored. You should also attempt to compare the vegetation that would be affected by regrading with what you feel can be reestablished.

The more recently the filling occurred (optimally, no more than a season or two), the more likely that removing the fill will save existing plants that were not cleared or severely damaged. Pull some fill away from the remnant tree trunks to examine the condition of the bark. Some species, such as sycamore, are more tolerant than others and may survive years under partial fill, sending numerous adventitious roots out from the buried portion of their trunks. Others, like tulip poplar, are much more sensitive and may show signs of bark rotting after only months. If the trees still appear alive and vigorous and do not show signs like dieback at the tips, loss of limbs, or broad areas of rotting bark, consider removing the fill. Substantial hand labor may be necessary to keep from further damaging the soil with heavy equipment.

• *Sometimes fill removal is required just to reduce problematic grades and permit adequate slope stabilization.*
Typically, a slope steeper than 3 feet horizontal to 1 foot vertical (about a 33 percent slope) on fill soils will have persistent problems of stability over the long term. Where the major vegetation has been killed by filling, complete

regrading may be needed to restore the original terrain or at least reduce the slope's steepness. Use the lightest equipment possible to minimize damage to the soil. A small dragline operated from the top of the slope to pull the fill away from the site may be a feasible method. Once the fill is removed, restabilize the ground as soon as possible.

• *Reuse soil to the extent feasible and properly dispose of the remainder.*
This step will require some real follow-through on your part. Transport is costly and disposal often creates environmental problems elsewhere. Reuse as much of what is removed as possible, separating the rocks and soil in the fill and using them elsewhere.

• *When fill is left behind, it often causes long-term problems for vegetation management.*
If the fill cannot be entirely removed, or too little can be removed to save existing vegetation, it may be necessary to cut down the dead or dying trees along walkways and wherever else they pose a hazard. Otherwise, leave dead trees in place for their value to wildlife. In either case, keep the trunk and branches for use on-site to restore the ground layer.

Filled slopes and valleys are often easily colonized by exotic invasives; therefore, control of exotics and replanting of native species are likely to be required on a continuous basis. Many such urban sites contain tons of construction rubble that years after deposition still support only mugwort, common reed, and a few isolated tree-of-heaven. The fill material is usually of very poor composition and may be excessively drained, very poorly drained, or both in patches. Adding a costly layer of topsoil is not enough to address the long-term problems of these areas, although it may support plants for a while before they start to die off.

Improving these areas is challenging. It is usually necessary to undertake detailed investigations of what comprises the fill before making any final decisions. At the very least, remove junked automobiles and old appliances, tires, and other large rubbish very carefully to avoid undue disturbance to the site. Remove all the trash you can. Any rubbish left behind is only an incentive for further dumping. As with graffiti, the most effective approach to dealing with trash is to promptly remove it.

In some places, construction rubble may be so mixed in with the fill soil that separation and removal of the rubble component is extremely difficult and removal of the entire layer too extreme. Do not expect such sites to support more than the toughest plant communities, at least for a long time. The presence of continuous cover alone will ameliorate extremely bad conditions by loosening soil and adding organic matter over time.

Never simply dispose of large boulders recovered from fill; they are usually extremely valuable. They can be used in streams to enhance habitat and make artificial "bedrock" surfaces to create single-file adventure trails. A boulder trail for climbing on or a path edged with boulders provides access and yet serves to protect the more sensitive areas of the landscape, such as the water's edge.

Stabilizing Slopes and Gullies

Not all erosion problems can be solved by vegetation alone. Where erosion is severe and has created gullies, soil replacement and additional support are necessary to hold the terrain in place until vegetation can become reestablished. Where runoff rates are so rapid that even established vegetation is damaged by stormwater, replanting alone will be ineffective, at least until better watershed management reduces flow volumes and velocities. Where current rates of runoff cannot be reduced, it may be necessary to enlarge and/or reinforce the channel using techniques such as bioengineering.

Biotechnical Slope Stabilization

"Biotechnical slope stabilization," also called "soil bioengineering," uses living and dead vegetation as well as various soil amendments to provide a greater degree of stability than that afforded by conventional grading and subsequent replanting. Under forested conditions, however, biotechnical slope stabilization may be of limited use because there is rarely adequate light to stimulate adequate shoot development. More important, many of the techniques entail significant soil moving, which is often not possible under forested conditions without causing undue damage. In general, it is best used in extreme circumstances, where there has been severe disturbance and there is no need to protect any remaining vegetation.

Live Staking

Live stakes are living pieces of stems or branches taken from trees or shrubs with the ability to root vigorously from cuttings that are driven into the soil and eventually root in place (Figure 22.9). In one bioengineering technique, they are used as vegetative stock for new plants and, like reinforcing bars in concrete, give added structural support to the soil. Soon after planting, the brush begins to develop roots, which further stabilize the soil column, often to a great depth, while the plants rapidly develop woody shoots above ground. Sometimes herbaceous cuttings are used as well as freshly cut branches that are still unrooted when planted.

Many native floodplain and wetland species—including willows of all kinds, many elderberries, shrub dogwoods, box elder, and, less reliably, sycamore, alder, and viburnum—are especially suited to this technique because of their tendency to root easily from cuttings, especially when covered with soil. Usually at least 40 percent willow is incorporated because it serves to stimulate rapid rooting of other species.

Push live stakes into ground at a 45-degree angle so that stems can develop roots along length.

Figure 22.9. Live stakes will root deeply if they are driven in right side up and at a 45-degree angle.

Beyond its remarkable suitability for bank and slope stabilization, live staking is a useful planting method that employs lightweight materials that can be harvested locally. Although only a handful of species can be established this way, the rapid establishment and effective stabilization afforded by this method make it appealing in special circumstances.

Some hints, learned through experience:

- *Live staking works best with vigorous cuttings.*

Look for firm green wood. You can harvest the same site repeatedly because species like willow, alder, and other plants used for live stakes sprout readily after cutting. This practice, called "coppicing," has been practiced for millennia, especially for basketry, and uses the same materials that are ideal for bioengineering.

When unrooted cuttings are used, collect and plant them on the same day if possible and no more than twenty-four hours after harvesting. Keep them moist during that time; do not let them dry out completely. The window for installation is also very narrow, confined to the coldest part of the winter, when the plants are completely dormant, which is typically during the month of February and a bit of March. The season must be advanced enough that the ground can be worked but not so late that buds are swelling.

Each cutting must have at least two bud scars near the top of the shoot. Cut the larger or butt end of the cutting at a 45-degree angle to help drive it into the ground as well as to indicate which end goes into the ground. Upside-down stakes do not grow. Cut the top end blunt or square. Tamp the cuttings into the ground at a 45-degree angle to maximize rooting. Use a dead blow hammer that has sand or shot in the hammerhead to absorb the shock of the blow rather than splitting the stake. Discard any stakes that split. In hard soil you can use an iron bar to create a pilot hole before planting. You can also drive live stakes between rip-rap and gabions (rock-filled baskets) to help establish vegetation along armored streambanks.

Fascines

Fascines are fresh, long cuttings secured every foot and a half with wire or twine to create a long roll about 6 to 8 inches thick. A fascine works like a cable to reinforce a stream edge or steep slope. The cuttings can be from 5 to 20 feet long and are easily constructed on-site (Figure 22.10).

Place the fascine in a freshly dug trench nearly the depth of the fascine and cover with soil (Figure 22.11). Use your gloved hand or a wooden stake to work the soil into air pockets in the fascine. Drive 18-inch-long split wooden stakes through the fascine about every 3 feet to secure it in the soil.

Fiber Roll

A fiber roll, or "fiberschine," is a long, sausagelike roll of coir (coconut) fiber or other material that reinforces banks and serves as the planting medium for

Figure 22.10. Workers easily construct fascines on-site.

Figure 22.11. The upper surface of the fascine should remain exposed to light after the trench is refilled.

vegetative cuttings and small seedlings planted after installation. The roughly textured coconut fibers, protected in a coir or jute mesh, take about five to seven years to disintegrate, and until then slow the flow of water, thereby causing sediment to drop and become trapped around the new plants. The living roots and shoots of growing plants in the fiber roll as well as the accumulated sediment gradually replace the dead coir fiber, which serves as a durable but temporary aid in stabilization (Figures 22.12–22.15).

Where less durability is required, you can use cheaper materials, such as straw, for the rolls. Fiber rolls help to avoid the use of stone gabions or sheet piling, and can be used with live staking.

To effectively apply these techniques, you should consult manuals available on biotechnical slope stabilization. You might start with these excellent references:

Bioengineering for Land Reclamation and Conservation, by Hugo Schiechtl (1980). University of Alberta Press, 141 Athabasca Hall, Edmonton, Alberta, Canada T6G 2E8.

A Streambank Stabilization and Management Guide for Pennsylvania Landowners, published by Commonwealth of Pennsylvania Department of Environmental Resources, Office of Resources Management, Bureau of Water Resources Management, and Division of Scenic Waters (1986). State Bookstore, P.O. Box 1365, Harrisburg, PA 17105.

Water Resources Protection Technology, A Handbook of Measures to Protect Water Resources in Land Development, by J. Toby Tourbier and Richard Westmacott (1981). The Urban Land Institute, 1090 Vermont Avenue NW, Washington, DC 20005.

Figure 22.12. The stream in Trexler Park in Allentown, Pennsylvania, before restoration.

Figure 22.13. Creative Habitats in White Plains, New York, propagated sixty species of native plants and installed coir fiber rolls (fiberschines) to stabilize the new shoreline until the vegetation became established.

Figure 22.14. Close-up of a fiber roll.

Figure 22.15. Within a single season the corridor was transformed from an actively eroding channel to a richly vegetated wet meadow that stood up to several major storm events within weeks of installation.

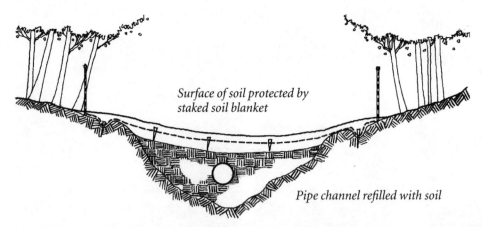

Figure 22.16. *Where a pipe disgorges partway down a slope and gouges a gully, it is sometimes useful to extend the pipe and refill the gully. Evaluate where you can reduce the volume of water or provide retention for runoff without undue damage.*

Biotechnical Slope Protection and Erosion Control, by Donald H. Gray and Andrew T. Leiser (1982). Van Nostrand Reinhold, 135 West 50th Street, New York, NY 10020.

Workshops devoted to these techniques are held throughout the year across the country. Watch your local environmental newsletter for announcements.

Stabilizing Gullies and Erosion Channels

Where runoff has increased dramatically and has created deep gullies, the eroded gully is usually far larger than the actual volume needed to carry the runoff. This occurs because the increased velocity of runoff erodes the gully bottom and incises a deeper channel from which floodwaters cannot escape, which in turn increases the volume and velocity of the flow. In such cases, consider partially refilling and reinforcing the channel to allow it to carry the flow without eroding a deeper gully (Figure 22.16). Where the gully has been created by an outfall that conveys water partway downslope in a pipe but creates erosion where it emerges unpiped midway down the slope, consider extending the pipe to a point closer to the base of the slope where and if grades are shallower. You can then fill over the pipe and partially refill the rest of the ravine, leaving a large enough swale to carry surface runoff. The ground surface may not require further reinforcement if the outfall was the primary cause of the gully formation.

This slope protection method does not provide any retention and may aggravate storm surge-related problems, however. Ideally, retention should then occur at the base of the slope, although in many steep terrains that is difficult or not possible.

Where shallow gullies have formed, a series of low, wooden log checkdams or fiber rolls staked in place reduce the velocity of runoff and encourage the deposition of sediment within the gullies (Figure 22.17). Reducing the velocity of

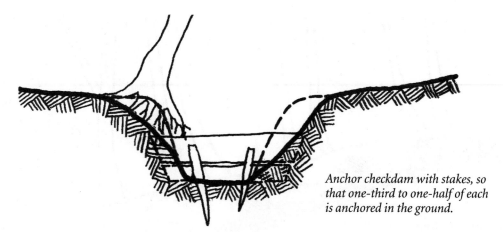

Anchor checkdam with stakes, so that one-third to one-half of each is anchored in the ground.

Figure 22.17. A stepped sequence of checkdams, designed as still pools and cascades, can reduce the erosive force of runoff in a channel. Checkdams slow the water enough for the sediment to settle out and accumulate inside on the slope above the dam.

stormwater in turn reduces its capacity for erosion. Checkdams also foster sediment deposition all along the length of the gully rather than only at the bottom of the slope.

Keep checkdams small and place them at frequent intervals. Start at the bottom of the slope and work upward (Figure 22.18). Locate the dams so that the top elevation of the downstream dam is no lower than the bottom elevation of the upstream dam or scour will occur. They also must be low enough to confine the flow of water within the gully or new channels will be cut at the sides in a major storm.

You can build checkdams from tree trunks and large branches found on the site. Checkdams made from wood typically do not last more than three or four years, by which time the site should be stabilized if runoff has been adequately controlled. You will not be able to complete gully restoration until the velocity of the runoff is reduced to a rate that does not adversely affect new vegetation once it is well established.

Cribbing

Cribbing is one of the most common techniques for repairing severely eroded slopes. Cribbing, or cribwork, generally consists of rigid supports, such as boards, logs, fiber rolls, or fascines, that are anchored in place like stair risers along the slope's contours to slow the flow of water. They reduce the erosive force of runoff and encourage deposition and the gradual buildup of the slope.

You install cribbing like checkdams, starting at the bottom of the slope and working upward. Be sure to place the logs exactly parallel to the contours to ensure that you do not direct water downslope but rather hold it behind the log. Use a level to align the logs (as shown in Figure 22.18); then stake them to hold them in place. Drive the stakes to anchor the log at right angles to the slope and at 3-foot intervals along the logs.

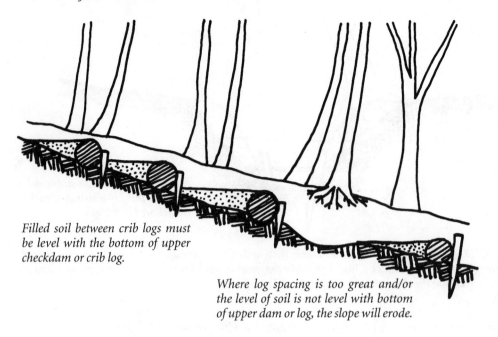

Filled soil between crib logs must be level with the bottom of upper checkdam or crib log.

Where log spacing is too great and/or the level of soil is not level with bottom of upper dam or log, the slope will erode.

Figure 22.18. The spacing of checkdams and cribbing is critical to erosion control.

Setting the log in a trench gives greater stability than placing it on the soil surface, but a trench is only appropriate where excavation will not unacceptably affect adjacent vegetation.

You can use untreated waste wood for cribwork. Where wood or logs are not available, you can use fascines made from brush. While they will not develop shoots under the shade of woodland conditions, they still serve to reinforce and retain soil and can be fashioned from locally available materials. Fiber rolls are also suitable. Place branches over the restored slope to help discourage trampling (Figure 22.19).

The two most common problems with cribwork are associated with commonly used, inappropriate cost-cutting measures. First, the "steps" may be spaced too widely, resulting in erosion between them and retarding permanent stabilization. Second, the size of each step may be increased to the point where no amount of deposition would ever create a natural terrain. Keep cribbing in small increments. You can always keep adding if you detect problems.

Evaluate the amount of sediment available to the site before deciding upon cribwork for long-term soil replenishment. In some places the soil available from runoff may never be enough to rebuild the slope, in which case it will indefinitely remain a stepped slope with exposed cribwork. The visible cribwork itself sometimes attracts trampling and even routine use because it provides a stairlike structure for climbing a slope.

These problems can be eliminated by immediately refilling the lost volume of soil to reestablish the desired grades rather than waiting for the process to

Figure 22.19. A sequence of fascines with jute matting restabilizes this gully until new vegetation becomes established. Brush on the surface discourages trampling.

occur incrementally over time. Although more costly, this "modified cribbing" is often well worth the effort because you can initiate the process of building a living soil surface and develop the landscape more rapidly.

You can use a similar method to improve drainage on short segments of a path or slope: simply place a log or fascine at a slight angle to deflect water off the path. These "water bars" can also be constructed entirely of soil or soil and stone, depending on site needs. Remember to use many shallowly placed diversions rather than fewer larger ones. The water bar is sometimes augmented with a small, shallow trench just upslope of the water bar to direct the runoff.

Check Logs

Where erosion is very limited and dispersed, you can use logs to check the flow of stormwater. Place lengths of long limbs and logs smaller than 3 inches in diameter along slope contours and stake them in place. Such "check logs" (Figures 22.20 and 22.21) are also an excellent way to dispose of woody debris in the landscape. Be careful not to create a channel by gradually concentrating the flow of water. The object is multiple redirectings and small impoundments to hold the water over the soil surface long enough for the maximum amount of infiltration to occur.

Logs can be staked in place to stabilize a slope for planting.

Figure 22.20. Check logs help retain small amounts of rainwater in the upper surface until ground litter can develop.

Figure 22.21. A check log along a former outlaw trail in Central Park. (Photo by Dennis Burton/Central Park Conservancy)

Replacing and Amending Soils

As a rule, the soil native to the site is preferred, for so many reasons, including cost, so you will want to preserve it wherever possible. The less disturbed the soil is, the less likely that any soil amendments will be needed, regardless of what conventional soil tests suggest. Where no major soil changes have occurred, except the loss of the upper horizons, you have an opportunity to make new soil and replenish damaged soils. The object of soil restoration is to mimic the natural soil of the site as closely as possible while minimizing the addition of outside inputs. Therefore, a basic requirement of soil restoration is not to consider adding any plants to the site that would require alterations to the native soil.

Most sites have a long history of uncontrolled erosion, so you will almost always need additional soil for restoration purposes. Importing large quantities of topsoil is not the best solution, however; it's more costly than adding and mixing soil amendments and has serious local environmental impacts where it is excavated. When replacing soil lost to erosion, do not overlook chances to recover lost soil from the site. Soil may accumulate at the base of slopes and gullies and can be reharvested in some cases, especially for small patch repairs. To improve locally available soil, choose appropriate amendments, which may include woodchips, humus, leaf mold, compost, gravel, sand, and expanded slate, depending on the conditions you wish to re-create. In the most disturbed conditions, waste products, such as dredge spoil, may also be suitable.

Changing Soil Structure

In some cases you may wish to modify the soil structure during the restabilization process. In very extreme cases of compaction, for example, it may be advisable to modify the upper soil layers to minimize the likelihood of recurrence, such as adding sand. These measures, however, are generally more useful in turf areas intended for heavy traffic. In woodlands and natural areas, controlling off-trail use and other causes of compaction is almost always more appropriate than modifying soil. Always emulate native soils to the extent feasible.

Where the soil is excessively clayey or silty, the addition of coarse material can make the soil less prone to compaction as well as drought because water movement is restricted in heavy soil. One practice is to add sand, in proportions of 5 to 30 percent; however, it may also excessively drain the soil during periods of drought. Or you can add gravel, which is less prone to droughtiness in the soil.

Where both drainage and water-holding capacity need to be enhanced you can add small particles of stone, such as shale or slate, that have been sintered, or heat-treated, to make them porous. Because these stone products, called "expanded stone," are permeable, they absorb moisture, up to 10 percent by volume or more. Although more costly than sand or gravel, expanded stone has the advantage of improving moisture retention. Today, fly ash, a waste material collected in smokestack scrubbers, is used in reclaiming soils. In fact, it is being heat-treated to improve its texture for use in reclaiming soils contaminated by a zinc smelter in Blue Lick, Pennsylvania. Densely porous ceramics are also now available and, though costly, have the advantage of being more uniform and easier to specify than sintered stone.

Another amendment is diatomaceous earth, composed of the skeletal remains of single-celled algae and nearly entirely silica, which has a very high surface area. It is used as a soil additive to improve moisture retention and to reduce compactibility to some extent. Diatomaceous earth is also easy to specify because it is more uniform than sintered shales and slates.

All of these amendments are suitable only for application in limited areas of heavy use in more urban, redeveloped landscapes and should not be widely used in natural areas. Diatomaceous earth is a known hazard to miners and should be handled with caution.

In sandy, gravelly, or rocky soils, excessive drainage, rather than poor drainage, may be the issue, especially where erosion has removed the more organic upper soil layers as well as the vegetation. Finer-grained material can be worked into the soil as well as organic matter to improve tilth and water retention. In general, however, deviate as little as possible from the native soil condition. The object is not to make all soils suitable for turf but to reestablish conditions as nearly as possible to those preceding the disturbance.

Adding Organic Matter

The addition of organic matter, probably the most common and trusted form of amendment, can make a great difference in the survival of new plantings. There are almost always a variety of local sources of organic matter that can be obtained at little or no cost. Many municipalities, for example, now operate compost programs for yard waste. In more remote areas, adding organic matter may be as simple as adding leaves that have accumulated nearby. But take care to exercise restraint in the use of organic matter. The increased fertility and added nitrogen may in turn restrict the natural germination and survival of indigenous plants in favor of invasives.

In many cases, you can eliminate the need to import nutrients or soil simply by using the site's own resources more efficiently. Think in terms of how nutrients are or can be cycled on the site and emulate the historic patterns that are likely to be less wasteful. Keep all dead wood on site, for example, instead of carting it away. Do not collect leaves from a site. Reduce the losses of nutrients and topsoil through better stormwater management. Where soil is more impoverished and an organic amendment is desirable, consider using raw wood or woodchips as the primary additives and add raw leaves and a few twigs to mimic the natural forest floor and provide a thicker layer of litter.

Save organic amendments for those sites where only subsoils remain, such as sand and gravel pits. In places like these, organic matter is more crucial to recovery.

Powdered rock dust is used widely for agricultural crops and gardens as an alternative to synthetic mineral fertilizers. Rock phosphate is a means of adding phosphorus. Granite is high in potassium. Greensand or glauconite is high in iron, potassium, and silica. Used as a mineral amendment, powdered rock dust can be sprinkled on top of or worked into the soil where restabilization is undertaken.

Although information on the impacts of rock dust on soil food webs, for example, is inadequate, its use in forests merits further evaluation. Consider a surface application in the fall, rather than in the spring. Forest applications are not worked into the soil, which would damage roots. Use lower rates of applications, one-half to one-third, than are recommended for croplands.

pH Modifications

Unless the native pH of the soil has been altered there is probably no need to modify pH. If you use a commercial soil test, remember that the accompanying recommendations will be geared toward adding lime to raise pH because turf and most vegetables are favored by a circumneutral (neither acid nor alkaline) pH. Many native soils, including extensive areas of the Northeast, are naturally somewhat acid and therefore support acid-tolerant communities, which are negatively affected when the pH is raised. This problem is further clouded by atmospheric acid deposition in the Northeast, which has lowered the pH of many soils and led to calls for liming of forests. Adding lime to make nutrients more available, however, does not necessarily help restore historic conditions of ammonium- versus nitrate-dominated soils. To answer questions about pH modifications in woodlands, we need far more study and many more field trials.

The most common method for raising pH is to add ground limestone to the soil. Where pH has been artificially lowered or where a temporary increase in pH is desired to increase germination of some species, the rule of thumb is to use 80 pounds per 1,000 square feet, or one and a half tons of ground limestone per acre, to raise the pH one point. Use pelletized forms of limestone, which produce less dust than simple ground limestone.

Prescribed burning offers a more natural method to temporarily elevate and re-create the pH of surface soils suitable for the germination of some species. Soil covered by fresh ash is an ideal seedbed for many species, such as oaks, as well as a variety of other less-expected species. Where burning is not an option, you can mimic similar conditions by raking small plots intended for reseeding with a straw rake to expose patches of the soil surface and sprinkling a little lime, possibly mixed with sand. Create small patches only to minimize erosion and the overall extent of modifications.

Landscape maintenance typically centers more often on raising rather than lowering pH, but those interested in habitat restoration in the Northeast face conditions where pH has been artificially elevated. While many natural landscapes are becoming more acidic, more neutral conditions often prevail in urban areas and on former agricultural lands. Soil pH is often elevated by the presence of concrete and mortar rubble from construction or the addition of lime for crops. In these cases, it is usually advisable to reduce the pH to reestablish the more acidic, native soil conditions that favor indigenous species and reduce competition from exotics. Powdered sulfur can be used for this purpose, although its application requires protective gear. A pelletized version that is reasonably dust-free is now available and preferable to powdered sulfur. Acidifying fertilizers that are commonly used for azaleas and rhododendrons and other evergreens are also suitable for individual woody plants that require lower pH levels. A general rule of thumb is to add 7 pounds of sulfur per 1,000 square feet, or 300 pounds per acre, to lower the pH one point. Where there are established plants, lower the amount of sulfur used by half and apply sulfur in the dormant season. Restoring the natural, lower pH typically favors native species over many exotic invasives.

Aluminum sulfate is also used to lower pH, although greater volumes are necessary (50 pounds per 1,000 square feet, or 1 ton per acre) and the added aluminum may be a problem because it can be toxic to many plants at elevated levels.

Charcoal is a highly absorbent material that can effectively immobilize many problem materials present in the soil. It has a long history of use as a soil amendment to "sweeten" soil and as an alternative to fertilizer, but relatively few studies document its effects. Several studies show its effectiveness even in such extreme cases as soils contaminated by herbicide spills. In laboratory conditions charcoal is commonly used in media for cloning plants or germinating seeds to enhance the survival of delicate, early plantlets. Consider the use and evaluation of charcoal as a soil amendment where contamination is present. Activated charcoal, a more expensive material, is more effective than ordinary charcoal and may be warranted in more problematic circumstances. Like wood, charcoal and partially carbonized woody material, which is sometimes referred to as "torreyized wood," appear to strongly stimulate soil fungal growth.

You can purchase fungi and other microorganisms for use in a wide array of extreme environmental conditions. They have proven enormously successful under the most degraded of conditions, such as strip-mine reclamation. New mycorrhizal products enter the market each year and may prove invaluable in the most drastically altered soils. Their role in habitat restoration may be more limited and should be evaluated carefully, however, since the widespread use of a single species or strain can have devastating effects on native diversity. If possible, obtain local strains. You may find what you need in your area. Now that the importance of microorganisms is more widely known, people are collecting and propagating local fungi in the same manner as they are collecting and propagating other native plants. Seek out a local mycologist and make every effort to develop indigenous communities of soil microflora. To learn more, try to attend a workshop on soil fungi.

Replacing Soil

Take care when refilling even a small amount of soil in an area where large roots have been uncovered by erosion for an extended period of time. If the root has been exposed long enough to form a protective layer of bark along its surface, the surface will rot if reburied. No soil should be added over exposed roots although the elevations of the soil can be restored between them.

When adding any new soil, be sure to scarify the surface of the subsoil for better adhesion. Where the potential for erosion will not increase, incorporate any new soil material well into the surface of the existing subsoil. When refilling over cribwork or fascines, or where the depth to be refilled is more than 4 inches, the first 4-inch layer should be lightly compacted before a second layer is placed. Each layer should be compacted to a density that is as close as possible to the adjacent undisturbed soil. Do not shortcut the process by applying a thicker layer all at once and trying to compact it from the top. It won't work, and the fill will be less stable. Additions of greater amounts of soil will require professional advice and engineering assistance.

Do not be overly concerned about evening out the terrain. You are not seeding a lawn where a smooth surface is advantageous; you are trying to re-create more historic conditions, which once were a terrain of pit and mound. The need to work around roots and other plants will help create a microtopography that will become more pronounced with time.

Except in the flattest terrain, blanket the soil surface to protect the area from erosion. Woody plants take a long time to become established. New plantings are typically very small when installed under woodland conditions and suffer high rates of mortality. (See "Stabilizing the Surface," later in this chapter, for more information.)

Repairing Compacted Soils

Where compaction has occurred, repair the site as quickly as possible or the compacted surface will continue to serve as a barrier to root growth, inhibit the exchange of atmospheric gases, and restrict the infiltration of water. In combination with airborne pollutants and hydrophobic substances, an impermeable surface crust is formed on compacted soil. When fill is added over compacted soil, the crust acts as an impermeable membrane preventing roots from growing upward into the new soil and leaving the site permanently less stable, unless and until the compaction is corrected.

Hand Excavating to Correct Soil Compaction

Where complete replanting is anticipated, you can disrupt the soil surface more extensively than otherwise in order to decompact it. Work carefully, using hand rakes to avoid damaging tree roots that are alive below the zone of compaction. Try using mattocks where compaction is severe. There will be few, if any, living roots in this layer, so you are not likely to damage healthy roots if care is taken

(Figure 22.22). Erosion, however, is a more difficult problem. Ironically, the soil may be quite stable when completely compacted, but as soon as it is loosened it is vulnerable to erosion and must be stabilized at the surface very quickly.

After the soil is loosened, add additional soil if necessary (Figure 22.23). You must then somewhat recompact the soil to restabilize the surface. Tamping by foot is adequate in most cases because the depth of loosening rarely exceeds a few inches. Then mulch with a thin layer of woodchips or raw leaves where woody cover is desired (Figure 22.24). Leave about 20 percent of the soil exposed. Depending on the extent of disturbance, a soil blanket, discussed later in the chapter, may also be necessary except in the flatter terrains (Figure 22.25).

Where compaction is locally extreme, an air gun is sometimes used to break up soils. It is reputed to minimize root damage under woodland conditions. The soil surface is, however, thoroughly disturbed, as it is with raking, and the process is very dusty. Protective gear is recommended.

Vertical Staking

Vertical staking (Figure 22.26) offers an alternative to thoroughly disturbing the upper soil to restore a compacted area and is especially appropriate where erosion is a serious threat and where vegetation should not be disturbed. Instead of loosening the entire surface area, try driving stakes made from cut branches vertically into the soil. Adjust the length of the cut branch depending on the depth of compaction and the amount of the stake you wish to leave above ground. This method loosens the ground at multiple sites to better convey water and air downward; at the same time it adds lignin to the soil.

In forested areas where raking or excavation could damage existing vegetation, vertical staking mimics the process of "piping" along the rotting roots of dead trees. After a tree dies, the decomposing roots leave long, continuous channels that are like pipes in the upper soil horizons, conveying water, nutrients, and gases vertically or laterally and serving to counteract compaction. Because only small areas are disturbed, numerous stakes can be driven into the ground around existing vegetation and no additional replanting is required for erosion control. In fact, the stakes themselves provide an added level of soil stability.

In a similar method, "vertical mulching," 2-inch holes are drilled about 1 foot deep and then filled with composted woodchips. Vary the spacing of the holes based on the severity and extent of compaction and take care to avoid major tree roots.

Subsoiling

Where compaction is deep rather than at the surface and native plant communities are absent or reduced, the deep compacted layer can be broken with only minimal surface disturbance with a vibratory subsoil mole plow. This method

Figure 22.22. At the Richmond National Battlefield Park in Virginia, an unofficial trail had seriously compacted the ground. Workers used rakes and trowels to loosen the soil, taking care to leave exposed roots uncovered during the work process.

Figure 22.23. Loose soil gathered from the bottom of the slope was returned to the slope to reestablish the original grade, except where roots were exposed.

Figure 22.24. Raw leaves added as mulch enrich the soil and are anchored in place by soil blanket.

Figure 22.25. Jute extending up and down the slope in the direction of the flow of water was anchored with split wooden stakes at 4-foot intervals with 6 inches of overlap and stabilized the surface. This project was completed at a National Park Service training workshop under the direction of Robin B. Sotir and Associates of Marietta, Georgia.

Figure 22.26. Vertical stakes made from cut branches driven into compacted ground in a dense pattern convey water and moisture downward into the root zone and loosen the surface as they decompose, without disturbing surface stability.

damages existing trees and cannot be used in woodlands; however, in open landscapes where turf is being converted to meadow or where forest or a grassland is being enhanced to improve rainwater infiltration, it is very useful. Be sure to keep an adequate distance from the roots of existing woody plants, which extend, in many cases, beyond their canopies.

Stabilizing the Surface

Surface stabilization in the forest presents a different range of concerns than those encountered in more open, sunny landscapes. Because quick seeding grasses do not persist in woodlands, conventional stabilization methods that rely on seeding do not work well, nor are they necessarily desirable under shady forest conditions. In forests it is the multilayered structure of trees and shrubs, as well as the herb and litter layers, that provides stability, not a dense carpet of herbaceous stems. The woody plants that are vital to forest stabilization may take several years or even decades, rather than months, to become established. Therefore mulch and a soil blanket, described later in this section, or some other surface protection are often required until a multilayered forest structure has developed.

The final mulch treatment to the soil is directed toward restoring fungi and other soil organisms. After the final grading and soil working are completed, spread on the soil surface a thin layer of woodchips, from 1/2 to 1 inch thick, to encourage the development of soil fungi. Scatter the woodchips, leaving about 20 percent of the ground uncovered, so that the soil remains bare and visible. This lignin-rich ground layer adds organic matter without enriching bacteria or earthworm populations. Fungi should develop rapidly. The thick, webby mat of growing mycelia adds structural support by adhering the mulch and soil particles together. Do not apply mulch to wet ground. Let the soil dry out beforehand, or you may encourage rotting of the newly installed plant material. Reapply mulch as necessary, but do not place a thick layer to save time later.

Where soils are especially sterile because of texture, such as waste piles or sites excavated deep into subsoil, for example, or damage from stockpiling, grading, contamination, or other causes, inoculating the soil with locally collected fungi may be advisable. You can colonize fungi from nearby landscapes with similar soils and vegetation that are less disturbed. The edges and less-disturbed areas of the site often offer good opportunities.

To prepare the transplants, place a layer of woodchips on the surface of the fungi-rich soil in small strips. The webby mycelia knit the woodchips into a delicate mat that can be harvested weeks later and moved to the more sterile site. Add raw leaves and a few twigs to mimic the natural forest floor and create a continuous litter-layer surface. Even simply adding a small amount of soil from a less-disturbed but analogous environment is an excellent way to introduce microorganisms. If the conditions are right, the fungi will spread rapidly from the inoculation area.

Please note, however, that you should not casually dispose of woodchips in a forest. Avoid using woodchips that might include treated wood. Chips from tree-of-heaven, walnut, and Norway and sycamore maples may inhibit growth through allelopathy. Where native communities are well established, any type of woodchip mulch can form a suppressive layer that inhibits herbaceous growth and the reproduction of many species. Where woodchips are used in a soil restabilization effort, a thick layer of mulch should be avoided for the same reason. Keep the layer thin and replenish periodically as necessary.

A soil blanket, also called an erosion blanket, provides additional protection for the soil surface on steep sites that are vulnerable to erosion or where runoff moves at high velocities. An erosion blanket is also useful where revegetation may be hindered, such as in deep shade. The blanket treatment is not a cure-all that can be used everywhere and should be confined to those areas where remnant vegetation is minimal. Where existing vegetation is dense it will simply grow up under the blanket and dislodge it.

There has been a veritable explosion of soil blankets and other site stabilization fabrics. Here are several overall guidelines to assist you in choosing and installing appropriate material.

• *Avoid products bound with netting, plastic, or fine string that can trap small animals.*
The netting will break down with time, but it can have unacceptable impacts for a season or longer.

• *Avoid nonbiodegradable materials that will not readily decompose, such as plastic or metal.*
You can make a simple and effective blanket on-site that will hold up until new plants are adequately established by placing jute or coir matting over the wood-chips. Where erosion potential is severe, erosion blankets made of coir, coconut husk fiber, provide a more durable and long-lasting surface protection than jute. Both are installed in the same manner, directly over the thin layer of wood-chips. The matting comes in 6-foot-wide rolls and should be laid from the top to the bottom of the slope (rather than along the contour) with about 6 inches of overlap.

Secure the matting with wooden stakes at 3-foot intervals; do not use metal staples. You can make the stakes by splitting 12-to 18-inch-long two-by-fours diagonally to make two long wedges. Then cover the mat with leaves and light brush to disguise all traces of the repair work. Uncomposted leaves may be used on the ground surface as well as woodchips because the blanket secures them and does not allow slippage. These materials are readily obtainable and can be easily carried into the forest landscape; they're light enough that equipment and vehicles are not needed.

Bonded fiber mulch, a cheaper alternative, may be useful in some severely disturbed forest applications. The mulch is made of gypsum mixed with wood fiber and sprayed onto the surface. It is relatively durable and does not tend to dry out as rapidly as cellulose fiber mulches. The gypsum breaks down with water, usually just as seedlings are developing. Its disadvantage is the vehicular access needed to bring in the required pneumatic sprayers used to install it; however, it may be useful on severely disturbed slopes bordering roadways, even in natural areas. The sprayers' reach is about 300 feet. Although the potential effects of bonded fiber mulch on soil food webs are not known, this approach merits further investigation and trials on disturbed fringes of natural areas, since gypsum improves the structure of heavy clay soils and remediates some salt damage along roadsides.

More severe situations in which tensile materials and petroleum-based fibers are presumed to be necessary go beyond the scope of this guidebook. Before you decide to use them, however, consider that in many such situations, addressing comprehensive stormwater management would be far better than resorting to so artificial a solution.

Managing Dead Wood and Brush

As a rule, individual dead trees should be left in the landscape as "snags" wherever possible. They are used as dens by many animal species and harbor insects and microorganisms that provide food for many other animal species. Woodpecker populations, for example, have increased dramatically in some places where gypsy moths have killed large numbers of oak trees.

A useful guideline is to leave at least three to five standing dead trees per acre for wildlife. Fallen logs and branches are also important to leave in place because they absorb and hold moisture like a sponge. Large logs are especially valuable to forest-floor creatures like salamanders. Two large and sound logs, in excess of 1 foot in diameter and 20 feet in length, and rotting logs, are recommended. Where logs are abundant, some can be moved to other locations where there is too little dead wood. The logs can also be along slopes placed to help control erosion. Partially submerged logs can be placed along shorelines to benefit fish, birds, and amphibious organisms. Logs in a stream both aerate water and provide additional habitat opportunities. Leaf litter and woody debris also can be reused elsewhere to add organic matter to eroded sites and to foster the restoration of important soil fungi and insects.

You can also use stumps, trunks, and limbs to construct checkdams, check logs, and soil stakes, as described earlier in this chapter. Where access is limited and chipping wood is not feasible, you can use the fine branches to build the litter layer. Brush may be temporarily effective in limiting access and discouraging trampling. When depositing brush on a slope to help control erosion, seek to create as natural an appearance as possible, mimicking the appearance of fallen limbs.

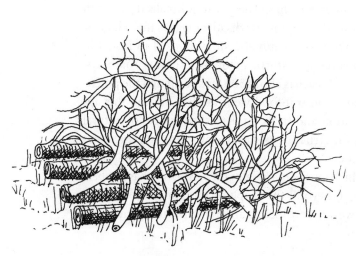

Figure 22.27. A brush pile oriented to receive some direct sunlight provides shelter for small creatures.

Figure 22.28. Logs laid on the ground disappear quickly and are excellent seedbeds for planting.

Figure 22.29. The ancient forest is filled with dead wood. Downed trees can be important assets for re-creating historic mixed-age, mixed-species forests.

A brush pile, if well sited with a sunny exposure, provides attractive and relatively safe shelter to wildlife in a small fragment of natural habitat, where small mammals and reptiles are often more visible and easily attacked. Such a shelter is also valuable in reducing mortality in winter and from vandalism. Brush piles also improve long-term soil quality and provide habitat for soil organisms.

To make a brush pile, select a sunny site, preferably away from human activity. Make the pile as compact as possible, placing logs at the bottom on the ground and laying larger limbs across the top to minimize wind damage.

Where there is a blowdown or other dead tree and you are not constructing a brush pile (Figure 22.27), leave the trunk and root mass in place. You can also partially cut up the branches to provide a higher degree of soil and wood contact. Cut the branches into pieces about 12 to 20 inches long and place them in the vicinity directly on the ground to maximize contact with the soil (Figure 22.28). Leave the stump as a snag if it provides no hazard. The soil mound thrown up by a fallen tree as well as the large log are ideal seedbeds for delicate species. Leave large logs as well to provide another specialized microhabitat (Figure 22.29).

CHAPTER 23

Controlling Invasives

The invasive exotic species rapidly expanding their ranges and prevalence in the eastern United States compete so successfully with native communities that indigenous vegetation is reduced year after year. The uncontrolled spread of invasive exotic vegetation also destroys wildlife habitat, further impoverishing the landscape. In many of our remnant landscapes where invasives hold sway, wildlife and ground-layer diversity has collapsed while older trees still persist, giving the appearance of a conserved landscape.

Two common rationales for not worrying about a particular exotic are "Oh, that's not a problem here yet" and "Oh, it's too late to stop that." In the meantime, kudzu is now reaching South Florida, East Texas, and Massachusetts—simultaneously. The more abundant an exotic becomes, the more difficult it is to control. Management usually does not take exotics seriously soon enough, even though keeping a new plant from becoming established is easier than removing it once it is entrenched. To do so, vigilance is needed, and a lot of us to act as monitors.

Every manager will be faced with conflicting demands when selecting among plant and wildlife populations. With restoration as a goal, you will need to give first priority to protecting those areas that still support mostly native communities from large-scale invasion. This chapter discusses strategies and techniques in the following areas:

- Identifying species of concern and patterns of infestation
- Developing an invasives management strategy
- Choosing control techniques
- Removing specific plants
- Broad-scale actions

Identifying Species of Concern and Patterns of Infestation

The first step in dealing with invasives is to document the species that are invasive in the landscape you hope to restore and the mechanisms and patterns of their dispersal. Many agencies and organizations are now compiling lists of problematic species. The list of invasive plants in the Appendix is not meant to be definitive but rather to alert the reader to many of the most commonly cited species. Not all the plants included on this list are equally invasive, but state agencies as well as environmental organizations have described all of them as severely invasive over a large area of the temperate forest landscape. All of these plants should be taken very seriously throughout the temperate landscape.

Also study your own site. An exotic species that is uncommon elsewhere may have severe impacts in your local natural area. A few species are likely to present the greatest threats. Develop your own list and update it regularly. Obviously, too, you should alert local nurseries about any potentially problematic species available in the marketplace. For further information, contact the National Coalition of Exotic Pest Plant Councils, 8208 Dabney Avenue, Springfield, VA 22152, 202-682-9400, ext. 230.

Developing a Management Strategy for Invasive Exotics

Do not underestimate the task of controlling invasive exotics. All of these plants are difficult to manage, if only because they reproduce so prolifically. These are plants that are spread by disturbance, that thrive under the same conditions that doom many other species. When dealing with invasives it is wise to have a strategy and be prepared for long engagement. Here are several recommendations:

• *Study the whole site and map the patterns of invasion before taking action.*
One relatively simple method is to designate the levels of invasion—"stable," "declining," and "degraded"—present on your site. A stable landscape is characterized by predominantly native species that are reproducing. A declining landscape shows signs of exotic invasion but natives are still present although declining. A degraded landscape may have remnant native elements, but all or most reproduction is exotic. These categories may be refined based on the relative levels of disturbance in your landscape. Study your map like the board in a game of Go or chess, carefully strategizing where you will make your next move. Mapping is especially useful in prioritizing where to manage. Update your map periodically to evaluate the success of the management effort and the effectiveness of the techniques you are using (Figure 23.1).

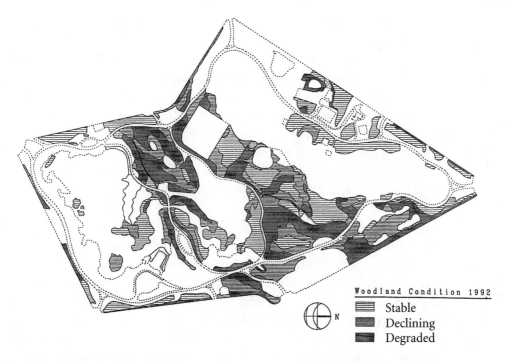

Figure 23.1. The staff of Prospect Park, under the supervision of Ed Toth, the director of landscape management, mapped the extent to which native communities were stable and reproducing, in decline, or degraded and replaced by exotic vegetation.

- *Give first priority to sustaining the richest sites and most valuable habitats.*

In general, initiating management in the most stable landscapes is advisable, removing exotics where they are just beginning to get established. Native species in the immediate vicinity will be the most likely candidates to fill the newly created gaps. The next priority might be to concentrate on declining areas that would serve to consolidate smaller stable patches into larger and more defensible exotics-free zones.

- *Prioritize which species are most destructive and active on your site.*

Concentrating on one or two species at the outset—those that pose the greatest threat and are spreading the most rapidly—is usually advisable. Make a least-wanted list and evaluate the extent to which native species and/or other exotics appear in the landscape.

- *Study the reproductive patterns of the species of concern.*

The reproductive strategy of the invasive plant is usually one of the most important pieces of information needed to control it. For instance, young trees not yet producing seed may be less problematic than older trees annually making an abundance of seed that is widely distributed across the landscape. Assess what factors are contributing to the invasion and the extent to which you can treat the root causes, such as seed-source trees, as well as the symptoms. Until Jean Marie Hartman began work on ti tree control in Florida, no one had identified the tree's seedling phase, which looks quite different from the adult form. Once recognized, the plant proved to be ubiquitous. Many clonal species are stimulated, rather than controlled, by fire, which may set off a rapid sprouting phase. Similarly, cutting before seed develops may be useful with annuals but would be relatively futile with plants, which spread vegetatively. Sometimes the propagule is inconspicuous. For instance, any attack on the lesser celandine should be timed before maturation of the tiny nodules that develop in the leaf axil, the plant's primary means of propagule distribution.

• *Assess the impacts of any plant removal before taking action.*
The activities associated with removing some plants may damage desirable plants or adversely affect wildlife or the site. For example, erosion may be a consequence where large numbers of plants are removed. Removing vegetation that provides the only cover for birds or serves other important functions also may be inappropriate. In such cases, replanting may be a necessary complement to invasives removal. A newly opened landscape also may invite trampling or serve as a pathway for yet another incoming invasive. Expect another exotic to rush in and fill the gaps you have created with removals, and another, and another.

One of the most important questions to ask is what will benefit from the reduced competition. Removing a large seed-bearing Norway maple may be important to controlling the future spread of that species, but if the understory is dominated by Japanese honeysuckle and juvenile Norway maples, they will be released and invigorated without competition from the old tree. Assess what will be needed to deal with the plants that will be favored by each action and evaluate alternatives.

• *Start with field trials and monitor them.*
After researching methods of control for the target species, establish a program that evaluates several methods before you proceed on a larger scale. Evaluate different conditions under which the invasive species occurs. Controlling established plants may be quite different from preventing the establishment of new plants. In follow-up monitoring be sure to look for what other exotics may have appeared as a result of your actions as well as any resurgence of the original target species. Evaluate the accuracy of your predictions, record current conditions, and reassess your control tactics.

• *Persistence and patience, not weaponry, are the keys to controlling invasive plants.*
What is important is not how hard you hit the target, but when and how often. After all, we are talking about plants that have shown themselves to be favored

and spread by disturbance. They are all resilient, prolific, and ubiquitous. Exotics control requires a kind of continuous attack policy.

When planning a control program, determine the rate at which to remove exotics by the amount of ground that you can replant and adequately maintain in subsequent years if natural regeneration is inadequate. Do not take on a larger area at a time than you can appropriately manage, or you will waste a lot of effort. Worse, you may expose an area to increased erosion by removing plants and then failing to establish new vegetation. Be persistent. The longer-lived clonal perennials can be remarkably tenacious. Where kudzu or knotweed is fully entrenched, it may easily take ten years or more to remove. Not all invasives show such an inexhaustible ability to resprout, but most show resilience for years.

• *Cease management activities, such as clearing and fertilizing, that foster the growth of invasive exotics.*
Clichés like "Create more edge to benefit wildlife" or "Open the canopy to stimulate understory reproduction," when applied to landscape management, have often favored exotic invaders. Bushwhacking, fertilizing, and other negative management activities should be halted; otherwise, we are actively creating the kinds of landscapes in which invasives thrive. Until we stop all the actions that continuously favor a few exotics, we simply will not know the recoverability of our natural landscapes.

• *Initiate management activities, such as prescribed burning and soil stabilization, that foster the regeneration of indigenous communities.*
The rapid invasion of exotics is as much a symptom as a problem. Simply removing the plant does not ensure that it will not return. The long-term prognosis is dependent on holistic changes across the whole system, such as reduced disturbance and the restoration of natural processes, like traditional fire patterns, to the site. Accelerate the natural process of "sealing" the edge by planting multilayered vegetation to reduce the penetration of light and wind into the forest interior. Under natural conditions, a newly created edge seals itself over time as existing plants extend new shoots and branches from the edge to fill the gap.

• *Reevaluate management objectives and strategies completely and regularly.*
Invasives management is an exercise fraught with uncertainty. Many of your actions will seem futile at the outset but will be rewarding in the end. Others will be the opposite, frustrating in the long run even if promising at the outset. There is a special problem associated with removing exotics: in our zeal to remove exotics we sometimes encourage them. To keep removing invasives from a site that is repeatedly disturbed is also somewhat pointless. Many nonnative species are going to be with us for a very long time. It is all too easy to imagine a

voice in the not-too-distant future championing the Norway and sycamore maples in the Northeast simply because they survive at all. Evaluate where your efforts will be the most effective in terms of sustaining indigenous systems. Do not focus so much on exotics that you lose sight of the larger goal, the conservation and restoration of native communities, not the serial removal of a sequence of exotic invasives.

· *Before removing plants, decide how to dispose of them.*
Many exotics are easily spread over even larger areas if appropriate methods are not used to dispose of them. Small slips of honeysuckle and many other vines can root if dropped on the ground. Do not, for example, transport plants such as multiflora rose, honeysuckles, and privet with ripe fruit on them to another location. Remove them before any seed has matured to eliminate this possibility. Plants that vigorously spread vegetatively, such as knotweed, can easily spread in mulch even after being chipped if they have not been fully composted. While most woody twigs and limbs and many leaves can be dispersed across the ground for gradual soil amendment, some species, such as tree-of-heaven and Norway maple, may be relatively toxic to seedlings of other species. Avoid chipping these species for mulch and carefully assess the use of their branches and limbs on your site. Large limbs and logs, however, may pose fewer problems than chips. Evaluate where it is important to remove all or parts of nonnative species from the site or where special disposal options are needed.

Control Techniques

You can control invasive plants in several different ways. Good judgment requires careful observation and monitoring. Garlic mustard, for example, a rapidly spreading biennial, dominates the forest floor once established, replacing a variety of native wildflowers. Garlic mustard was first found in the United States in 1868 and is now widespread. It invades woodlands from Canada to Virginia, from Missouri to Minnesota, and all points between, but its actual range may be even larger. Researchers note that populations can double every three years. The following description from the "Illinois Garlic Mustard Alert" flyer, prepared by John Schwegman, illustrates the kind of options that must be considered and just how complex control efforts can be:

> Minor infestations can be eradicated by hand pulling at or before the onset of flowering or by cutting it at or within a few inches of the soil surface just as flowering begins. If flowering has progressed so that viable seed may exist in the cut or pulled plants, remove them from the area.
>
> For larger infestations, fall or early spring burning is effective. The evergreen first year plants are killed by fire, however dense stands of the plants will not burn without additional fuel. Dense populations may best be burned in fall when new leaf fall provides adequate fuel. Spring burns should be early to minimize possible injury

to surviving spring wildflowers. Severe infestations will require several years of burning and should be followed by hand pulling or cutting of remnant populations.

Application of 2 percent Roundup herbicide (a formulation of Glyphosate) to the foliage of individual plants and dense clones is effective in fall and spring. At this time, most native plants are dormant but garlic mustard is green and vulnerable. Be sure to avoid native species with green leaves and remember that herbicides must be applied per label instructions.

The Missouri Chapter of The Nature Conservancy and the Kansas City chapter of the Missouri Native Plant Society have successfully controlled garlic mustard with annual hand-weeding and controlled burning in selected areas. The native groundcover rebounded, and garlic mustard has declined in the managed areas.

There are two major approaches to exotics management: (1) taking actions directed toward controlling specific plants; and (2) seeking to modify the larger environmental conditions to favor natives over exotics. The most successful program is likely to involve a combination of both approaches and techniques that are revised over time. The approach taken in this manual is to concentrate on the most pernicious species by waging a continuous effort of monitoring and control. The broad goal is to keep undisturbed areas free of invasion.

Here are some general guidelines relating technique to the life cycle of the plant that can be helpful in developing a control strategy:

Annual Species

Annual species germinate, mature, set seed, and die in a single growing season. In the temperate forest landscape, annual species are typically a problem at the very earliest stages of succession, such as after cessation of agriculture, grading, and new seeding. Until recently, the invasive annuals would diminish rapidly as the landscape matured. Today, however, there are several notable exceptions, such as stilt grass, an annual grass that persists in shade and outcompetes spring ephemerals mile-a-minute and knotweed. Even though such species as ragweed, foxtail, and lamb's quarters may not persist on a maturing site, there are usually more than enough disturbed areas in the larger region at any one time to keep these species abundant and widespread. The presence of these early weedy species can provide severe competition to native seed, especially in meadows and prairies.

One management option in open fields is to repeatedly till to encourage rapid sprouting of the weed seeds in the soil until they are exhausted. This method is generally appropriate only where tilling cannot otherwise be avoided and where exotic infestations are high. Many annual exotics are agricultural or turf-related weeds that favor a circumneutral pH. Where native soils are naturally acid, consider lowering the pH, especially on sites where the current pH is higher than historic levels because of past land uses such as agriculture or construction.

Where germination has already occurred, it is advisable, whenever possible, to "deadhead" the plants, cutting off their seed-bearing structures, to prevent them from going to seed. The adage "One year's seeding, seven years' weeding" is no exaggeration. In some cases, it is possible to set a cutter bar high enough to cut the tops of the target species if it is taller than the other plants in the meadow, which is often the case, especially with annual invasives.

Of course, there is a herbicide option for every kind of plant, but it is not always the best one to exercise. For example, where tilling is not an option for weed control in early successional landscapes, the use of a pre-emergent herbicide may affect the seeds of both undesirable species and desirable native species that are present in the soil, although it may benefit seed that is later sown. Therefore, any use of herbicide should be carefully weighed against other possible management options. Herbicide use is discussed in more detail in "Removing Specific Plants," later in this chapter.

Biennial Species

Biennial species take two years to complete the cycle of flowering and setting seed. The first year's growth is typically a basal rosette of leaves followed the next year by a flowering stalk. The growth form is notably different each year, a characteristic that aids in selecting an appropriate management technique. For example, the ubiquitous Queen Anne's lace and mullein, common biennials and classic roadside weeds, are favored by continuous edge and the practice of erratic but infrequent mowing. As with annuals, pre-emergent herbicides and tilling may be useful controls for biennials during a site's earlier phases, such as the first season after a field is released from agriculture or on a bare soil after disturbance. Because many biennials are typical of early oldfields, they often disappear as the landscape matures, with or without management. Infrequent mowing, or none at all, hastens their demise. You may determine whether or not there is need for any management at all based on potential off-site impacts—that is, whether or not the plant is spreading locally—as well as the shorter-term consequences in your landscape. Mowing just before the onset of flowering and before any seed has set may be useful in some meadows where adjacent invasion is a concern.

Garlic mustard is a somewhat exceptional biennial in that it is a denizen of woodlands, whereas the bulk of biennial weeds, like annuals, favor grasslands and other, younger landscapes. Mowing in woodlands is not appropriate; however, hand-weeding in the first or second year (before flowering) is slowly effective. Where the infestation is especially dense and no other plants would be damaged, a swing blade may be useful for removing the flowering stalks before seeds mature. Prescribed burning, herbicides, weeding, and cutting—in combination—are common approaches, although some managers report increased infestations after fire, underscoring the need for local field trials. Because garlic mustard stays green out of season, it is easy to weed as long as the ground isn't frozen. This same trait, persistent green leaves, makes garlic mustard a relatively easy herbicide target in the late summer, when many ephemerals have largely disappeared. Delay spraying until this time to avoid nontarget species. Choose a warm day, however, to maximize uptake by the plant.

Perennial Herbaceous Species

Perennial species live three or more years. In fields and other open landscapes, repeated mowing of the target species, up to three times a year, usually exhausts the rootstocks in two years or less, except for a few exceptionally persistent species such as knotweed. In some cases, such as mugwort, the addition of several inches of soil over existing plants can be effective as a control. Fill, available, for example, when a previously filled site is being regraded, sometimes works with common reed and can later be removed with or without the rootstocks and reused on other patches.

Japanese knotweed is one of the most difficult perennials to control. Your best chance of eradicating it is when it is first becoming established. Once it is entrenched, knotweed shows astounding resilience. Weeding is not always effective because the plant sprouts easily from even small pieces of root or shoot that are left behind. This weed and other herbaceous perennials that are capable of spreading long distances by cloning tend to do very well under conditions of disturbance. To control them more or less effectively, consider the use of herbicides in combination with weeding. Carefully dispose of knotweed and other ready sprouters, making sure you do not inadvertently disseminate the plants in the course of transporting them from the site or by using them as a mulch. Long, hot composting is one good disposal method. If you have large amounts of cut brush, make a simple pile of the debris and place the invasives inside. The pile at least contains the plants somewhat and is a good way to handle large amounts of cut brush.

Purple loosestrife is another difficult-to-eradicate perennial. Hand-pulling in patches works well if you are careful not to scatter seed, but getting all the root is extremely difficult. Do not dig loosestrife mechanically, which tends to spread the plant. You can burn loosestrife, but that will not kill the roots. Herbicides in combination with weeding and cutting are commonly used for purple loosestrife.

Vines

Again, a practical way to eradicate exotic vines is to vary your strategy according to the life stages of the plants. The same species may appear in a variety of forms at different ages and in different environments. Japanese honeysuckle is typical of many woody vines. It may heap and mound when young and climb in trees later. When mown, it can become a dense, low, almost continuous groundcover. Under light canopy Japanese honeysuckle may cloak the ground and hug the bases of trees.

Vines climbing in trees can generally be controlled by cutting a segment from the stem and leaving the top growth stranded in the canopies. Manual weeding can be very effective as well. If you are dealing with an extremely aggressive species, an herbicide directly applied to the stump or cut stems may help minimize sprouting from the base. The nearby evergreen Japanese honeysuckle is especially vulnerable and easy to target on a hot fall day. Where the vine blankets the ground, consider a suppressive mulch. Hand-weeding is also useful. Kudzu, when growing low on the ground, is vulnerable to fire.

Clonal Woody Species

Clonal woody species, such as black locust, autumn olive, and honeysuckles, can be especially difficult to manage. Weeding may be impossible because these plants have many ancillary stems. Repeated brush clearing or burning actually seems to invigorate some of these plants or at least encourages rapid sprouting so they become widely entrenched. Nevertheless, they cannot sprout indefinitely and will eventually succumb. Try repeated cuttings and, in the most severe cases, evaluate the use of spot herbicide treatments in combination with cutting.

Trees and Shrubs

For those tree or shrub species that do not sprout after cutting, simply removing the mature plant is adequate and can be done all at once or in sections over time. Most invasive tree species sprout vigorously, however, and if they are too large to be weeded or weed wrenched (see "Removing Specific Plants," later in this chapter) and removed entirely, you may need to grind the stump or treat it with herbicides. As with other invasive plants, the real key to control may lie in understanding the mechanisms of reproduction and where in its life cycle the plant is vulnerable. Light ground fire, for example, which usually does not affect mature trees, can prevent species with very fire-sensitive seedlings, such as Norway and sycamore maples, from becoming established despite the presence of mature seed trees.

Avoid unnecessary soil disturbance, such as grubbing and rototilling. These methods may be appropriate for renovating a horticultural bed but not woodlands, except where existing vegetation entirely consists of invasives you wish to eradicate and full replanting is to follow. If the soil surface is stable, even if it is supporting only exotic invasives, beneath the soil there may be roots of adjacent plants, often from some distance away. Remember, roots are opportunistic and go where the going is easy. A tree's roots may easily extend well beyond twice the limit of the branches. Grubbing and rototilling also disrupt fragile soil microorganisms upon which good forest growth depends.

The following methods are recommended for the control of invasive vegetation. But removal of these plants is only the beginning. Regular monitoring is a

crucial part of any invasive control program combined with continuous experimentation and evaluation.

Removing Specific Plants

Hand-pulling, or weeding by hand, is one of the most important and effective
management techniques in fragile landscapes such as forests. Real diligence is
required, but hand-pulling often provides excellent control over time. Where
invasives are mixed with desirable native species, repeated weeding of the invasives may give a competitive edge to the native plants.

The primary limitations of hand-pulling as a control is the size of area that
can be feasibly controlled on a routine basis and the trampling that usually occurs in the course of the work. In theory, if you get all the roots, the plant is
gone. Getting all the roots can be very difficult, however. It usually helps if the
ground is wet but then soils are more easily compacted and damaged by trampling. Consider using boards for walking on the soil surface and/or have the
team work as a bucket brigade to reduce damage to the ground.

Where the landscape is relatively pristine, hand-weeding is probably the best
control of invasives, even at a very large scale, and it can be undertaken continuously by all those who monitor, manage, and use the site. Hand-pulling is effective where plant stem diameter is less than 1/2 inch. Beyond that, some mechanical assistance, such as a weed puller, may be useful,

A weed puller is a long lever with a fulcrum plate and jaws at one end and a
handle at the other, specifically developed for removal of trees and shrubs. The
fulcrum is placed flat on the ground, and the jaws grip the base of the plant. By
pulling on the long handle of the tool, the operator simply levers the entire tree
or shrub, including root system, out of the ground (Figure 23.2). Tough plants
sometimes require several pumps of the handle before they give way. A weed
puller is not effective with root sprouts, which break off at some point because
the larger clonal root system to which it is attached cannot be extracted. The
weed puller extracts almost no soil along with the roots, which in turn minimizes local soil disturbance.

When removing multistemmed trees and shrubs, first prune out stems at the
base of the plant so that the weed puller's jaws can reach around the main stem.
Weed pullers are effective for removing species such as shrub honeysuckles,
privet, oriental bittersweet, winged euonymus, and tree-of-heaven.

The Weed Wrench is an excellent weed puller that easily handles saplings up
to 3 inches in diameter. It is simple to use and comes with a set of instructions
and safety cautions. Four sizes are available, all of which are portable, although
the largest size is a heavy load to carry on foot all day. None requires exceptional
strength to use and can be effectively wielded by a relatively slight person. They
are made by New Tribe, 5517 Riverbanks Road, Grants Pass, OR 97527,

Figure 23.2. Joe D'lugash, a volunteer who spearheaded landscape restoration in Philadelphia's Wissahickon Park, wields a weed puller.

503-476-9492. The Root Jack, made by Michael Giacomini (P.O. Box 726, Ross, CA 94957, 415-454-0849) is a similar device.

For the most degraded landscapes there is the Brush Brute that replaces the bucket of a front-end loader. Its steel frame has 18-inch teeth that are driven into an undesirable plant, which is then lifted out of the ground.

Herbicide use is always hotly debated and often more so in restoration projects. For some people, it is difficult to defend even responsible and selective use because chemical herbicides and pesticides have been overrelied upon or applied excessively in the past. If landscape convention and practices used less herbicide on gravel and mulch beds, roadsides, and building margins, the strategic herbicide control of invasives would entail no net increase in pesticide use, and might even result in a reduction.

Many practitioners are having good results with herbicides and would be very reluctant to eliminate that option for exotics management. One justification is that they are labor saving. Another argument is that the most intractable aliens, such as Norway and sycamore maple and Japanese knotweed, seem to be simply uncontrollable without herbicide usage. Managers in Rock Creek Park in Maryland, for example, are reluctantly turning to herbicide in an effort to stop the spread of lesser celandine, which is filling the floodplain and has already obliterated a once-notable colony of trout lily. Nothing else they have tried has yet slowed it. In some cases, the use of herbicides also eliminates or reduces soil damage from the foot traffic accompanying weeding.

Because arguments on each side of the herbicide issue are strong, the following recommendations provide techniques for both nonchemical and herbicide use. The guidelines for herbicide use seek to minimize the volume of chemical used and to maximize the selectivity of the application.

When herbicides are used, we recommend that you contact an Integrated Pest Management (IPM) practitioner in your region. IPM seeks to achieve effective control rather than total eradication and minimizes the use of chemicals by employing a combination of methods, including mechanical and environmental controls, such as prescribed burning. IPM practitioners will provide specific guidelines for your site that you should consider in your overall planning and monitoring.

• *Make every effort to avoid the use of herbicides.*
In some areas, any herbicide use at all may be inappropriate, particularly where levels of biodiversity are high. Study labels and manufacturers' guidelines to make sure the herbicide is listed for use on the species of concern and is appropriate for your site's conditions. Learn to recognize where the damage to native plants from exotics would be greater than that likely to result from herbicide management, and monitor your results.

• *Minimize herbicide use by confining its application only to the most intractable and severely invasive exotic species.*
Use methods of application that minimize the volume of herbicide you use, including focused spot treatments using adjustable nozzles and wand applicators to reduce drift. Apply only on warm, sunny, calm days and use a shield such as aluminum flashing or sheet plastic to protect neighboring plants. Be especially cautious about using herbicides like picloram and dicamba, which have residual effects in the soil.

• *Wherever herbicides are used, independently evaluate mechanical, nonchemical procedures and their combined use with herbicides in field trials to compare their effectiveness.*
Examine other factors beyond the control of the target species, especially impacts to nontarget species and variability in subsequent species recruitment. Try using a commercial dye in the herbicide to determine where the spray actually goes for training and evaluation purposes.

• *Herbicide use should diminish over time on each site, as initial control is accomplished and more desirable plant communities become established.*

Where herbicide use has remained high for several years, reevaluate the management program. Herbicide management should not become routine.

• *Be critical and conservative, and do not underestimate the potential hazards.*
Ultimately, toxins are not the solution and at best are an interim tool. At worst, they should not be used at all.

• *Anyone applying herbicides should be certified for their use, even for herbicides that do not require certification.*
The issue of safety is especially important where volunteers and other nonprofessionals may be involved. The training program is extremely useful and serves an invaluable role in establishing appropriate safety procedures. After spraying an area, for example, you should move away backwards to avoid walking through the sprayed foliage.

Suppressive Mulches

Mulch may benefit plants or suppress their growth, whether they are weeds in forests or in horticultural beds and vegetable gardens. A deep layer of mulch is an effective growth suppressant; a thin layer with lots of gaps is often an excellent ground layer for recruitment. Neighbors who dump their leaves and garden waste in nearby natural areas have been practicing suppressive mulching, often to the detriment of the natural communities, without knowing it.

The biggest problem with a suppressive mulch to control exotics is that it is typically nonselective. Very few plants, mushrooms aside, find habitat in thick mulch. Therefore, mulch is of limited use in many restorations where increased native recruitment is the goal, except where individual plants or clones can be successfully targeted. The most common mulch consists of organic waste such as woodchips or shredded bark, but plastic mulches are increasingly available. Instances where mulch might be used to minimize exotic invasion include the following:

- To protect selected new plants from competition from exotic invasives and aggressive native species until they become established
- To eliminate local patches where invasive species are entrenched
- To mimic natural conditions under which allelopathic suppression occurs. For example, chips from tree-of-heaven may minimize resprouting from a cut stump.

Organic Mulches

Organic materials are most commonly used as mulch although an increasing number of synthetic mulches are appearing on the scene. In general, only organic mulches should be used except where the mulch is intended to be temporary and will be removed.

An astounding array of materials can and have been used as organic mulches. Remember, however, that they may not be equally appropriate in the context of restoration. To the extent feasible, stick with materials that would be part of the landscape system anyway, such as leaves and twigs and logs. These are all soil builders in the forest as well as potential suppressive mulches under selected conditions. Woodchips and shredded bark are also excellent and may be readily available in many circumstances. Straw may also be available; sometimes it is mixed with a little manure, which is often referred to as "horse-blanket." Mushroom soil, spent mushroom-growing medium, is also locally available. Like any highly enriched materials, they may be of limited utility as mulch in forest conditions.

Rags, paper, waxed paper, and other waste products are effective mulches that do not flood the system with nutrients. They have been useful in reducing competition to newly planted trees in reforested areas. Avoid the use of any product you judge may be inappropriate in more natural areas.

Plastic Mulches
Both black and clear plastic sheeting mulches are useful where use of an organic mulch is undesirable or has been ineffective. Although plastic is unsightly, it is later removed and involves no herbicides (Figure 23.3). As a general rule of thumb, use black plastic under shaded conditions to starve exotics of light energy. Use clear plastic in the open to create overly hot conditions.

23.3. Temporary plastic covering is field-tested in Wissahickon Park as a control for Japanese knotweed. (Photo by Clare Billett)

Before applying plastic, cut and remove the existing invasive plants. A weed whipper with a blade, rather than a plastic string, can be especially useful for cutting thicker stems. Tack the polyethylene tightly in place on the soil surface, using metal staples or wooden stakes. Where appearance is an issue, use an open mesh fiber to secure a leaf mulch over the slippery plastic. It may take at least eighteen months to adequately control species such as Japanese honeysuckle.

When the plastic is removed, complete restabilization of the site is necessary. Jute matting and additional replanting may be required as well. Like other methods, plastic is variably effective.

Root barriers also have some limited applications in controlling the spread of exotics where runners are a problem. Long-term effectiveness is limited. They are generally plastic or other synthetic fabrics and may not be suitable as permanent installations.

Removing Parts of Specific Plants

Cutting roots or shoots rather than removing the entire plant is another control option. Several successive cuttings can be extremely effective in exhausting the rootstocks of many perennial species or preventing the formation of seed in annuals and biennials. Typically the aboveground shoots are cut and either left on the site or removed. In many instances, you may also use an herbicide for stump treatment of vigorous sprouters and clonal spreaders.

Ideally, try to recycle most plant material on-site. Under forest conditions, every bit of woody debris may be vital to rebuilding soil flora.

Note that in grasslands, the removal of cuttings generally benefits shorter-lived (or mower-dependent) species more than longer-lived perennials.

Hand Tools

A variety of tools are available to tackle different kinds of plants at different scales of intervention. A weed whipper with brush blades is very useful for removing top growth in heavy brush such as multiflora rose but is very hazardous and requires a skilled operator (Figure 23.4). If you use hand-held motorized tools, consider the new four-cycle models, which pollute substantially less than the two-cycle models. Weed hooks and brush hooks offer nonmotorized alternatives that are also very portable. Small one-handed chain saws are also quite handy but potentially dangerous in inexperienced hands.

Be especially careful when working in tangled vines. A ratchet lopper deals with vine growth well because it allows you to cut out a section of the main stem, leaving the stranded top growth to rot in the trees. Also try a three-pronged claw, but be sure not to dig into the soil. Hand-held pruning shears are very versatile and easy for inexperienced workers to manage.

A flame-wilter has much the same effect as cutting a vine shoot; however, some exotics such as Japanese honeysuckle can be quite difficult to damage

Figure 23.4. A weed whipper with brush blades is useful for severing vine connections to the ground.

sufficiently, although evergreen vines such as English ivy are usually easily killed this way.

We recommended the following schedule for removing shoots of hard-to-eradicate invasive perennial exotics:

- **Spring:** Just after plant has fully emerged from dormancy, complete hand cutting and removal of top growth.
- **Summer:** After full regrowth and/or when flowering, complete hand cutting and removal of top growth.
- **Fall:** Continue follow-up monitoring and remove fall flowers if necessary.

Note the need to hit the plant repeatedly at times of its greatest energy demand in order to exhaust its root system.

Girdling

Girdling is a commonly used technique for killing woody plants. It consists of stripping away the inner bark of the trunk, the "phloem," and leaving the sapwood, or "xylem." The idea is to starve the roots, which continue to send nourishment up into the trees while nothing comes down to the roots from the canopy.

To find the right depth to cut, make a slash in the bark with a knife or ax. The outer bark separates relatively easily at the weak cambium layer separating the xylem and phloem. Make several gashes until you know how deep to cut. Do not worry about damaging the plant; you are trying to kill it anyway. You do, however, want to leave enough sapwood intact to drain the roots. Girdle the plant as close to the ground as reasonable to minimize sprouting.

You may need to reopen the wound once or twice if bark starts to regrow, so check the site periodically. Remove sucker growth as well. The width of the girdled band of bark should be about 3 to 6 inches.

Vigorous clonal sprouters such as black locust are stimulated by girdling in many instances, so evaluate your site's conditions carefully. You can also girdle a major root to weaken a tree to kill it over time. Girdle the root close to the trunk.

Sequential Removal

Where it is not desirable to remove a major tree at once, you can effect a more gradual transition. This may be preferable to outright removal under several conditions. In some cases, the too-sudden loss of canopy may be problematic because no native tree is there to replace it. Successive severe prunings that reduce the canopy over time may allow a native replacement tree to become established beneath the target tree before it is completely removed.

Several methods can be used to damage a tree and send it into decline where that is preferable to complete removal. Where there is no risk of damage to adjacent plants in the event the tree falls, you can cut several larger roots to substantially reduce the tree's capacity to feed itself. Similarly, when large limbs have been removed in a sequential removal, the wounds left behind may encourage fungi and rot in the tree as well. Bad pruning—cutting the limbs too close to the trunk and/or leaving behind lopped-off limbs—can increase the likelihood of infection. A damaged tree also provides excellent habitat for many insects and soil organisms.

Where root cutting might create a hazard, remove large limbs to eliminate the tree in pieces over time. Take down the tree eventually, or leave it as a standing dead tree for wildlife habitat. Where a tall dead tree presents a hazard, leave it behind as a shorter snag or as large logs, both sound and rotting, on the ground. Remember that old growth is characterized by extensive accumulated dead wood on or near the ground. Consider establishing seedlings and saplings of native replacement species in the new gaps during this time.

Broad-Scale Actions Directed Toward the Whole Landscape

The problem of exotic invasion is usually associated with broad-ranging environmental and land-use changes that foster the spread of weedy vegetation. The very conditions that once selected in favor of specialized vegetation have been replaced by those that favor the appearance of the hardy generalists. Evaluating what processes are directing these changes, such as altered hydrological regimes or disruption of historic processes such as fire, and seeking to reestablish healthier ecosystems overall are as important as removing exotics as they occur.

Alterations to Hydrology

Many exotics and aggressive natives become established when the natural hydrologic regime is suddenly modified. A collapsed pipe or blocked drainage in a tidal area can invite the invasion of common reed very rapidly. A new highway through a lowland forest almost always produces death of the trees, from flooding on one side of the road and drying out on the other. The site becomes an easy target for exotics. If action is taken quickly, the preexisting community may in some cases be able to reestablish itself or be replanted, especially where you can restore historic hydrologic conditions. Removing dikes or blocking drainage ditches to restore flow, including tidal conditions, may be more important factors to address than removing exotics.

In Kearny, New Jersey, the natural hydrologic regime was interrupted early in the region's history and, like thousands of acres in the Hackensack Meadowlands, much of this landscape was converted from saltwater cordgrass tidal marsh to a dense blanket of common reed broken only by thin strands of marsh creeks. Wildlife usage was severely altered with the change in landscape type. Then, in the middle of this century, the construction of the New Jersey Turnpike served to create a drainage barrier that impounded freshwater runoff over hundreds of acres of this common reed marsh. The increased water level and more prolonged flooding stressed the common reed severely and resulted in the death of about half the area of reed over time.

While the flooding did not restore historic conditions, it did create a pattern of open water and vegetation in roughly equal parts that greatly improved the value of the habitat to a wide array of waterfowl, including several rare species. The flooding triggered a chain-reaction sequence of changes in this wetland, many of which could be integrated into a successful restoration program if stormwater were adequately managed. Unfortunately, stormwater management policy and planning are rarely driven by the needs of natural areas, and management of this marsh has shifted with political directions.

The impacts of changes in hydrology are usually more difficult to see in more transitional areas where water is a less-dominant factor in determining vegetation. Where deep gullies and erosion channels drain away groundwater, for example, native communities in the immediate vicinity may be stressed and more vulnerable to exotic invasion.

Prescribed Burning

Prescribed burning can be an important tool for restoring historic conditions to favor indigenous species, although exotics will be variously affected. Garlic mustard and autumn olive, for instance, may benefit.

If you wish to explore this option, get as much information on the fire history of your site and evaluate how it may have been altered to favor or inhibit some species. You will also require knowledge about controlled-burning techniques. The Nature Conservancy is making a concerted effort to incorporate prescribed burning into its local landscape management programs and sponsors regional training. There is no better way to learn about managing a prescribed burn. Local fire departments in rural areas are often willing to perform controlled burning in return for a contribution. Some state forestry departments have prescribed burning programs as well.

Mowing

Mowing is one of the commonest techniques applied in the destruction of native plant communities, but it can also be used as a tool for controlling exotics, especially in grasslands where burning is not an option. At the very least, major changes in conventional mowing regimens could foster a greater array of native communities, which in turn might make the landscape less suitable to weedy species. However, mowing and brushhogging should never be done in woodlands. The eastern temperate forest is often referred to as the sprout hardwoods, and in fact, almost all deciduous woody plants in the Northeast sprout after cutting, but this capacity is limited. Woodland ephemerals, for example, many of which spring from ancient slow-growing clones, are substantially reduced by even a single mowing. Mowing and brushhogging are just clearcutting on a smaller scale.

Mowing in general benefits short-lived species and those whose buds and other propagules largely develop underground, at least part of the season, because it removes the top growth. Woody species are especially intolerant of mowing because reserves in the root system must be drawn upon to regrow stems and leaves as well as buds. Repeated mowing or cutting simply exhausts the plant. Well-timed mowing at least three times in a season can help control many woody invasives such as multiflora rose. Vines such as Japanese honeysuckle, however, are stimulated to send up shoots after mowing and simply form a denser, lower carpet.

Alterations to pH and/or Fertility

Many native communities are adapted to low fertility. At the same time, many of the pervasive human impacts to natural systems entail a substantial increase or release of nutrients into the system, from clearcutting to fertilizing and air pollution. Similarly, many native communities, especially in the Northeast, occur naturally under relatively acid soil conditions, which are also associated with reduced fertility. Therefore, it should come as no surprise that reducing, rather than increasing, fertility and maintaining a low pH often fosters native species. Many opportunistic weedy species, on the other hand, are shallow rooted and nitrophilous. Do not, however, try to lower pH beyond local natural ranges. Sugar is also used as a treatment to lower nitrogen, but it is still in the very experimental stages.

Shading

"Closing the canopy" is a common strategy where blowdowns and other losses have left extensive gaps in the forest canopy. You may be able to use this strategy to inhibit the establishment and spread of some exotics of concern by closing the canopy with plantings of indigenous species. Some invasive exotics, such as giant nodding foxtail and mugwort, thrive only in full light conditions and will fail once woody species become well established. Others spread more slowly once the canopy has closed but are not controlled by shade. Knotweed, for example, colonizes shady locations more slowly, often gradually working its way in from the edge. It can, however, persist in the dense shade that may grow up about it as trees mature. A woodland gap filled with native plants may better resist this gradual colonization.

At the same time, however, consider the extent to which native recruitment would be restricted by heavily planting in gaps. Gaps provide the best conditions for adding propagules of native species if exotic recruitment can be adequately managed. In 1993 over 20 percent of the forest in Brooklyn's Prospect Park, for example, consisted of gaps wider than 20 feet across. With few native seed sources nearby and often high levels of disturbance, invasive species such as Norway maple often were taking full advantage of these opportunities. The current management includes both exotics removal and replanting historic species by associated soil type.

A similar strategy is called "sealing the edge." When a wooded landscape is partially cleared, the newly exposed edge is very vulnerable to invasion by exotics and other edge effects. With dense planting you can somewhat limit this, however. The idea is to intentionally plant the very places where natural reproduction is most likely and displacing other undesirable plants by shading and occupying the space available. Shade cloth may be useful to further inhibit selected vegetation or to establish selected species, although the effect is only temporary and the cloth must be removed as it deteriorates.

Increasing Competition

Much of management consists of attempting to shift the balance of competition in favor of native communities. Any time you remove rather than plant exotics you are putting your finger on the scales of competition. The same thing is true when you plant natives or add large amounts of seed and other propagules of indigenous species: You are acting to favor competition for natives.

Managing Invasive Wildlife

Invasive plants are difficult enough to manage, but invasive wildlife is much more so. Some simple pests can be removed by hand at a small scale, but most wildlife problems will take all the expert help your team can find. Populations of wildlife can shift with far greater suddenness and speed than plants. Relatively few individuals are needed for an invasion. The habitat does not even have to be disturbed for it to be invaded, but that often helps.

Because animals are mobile, controlling their movements at a small scale with much effect is often impossible. Their impacts may be pervasive beyond our current imaginings. For instance, the extent to which exotic earthworms have colonized is poorly understood. Lack of knowledge is a huge obstacle. Recognizing quickly enough that an invasion is already under way is also almost impossible because so few individuals are required. Prevention is the most important strategy but becomes ever more difficult as global commerce continues to increase and new species enter our ecosystems.

The problems associated with introductions of exotic animal species tend to occur over a very large scale, which compounds the effects of questionable actions taken in the name of controlling them. Take, for example, the gypsy moth. At the outset, taking strong measures to eradicate it when it was occurring locally might have checked its spread, but now that the species is widely distributed, we have no easy way to control it. For decades communities have saturated themselves with pesticides to escape the consequences of a gypsy moth year. Nevertheless, there is no evidence that long-term damage to vegetation by the gypsy moth is any less in treated areas than untreated areas, because treated areas experience more frequent cycles of population explosions.

Today, many communities spray rivers and streams to control gnats as well as forested areas for the gypsy moth, with the bacterium *Bacillus thuringiensis*, Bt, which is touted as safe to humans. Bt is, however, death to any larvae that happen to be alive at the time of spraying. All caterpillars and many kinds of larvae are decimated, not just those of the gypsy moth or gnat. All species dependent upon the cyclical abundance of caterpillars and larvae for food are affected, migratory birds and fish in particular.

If you are in an area where Bt is used as a control for caterpillars, make every effort to minimize its impacts. Be sure that headwaters and tributaries are not sprayed because the gnat larvae are not there anyway; they are in the large, warm riffles of rivers. Be sure to concentrate eradication efforts on the first generation of larvae, because it spawns all the rest. If you don't skip at least two generations after that, the population will develop resistance very quickly.

Introductions of new species also sometimes occur deliberately in the effort to control other pests. As noted earlier, an insect introduced to help control purple loosestrife also appears to be affecting nontarget species. Be extremely careful about considering the use of an introduced species as a biological control. Most have unintended impacts, whether or not they are recognized, on other species. Make sure any such efforts are adequately monitored: misinformation abounds, and so do assurances that one control or another has no adverse impacts. At present, however, biological control may be the only alternative in development that can be expected to be successful with ultra-competitive invaders, including both plants, such as loosestrife, as well as animal species. In fact, a biological control is at present reducing populations of gypsy moth. A fungus, *Entomophaga maimaiga,* that is a natural control of gypsy moths was first observed in North America in 1989 and since has been widely introduced. Populations of the moth larvae crash as the fungus spreads, working most effectively in wet springs.

CHAPTER 24

Planting

Returning complex native plant communities to the forest—and with them native wildlife—is usually the primary goal of those interested in restoring the landscape. Yet simply replanting a diversity of species on a site in which they may have thrived in the past is often doomed to failure. Eager volunteers may plant hundreds of wildflowers in an urban woodland fragment, only to find that even with vigorous maintenance almost nothing is left after a few years. Elsewhere rare plants, rescued and relocated from a development site to a nature preserve, seem to thrive for a season or two and then suddenly disappear for reasons not understood.

The mortality rate for forest transplants can often be very high, above 90 percent, so assessing what is really feasible in any given situation is especially important. People often want to start replanting prematurely, before disturbance is controlled, planting woodland wildflowers, for example, before the litter layer is stable enough. In a relatively degraded area, efforts to reestablish conservative species may be futile until a stable landscape matrix is established. Seedling survival rates are also better where weedy competition is under control. Good managers learn to "read" the landscape, to take the cues about what approaches are likely to work well in the future, by monitoring natural analogs to the landscape on reference sites and recording site conditions, stabilizing erosion, and removing exotics in concert with replanting. If these tasks are done at the beginning, the chances of reestablishing indigenous species are certainly better.

Before embarking upon a planting program, the restoration team should address the issue of how to obtain plants. Any serious restoration will require establishing a propagation program, either independently or in cooperation with another agency, organization, or specialized nursery. Commercial nurseries may also be important to your effort. You must evaluate the sources of the plant propagules as well as the suitability of their production techniques—for example, local varieties produced by seed versus vegetative monocultures and grafted plants. If you are fortunate, you may find a local naturalist or botanist who has been collecting and propagating local flora and will be willing to provide propagules for your effort since no major seed producers grow local varieties of native meadow grasses and wildflowers of the eastern forest. All "regional" mixes of meadow seed, for example, by definition, consist of nonlocal subspecies, which may threaten the genetic identity of local ecotypes. Another key question is whether or not you will use material produced by seed or vegetative means, as well as what propagules should be locally collected and according to what protocols.

A variety of different planting techniques, each with different opportunities and drawbacks, can be used in wooded landscapes and should be considered in developing replanting strategies. Sometimes the most effective way to plant is to release the seeds that are already on the site, such as with prescribed burning and exotics removal or even by shaking the seedpods to distribute their contents, rather than to introduce propagules from elsewhere. For most sites a combination of approaches is the most useful. This chapter describes a variety of strategies:

- Planting methods and strategies
- Fertilizing and pruning
- Propagation, transplanting, and collecting
- Exclosures

Guidelines for Reestablishing a Forest's Structure and Composition

A forest often dies from the ground up. The smaller plants, both woody and herbaceous, are the most vulnerable. The canopy may be the last to go, declining for years above a trampled, compacted, and eroded ground.

Restoration too, begins on the ground, with new recruitment. Many herbaceous species, for example, will not find appropriate habitat until the forest floor is alive again. Where the soil has been severely disturbed, it may be a very long time before conditions suitable for many species develop. Here are several guidelines for reestablishing forest layers:

- *Watch where reproduction is occurring naturally—these are places of opportunity in the landscape.*

Under relatively stable conditions, the gaps in a forest created by the loss of a large tree (or trees) are usually the places where most of the expansion of existing plants and the recruitment of new species is occurring. In a more disturbed landscape this effect may be due to edges and clearings newly created by human activities. The most immediate beneficiaries of the space are the already-present smaller trees that are "released" to grow rapidly into the gap, but new species also appear. Some plants, such as tulip poplar and black birch, colonize so well under these conditions that they are called "gap-phase species." These same conditions offer the restorationist somewhat enhanced chances of establishing new plants in the wild. Exotics also often benefit from gaps, so expect some new problems to arise.

• *Look for analogous natural conditions and mimic their planting patterns.*
Plant in patterns you have observed on the site or in analogous habitats, which are not evenly spaced as in an orchard or plantation. Planting patterns found in nature reflect important functions; small trees growing in clumps, for example, protect one another and are less vulnerable than a single stem. Let the site inform you about how to plant, based on its soils, aspect, drainage, moisture, successional status, and associates—that is, species commonly found together under natural conditions.

• *The reintroduction of herbaceous plants is especially difficult, so expect some mortality and start with small-scale experiments.*
Species disseminated by birds and wind, including most trees and many shrubs, reestablish relatively rapidly. Other means of dissemination, such as ants, are far less effective in the fragmented forest. Seek help from an invertebrate specialist to develop an ant and/or beetle restoration program. Consider planting on sites that will get ongoing maintenance, and, of course, monitor what happens. You may need to establish fairly large populations before there are enough individual plants as food sources to support ant colonies, for example.

• *Periodically reassess the need to plant at all based on changes that have occurred over time and the rate at which the landscape is recovering or deteriorating.*
As a general rule, it is preferable to start with exotics removal and stabilization of bare soil and allow for some natural recruitment before determining which species to consider planting. There is little need to plant species that are reproducing locally; they will develop more rapidly than transplants anyway. Nevertheless, achieving diversity may still depend upon your introduction of an array of potential propagules. Use the restoration model site and historic records (see Chapter 21, "Monitoring and Management") to determine which species that are not now part of your landscape are most characteristic and under what conditions they would likely reoccur. Add propagules incrementally to restore some of the range of recruitment opportunities that prevailed in the past. Some species will require repeated introductions to establish themselves.

• *Protect and preserve the genetic raw materials of the local landscape.*
Learn how to propagate local subspecies responsibly. Study the plant population genetics of your landscape to determine local genotypes and ecotypes and
the level of local adaptation. Develop an approach and guidelines to reintroduce plants in the landscape that conserve genetic richness for adaptation and
evolution. Recognize the role we play as consumers and distributors of plants
and seeds and other organisms in creating the problems of blight and disease,
as well as by introducing hybrids and varieties that trigger competition for land
area that decimates local subspecies. The best way to produce sustainable landscapes is to patronize responsible producers and support local propagators and
regional nurseries that practice sustainable horticulture.

For a set of *Guidelines for Gene Conservation through Propagating* and
Procuring Plants for Restoration, contact Marylee Guinon at Sycamore Associates, 910 Mountain View Drive, Lafayette, CA 94549, 510-284-1766.

Planting Methods and Strategies

Techniques for reintroducing native plants are developing rapidly as more and
more people become involved in restoration. The developers of new tools and
techniques are moving away from large plants and heavy equipment to lightweight materials that can be carried in by backpack and to smaller, more agile
equipment that can pick its way across vulnerable terrain. New production
techniques offer smaller, readily transplantable plants whose survival is enhanced by mycorrhizal inoculations, water-holding gels, and vigorous root systems. New products can be used to protect vulnerable new plants until they become well established.

The production method and size of the plant determine what planting
methods are suitable. Large-rooted specimen plants are usually transplanted
with a large volume of soil; smaller specimens of many woody species can be
transplanted without soil.

Under woodland conditions consider planting woody seedlings in and
around stumps and logs. This approach allows the least possible soil disturbance and enhances seedling survival, especially during dry spells, because the
spongy log may hold water far longer than surface soils. In addition to stumps,
you can plant just above check logs to give the plant both protection as well as a
better supply of moisture and nutrients. These methods are the most appropriate overall for establishing the widest array of species under forested conditions. Other good sites are the clefts of roots at the base of trees, where there is
minimal competition from tree roots and additional moisture from stem flow,
that is, water flowing down the trunk.

When trying to establish forest in expansive, open conditions, try and evaluate several different sizes of plants to increase survival rates. Mixed-size plant-

ings are less vulnerable to exceptional climatic conditions, which might eliminate an entire season of work if only one method is employed. Use mixed species as well, including both canopy and understory species. Plant in loose clumps and gaps—this technique is called "nucleation"—rather than in rows plantation-style. More random placement makes the woodland look more natural in appearance and varies the conditions available for plant succession.

Recommended stocking rates—the density of planting—vary widely based on available budget for both plants and long-term maintenance. Seedlings are often spaced as close as 3 feet and as far as 8 feet apart with good results (Figures 24.1 and 24.2). Whips—young trees—may be from 6 to 12 feet apart. Consider these to be average densities rather than recommended spacing. The goal is to establish a dense and layered forest of mixed species rather than to space each tree for optimal specimen growth. Trees planted in mixed clusters in variably spaced patches have a higher survival rate than those planted on equally spaced centers or than monospecific plantings.

Mulch plants with a thin layer of woodchips. If possible, water newly planted trees during severe dry spells. An occasional deep soak is better than a frequent light sprinkling. Of course, the first watering may have to be the last, but your chances for success go up with limited supplemental watering during the first season.

Figure 24.1. Planting clusters of native trees was the first step in restoring this site.

Figure 24.2. Mixed-species groves now stabilize ground once eroded by trampling.

Balled-and-Burlapped Plants

Balled-and-burlapped (B & B) plants are transplanted in a ball of soil wrapped in burlap. Today, especially larger specimens may also be contained in a wire basket that holds the rootball intact. B & B is the most common planting method for larger trees and specimen shrubs in the Northeast. Indeed, many native species with taproots that do not transplant well by means of B & B are largely unavailable in the trade. Transplanting with this method is usually successful because it sustains enough of the root system for the plant to survive.

The Association of American Nurserymen has developed recommended minimum standards for B & B plants. Do not be tempted by bargain prices offered for a larger plant than specified if it comes with a smaller rootball. Do not accept a plant wrapped in plastic fabric and twine instead of biodegradable burlap and jute twine. The rootball should be completely intact. Cracks in the rootball mean damage to the root system. Reject any damaged rootballs. Dig plants when dormant. Late digging, after the buds have started to swell in the spring, substantially reduces the likelihood of survival, although some species can be dug when in full leaf if they are well maintained before and after planting.

Although many species are available in B & B form, especially in larger sizes, the size and weight of the rootball are major limiting factors in woodland

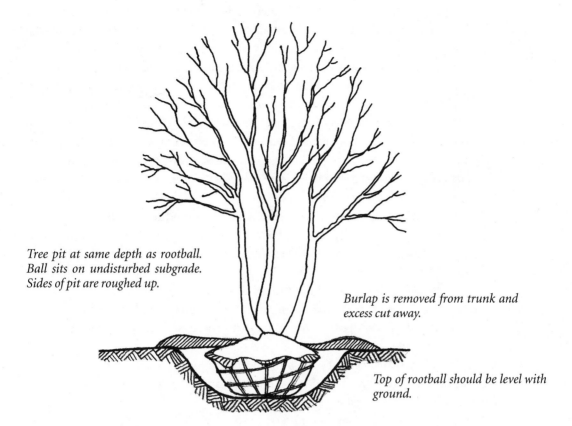

Tree pit at same depth as rootball. Ball sits on undisturbed subgrade. Sides of pit are roughed up.

Burlap is removed from trunk and excess cut away.

Top of rootball should be level with ground.

Figure 24.3. Balled-and-burlapped tree placed in planting pit.

restoration. Even a relatively small rootball is difficult for two people to carry any distance, which places severe restrictions on the use of B & B plant material in woodland areas where vehicular access is undesirable or unavailable. In addition, B & B plants are expensive, further limiting their use in restoration. The large excavation required for their planting adversely affects adjacent plants and entails extensive soil disturbance. Current best practice advocates as wide a hole as possible, especially in the first foot of soil, to encourage roots to expand outward. This convention, however, is inappropriate in all forests, except the most disturbed sites, where soil reworking is necessary for other reasons. Large, well-developed specimens rarely do well planted under canopy cover and may recover poorly from transplant shock in the shade.

When planting B & B trees (Figure 24.3), dig the hole slightly wider at the top than the bottom to ensure that you can backfill without leaving air pockets. Dig the hole no deeper than is required to accommodate the depth of the ball. A deeper hole does not benefit the plant and may jeopardize it should settlement occur. The top of the rootball should be at the same elevation as it was before the plant was dug, or slightly higher if the soils on site are wetter than in the nursery.

Position the tree so that it is generally perpendicular to the ground, or leaning in the same direction as adjacent trees. Leave the biodegradable burlap wrap and twine in place, cutting away only excess wrappings. Do not dislodge the rootball in any way. Where clay soils predominate, use a shovel or pick to scarify edges of the pit to break through the impermeable layer of soil to minimize the likelihood of glazing and ensure better root penetration beyond the soil of the rootball. Backfill around the root area and gently tamp the soil to prevent major air pockets, but do not overcompact. Do not use a backfill mix or modify the native soil to which the plant must adjust over time. Watering the plant with a hose at a slow dribble helps settle the soil. Prune only broken, damaged, or wilted branches after planting.

Some planting recommendations suggest mounding the soil at the outer edge of the planting ring to form a water-holding berm; however, this approach may encourage root growth to remain within the berm, close to the tree, as if the plant were still confined to its B & B state. In most instances a ring of woody mulch over the rootball is adequate to conserve moisture. Do not mound mulch up around the trunk of the tree. It will damage the bark and encourage insects and disease. This common landscape practice has shortened the life of many plants.

Movement is necessary for building the trunk's strength. Swaying actually stimulates root growth, so not staking the plant is really best. In extreme conditions, use a flexible stake so that the trunk will sway in the wind. Remove the stake and wire after one year or less. Leaving wire or string around the tree can kill it. Do not wrap the trunk with tree wrap or protective tape, even to discourage pests. Tree bark needs air and sunlight in order to build a healthy protective sheath. The wrap slows the tree's ability to adapt to the site and may in fact provide a cozy home for insects.

One exception to the no-wrap/no-stake guideline occurs in areas where theft or vandalism of smaller trees is predictable. To prevent theft, you can secure saplings with flexible adhesive wrap to a stake driven at least 12 to 18 inches into subsoil below the hole you've dug. The new plant may, however, need protection from browsing, such as deer fencing. Other methods for protecting new plants are described later in this chapter.

Container-Grown Plants

You can purchase many smaller plants, both herbaceous and woody, in containers ranging from plugs to several gallons in size. Although the plants may be smaller than those available in balled-and-burlapped form, many more species are commercially available. The soil in the containers may make them heavy; however, because the plants are often small, they are generally not difficult to transport to the planting site.

Container growing is ideal for many native species that are difficult to transplant in B & B form, such as hickories, sassafras, sumac, and black gum, as well

Figure 24.4. A wide range of species and sizes of plants may be purchased in containers from local propagators.

as herbaceous species (Figure 24.4). To add to your planting stock, you might consider having volunteers start and grow native plants in containers from seed. When planting in containers, repot frequently to allow healthy, rather than girdled or pot-bound, root systems to grow. Planting times are also broader because the plants are less subject to moisture stress from loss of roots in the transplanting process than with B & B.

The shoot growth and root and soil mass also contribute to site stabilizations on steep slopes (Figure 24.5). The most difficult task in transplanting container-grown plants is determining at what stage a new seedling is suitable for planting out in the field. Younger specimens are often better in woodlands than older ones; an older plant might have difficulty adapting to the sudden change from the nursery to the forest. Plant very small sizes only in the springtime; they are too vulnerable to frost-heaving in the fall.

When purchasing container-grown plants, inspect the bottoms of the containers to see if the plants have outgrown them, rejecting specimens with overgrown, kinked, coiled, or girdled roots; they will not develop properly after planting.

Transplanting container-grown plants into the ground needs to be done with care to avoid damaging roots. When planting, carefully remove the plant from its container by gently upending the container or cutting it away from the soil, which should be held as a mass by the dense roots of the plant. If the soil is not held well by roots, the plant probably has not yet developed adequately. Be extremely careful when unpotting herbaceous plants or when examining their roots because they are much more sensitive to handling than woody species

Figure 24.5. Containerized plants hold well on steep slopes.

and more likely to be damaged. If a plant is very sensitive to handling, try soaking the pot inside a bucket overnight. Examine the sides of the root mass for roots that are J-shaped, kinked, or circling the pot and that could develop into girdling roots over time. If they are, the plant has poor survival prospects because these problems do not correct themselves. Pruning the plant's roots in an attempt to correct these deficiencies in the field may expose the plant to soil-borne diseases.

Set container-grown plants in a hole no deeper than, and only slightly wider than, the container. Leave only enough room to backfill properly, and water and mulch the plant, if possible. Water lightly in very dry conditions during the first growing season, but don't overwater; the object is to ensure survival, not to weaken the plant for the long haul by accustoming it to a high level of maintenance.

Plugs

Plugs are very small seedlings often grown in plantable pots made of organic material. The containers biodegrade in the soil so you won't have to remove them before planting, reducing root damage and labor. You can use them to sprout most kinds of plants. Because plugs are small, they disturb the site less than larger plantings. Plugs can be planted with a tree planting bar, as close as 1 foot in open areas. Use plugs in the spring only, as they often frost heave if planted in the fall.

Rootstocks, Corms, Rhizomes, and Bulbs

Some plants, such as ferns and many wetland species, are available as rootstocks, corms, or rhizomes. They may have specific planting requirements, such as a specified depth, and it may not be obvious which side is up, so obtain planting instructions from the supplier.

You can wrap wetland rootstocks in squares of burlap weighted with a large stone or two and then drop them into shallow water. The burlap affords some protection from waterfowl grazing (Figure 24.6), and the weight of the stone holds the propagule in place without digging a planting hole.

Bareroot Plants

Bareroot trees and shrubs are less expensive and lighter than B & B or container-grown plants because they are sold without soil. Dug when dormant, usually late fall or early spring, bareroot plants are kept in cold storage at temperatures just above freezing with a very high humidity until they are shipped. The bareroot plant usually arrives with its roots wrapped in moist straw inside a plastic bag.

Bareroot planting is a convenient method for planting large numbers of small trees (Figure 24.7) because they are easy to handle and readily available from nurseries. Increasingly they are sold at larger sizes, but an individual can still carry them without difficulty. Bareroot planting also entails far less soil disturbance than container or B & B planting. Not all species, however, are

Figure 24.6. Wrapping wetland rootstocks in a square of burlap, weighted with a fist-size stone, allows you to simply drop them into shallow water, rather than trying to dig and plant them in muck.

Figure 24.7. Bareroot plants are lightweight and easily carried on-site.

suitable for bareroot planting. Soft-rooted plants like magnolias are extremely difficult to handle in bareroot form, but different peoples' experience and recommendations vary.

Bareroot planting times are restricted. Ideally, plant bareroot trees as soon as the ground can be worked to take advantage of cooler temperatures and damper soils. Do not plant bareroot material after the weather turns hot and dry. Expect high losses with bareroot trees if they are stored, shipped, or planted improperly. For best survival, try larger sizes (8 to 10 feet), planted early in the spring and watered through the first season, especially where browsing by deer is likely.

Upon arrival on the site, store bareroot trees in the shade and cover their roots with moist mulch, straw, or compost. Do not allow roots to dry out. Consider additional protective measures, such as a wetting solution that is mixed with water to form a gel thick enough to cling to the roots (Figure 24.8). Wetting agents retain many times their own weight of water. Before planting, clip only broken or damaged roots with a clean, sharp tool and dip the entire root system in the wetting solution to rehydrate the root system and to maintain an even moisture balance in the soil after planting.

For larger bareroot plants, greater than 1 inch in trunk diameter, dig a hole 6 inches larger than the size of the root system when it is spread out and about 18 inches deep (Figure 24.9). Leave rocks in the soil where they help improve drainage. Make a small, well-compacted mound of earth in the center of the

Figure 24.8. Keep bareroot plants from drying out. A wetting gel that coats the roots before planting is helpful.

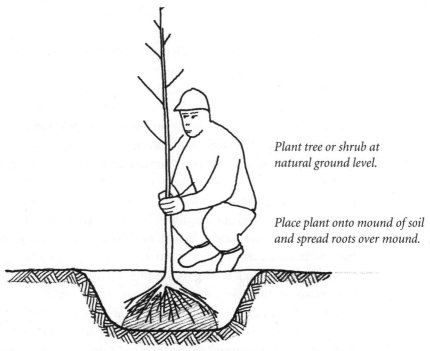

Plant tree or shrub at natural ground level.

Place plant onto mound of soil and spread roots over mound.

Figure 24.9. When planting bareroot, be sure to spread and press the roots around a mound of soil in the pit to ensure good soil and root contact and to eliminate air pockets.

hole and spread the roots over it. Position the tree so that its main stem is growing more or less straight up. Backfill soil around the roots. Tamp the soil evenly in layers to eliminate air pockets and poorly compacted areas.

Keep bareroot seedlings in a bag with wet straw or water until you can plant them; do not carry them to the site in your hand or they will dry out. To plant seedling- and sapling-size bareroot plants, use a planting bar or dibble bar to create a 12-inch V-shaped notch (Figure 24.10). Place the seedling upright about 1 inch deeper than in the nursery. After planting, tamp the soil firmly with your foot to prevent air pockets, but do not pound the soil. Water in well. One experienced volunteer can plant 500 to 700 seedlings a day. During dry spells, bareroot plants often develop slowly enough to survive conditions that would kill a B & B plant. Do not stake except in the most extreme conditions.

Cuttings

Cuttings are new plants developed from a vegetative part of the parent plant rather than from a seed. They are relatively small, lightweight, and easy to handle and may be planted before or after roots develop. Because they are produced by vegetative, rather than sexual, reproduction, their genetic diversity is limited to that of the parent stock. The plants produced are very uniform and therefore are often used in agriculture and in high-production horticulture to reproduce, for example, woody plants such as blueberries. Tissue culture, a form of cloning from selected cells, is a more intense vegetative process that utilizes parent material efficiently and generates high volumes of identical plants at relatively low cost.

Despite the limited genetic diversity, cuttings may be especially useful in situations where enough plants are needed to ensure a very dense cover very rapidly, such as for erosion control with soil bioengineering. Many woodland shrubs, such as blueberries and huckleberries, naturally spread vegetatively and may have very low seed viability, making them good candidates for establishment by cuttings. New technologies can speed up the process of establishment. For example, you can use vine cuttings that are rooted in growth chambers, then spread over the ground and watered temporarily in a rooting medium to create a bed of groundcover that otherwise would take years to establish. Variations on this method, although not yet perfected for forested conditions, hold promise for reintroducing finicky herbaceous species that also may benefit from additional care during an initial establishment period.

Many soil bioengineering techniques, including those described earlier for streambank stabilization, require cuttings to be planted directly in the field, before they have developed roots. These are useful planting techniques even where extra stabilization is not required and are easy to implement if source material is available nearby. They are most useful at edges and in open landscapes where rapid reforestation is desired. However, because these techniques rely on roots developing after planting, they must be used in areas of full sun exposure and so are of limited application under canopy cover. Currently, the

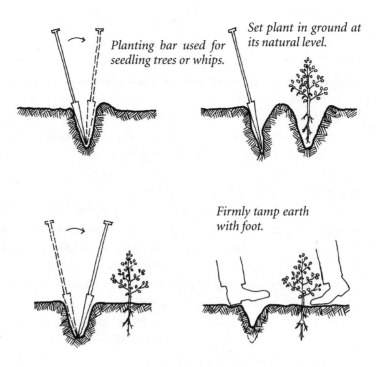

Planting bar used for seedling trees or whips.

Set plant in ground at its natural level.

Firmly tamp earth with foot.

Figure 24.10. A planting bar eases the task of planting seedlings.

diversity of species that may be used in this way is very restricted, although the list of suitable plants will certainly expand with more field trials. Typically, species that spread vegetatively as a common natural reproduction method, such as willows, shrub dogwoods, elderberries, and alders, are most suitable. You can harvest plant material from the site or nearby natural areas with minimal impact and with care.

Direct Seeding

Direct seeding of forest species is not a commonly used planting method, in part because erosion- and sediment-control regulations promulgated in the past few decades have served to shift the objective of many restoration projects from reforestation to erosion control, which in turn promoted the use of rapidly established grasses and eliminated necessary research on seeding of other species. Before then, there was a lot of interest in this approach because of its low cost. To dry up wetlands, "reclaimers" seeded the now-invasive pest ti tree into Florida wetlands from the air, and foresters dropped black locust seeds onto remote, abandoned mine sites on Appalachian hillsides in an effort to stabilize the slopes. Today, under the Conservation Reserve Program, farmers are planting oak forests from acorns on marginal soils withdrawn from agricultural production. To learn more about seeding techniques, consult journals such as *Restoration and Management Notes,* which describe advances in forest seeding techniques for restoring native mixed-species communities.

Establishing plants from seed in existing woodlands is difficult because of intense competition from established plants. At the same time, though, you will probably want to enrich the palette of plant material on your site, and it is easier to simply collect seed than to grow a new plant as well. You can bring in a rich array of mast species with larger nuts, such as oaks, hickories, beech, and walnut, from adjacent natural areas to enrich a forested site. We recommend two planting methods: squirrel-cache and stump planting—both of which give your precious seed the best chance possible.

Squirrel-Cache Planting

Squirrel-cache planting is very simple in concept and practice: just place several seeds in the same hole. The multiple seedlings have safety in numbers; they also benefit one another by sharing shade to conserve moisture. In severe environments the cache planting method may make the difference between surviving and dying. One example is the black soils of many abandoned coal mines, where the temperature gets so high that seedlings easily die. With squirrel-cache planting, only a portion of each tree's bark is exposed to heat damage, the outer edge rather than the entire circumference. In many cases all the seedlings survive where one would have failed.

Stump Planting

A seed that falls onto or gets taken to a rotting stump where there is the space and opportunity to grow as well as some protection from other already-established plants is lucky indeed. Mimic these conditions by directly planting seeds in stumps. Wait until the interior of the stump is spongy enough to hold moisture well and to embed the seed or seedling in the rotting wood.

Seedbed Preparation

Some seedbed preparation in woodlands is useful for fostering the germination of new seeds as well as seeds already in the soil. Seedbed preparation in woodlands is driven by the need to minimize damage to existing plants in the landscape. Plowing, disking, raking and other techniques often used for seeding are entirely inappropriate in woodlands.

Although minimizing soil disturbance is desirable, good soil and seed contact is necessary. Pull back small patches of the litter layer to expose areas of soil before seeding. Do not rototill or work the soil well. Simply press the seeds into the soil well and cover with only a minimal amount of litter.

Because sprouting may be improved with prescribed burning, research the fire history of your landscape to evaluate the possibility of incorporating prescribed burning into your seeding program. Where you cannot reintroduce fire as a large-scale management tool, you may be able to mimic the conditions that occur after a fire to some extent. For example, you can use small-scale spot burns in small patches to create more favorable conditions for the germination of oaks, sassafras, and many woodland ephemerals. Exposing small areas of bare soil beneath ground litter and sprinkling the surface with very small amounts of lime and sand mimics postfire conditions to a degree. The result may well be dozens of oak seedlings appearing in a single season in a square-yard patch. Some herbaceous species do not establish well unless methods like these are used.

Protecting New Plants

New plants may need protection from browsing by deer, rabbits, and other wildlife. Saplings and larger trees may require either fencing or protective collars to protect their lower trunks from damage. You can make individual temporary tree fences from a variety of materials, such as old snow or silt fence. Commercial collar—a coiled strip of plastic that can be loosely wrapped around the trunk—also is available. Sizes range from tall enough to protect from deer browse and rubbing to short collars for use where only smaller mammals such as rabbits hinder establishment. You must remove both fencing and protective collars after a season or two, before they inhibit tree growth or fall apart. As always, evaluate any new product or procedure in the field before applying it on a large scale.

One currently popular method is to completely enclose smaller-size plants in a tube to protect both leaves and bark from browsing while also somewhat conserving moisture by creating greenhouse-like conditions within the tube. This interim level of protection gives the plant an opportunity to establish under difficult conditions. There are varied tube-type protectors on the market, and new ones appear frequently. They typically vary in strength from hard plastic to a thin film, in height from a few inches to 8 feet, and, of course, in cost and weight. Opacity is also variable; use the most transparent products available if they are to be used in woodland conditions. You must cover the top with screening if the sapling is too small to allow birds to escape after entering in search of insects. Remove tube-type protectors after no more than one year because the tree develops no strength when it is confined in one. If left in too long, a plant may simply flop over when the support is removed, just as if it had been staked too long. Despite the early growth spurt, many managers report difficulty in weaning plants from the tubes and continued dependency on staking. Because unprotected trees catch up in height with the assisted plants within a few years, it is probably preferable to find other ways to protect plants from browsing.

Avoid all such products when they are not truly needed. All entail extra cost as well as labor, and thus disturbance in the landscape. You may be able to accomplish an adequate degree of protection more simply. One group working in Wissahickon Park in Philadelphia noticed abundant tree seedlings in a patch of the invasive and thorny devil's walkingstick in an area where deer had consumed most of them. The team removed some of the walkingstick to lessen competition with the seedlings but left enough to continue to deter deer. They then stuck the thorny stems they had removed upright in the ground around newly planted saplings to afford them better protection. In Central Park, one crew member "planted" dead brush upright in the landscape to discourage walking on top of new plants (Figure 24.11). In released meadows, herbaceous growth may be so dense that additional protection of woody plants is unnecessary. Plots at the Pennypark Ecological Restoration Trust, outside Philadelphia, suggest that woody seedings and saplings grown in sulfur-treated meadows may be less palatable to deer.

Fertilizing and Pruning

Fertilizing is not necessary or desirable except in the most extreme conditions. Do not consider fertilizing unless you have taken soil samples and have determined critical limiting factors that exceed the range of conditions to which local species are adapted.

Phosphorus is the nutrient most necessary during transplanting and may be somewhat limited where nitrogen loads are high. Nitrogen can have negative effects and should be avoided whenever possible. Where necessary, use a slow-release variety. Allow no fertilizer to come into contact with root systems. Mineral rock dust as an alternative to fertilizer is an option that merits further exploration.

Pruning should not be necessary except to remove broken and damaged or hazardous limbs. In general, wait at least a year before pruning to ensure that

Figure 24.11. Inconspicuous dead branches driven into the ground deter trampling around new plants in Central Park.

you remove only dead wood. The best time to prune trees is late winter or very early spring, when the plants are dormant and before buds swell and the sap begins to flow. Never leave stubs partway up the branch; cut off the whole branch instead. Never prune behind the branch's bark ridge, the folded area of bark where the branch leaves the tree trunk or a large branch; this practice wounds the tree. Never top trees by completely severing the leader, the main upright branch, midway up the tree. Where a major limb is to be removed, locate the branch bark ridge; then stub cut the branch about 12 inches above the branch bark ridge so that the weight of the branch does not rip the final cut. Next, locate a point (*A* on Figure 24.12) at the crotch of the branch but outside the branch bark ridge. Locate a second point above the swelling where the branch meets the branch collar (*B* on Figure 24.12). Make the final cut at line *AB*. If the limb is large enough that tearing is a risk, use the three-cuts methods to prune. The first cut goes under the branch about a foot or so from the trunk. The second cut goes just beyond the first cut and starts from above. If the limb falls and the branch tears, the rip will stop at the first cut. The stub of the branch can be cut safely to the desired point with cut three.

Alex Shigo has spent his life studying the nature of tree wounds and their effect over time on the plant. His books are singularly informative. A good introduction is *A New Tree Biology: Facts, Photos, and Philosophies on Trees and Their Problems and Proper Care.* It and other related publications are available from Shigo and Trees, Associates, 4 Denbow Road, Durham, NH 03824.

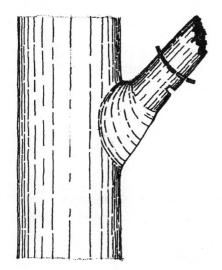

1. Notch underside of branch above branch collar.
2. Cut stem as shown by dashed line.

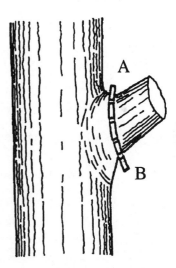

3. Final cut is just above branch collar between points A and B.

Figure 24.12. Pruning.

Propagation, Transplanting, and Collecting

Propagation, like other aspects of gardening, is part art, part science, and part joy. There is an incredible amount to be discovered about native species and their successful recruitment. The goal of these guidelines is only to give you some incentive to start. Start small. There is a lot of "feel" and experience to successful propagation. Take advantage of area horticulturists and spark their interest in the challenges of trying to propagate less-familiar native species. Volunteers with The Nature Conservancy in Chicago have demonstrated just how creative they can be at developing new methods suited to particular native species:

> "Everyone loves to gather Indian grass," says Steve Packard, "because it feels so good when it comes off. When it's ripe, you just pull your hand up the stalk, and you have this big handful of seeds, and you dump them in the bag. With rattlesnake master, on the other hand, the seed is spiny and tough. One way to do it is to wear gloves and wrench the heads off. Later on, we developed better methods. With rattlesnake master, for instance, we cut the heads off with garden snippers and throw them in a box with a screen on the bottom. Someone rubs the heads, and the seeds separate and run through the screen, and you catch them in another box" (Stevens 1995, 54).

Collecting Seed

The following guidelines for seed collecting were taken from "Growing Pains: The Ethics of Collecting from the Wild," by Sue McIninch (1993):

1. Identify the plant before you take the seed. Make sure the plant is not a rare or endangered species.

2. If the plant is on private property, ask permission from the owner of the property before collecting any seed from the plants. Landowners are generally responsive to such requests.

3. NEVER take all of the seed, always leave some for natural dispersion. If the seed is a fruit or a pome and considered food for wildlife make sure you have left some for this purpose too.

4. If the plant is common in the area, take samples from a variety of plants. Sometimes pollination, maturation and seed ripening can occur at different times even within the same community. Sun exposure, prevailing wind patterns all have an effect on the ripening process.

5. Identify or test soil conditions and observe the habitat that the plant grows in. Feel secure enough that you can replicate these condition. If you do not think you can do this, pick another species to propagate. It is important to not waste seed (5).

Collect seed only when it is ripe, which is likely to be in mid to late fall. In general, ripe seed is darker, harder, and larger than the unripe seed. The leaves of the plant may already appear brown and dead. Collect seeds inside fruit when they are ripe or just after. Gather seeds in hard capsules just before the capsule splits. Often, simply shaking the stem will release the seed, telling you when it is ready. The fresher the seed, the better. An added virtue of seed is that it is free of pathogens such as bacteria and viruses, so you are unlikely to spread disease via seed.

Cleaning Seed

Most seeds keep best if they are cleaned of all extraneous chaff and dust so dust-borne or chaff-borne mold and fungi are less likely to grow. But cleaning seed is sometimes easier said than done and may require inventiveness and experimentation if no effective method is described in your reference books. You can clean seed in many different ways, including shaking them in a bag or over newspaper, winnowing them with an electric or manual fan, or by blowing lightly over them in a shallow container. Soak fleshy fruits in water to ferment them for a few days and separate the seed easily. Allow the seed to dry for a few weeks in paper bags, envelopes, or open cardboard trays, but not in plastic.

Storing Seed

Some seeds require scarification—slitting or softening of the seed coat in water—and some need a period of rest before it is time for them to sprout. Dormancy, temperature, and humidity requirements for germination vary widely among species. The cold and damp also trigger certain chemical changes that break the dormancy of seed. When in doubt, you can always mimic nature and simply plant seed outside where it might fall naturally if adult plants were around, although that option is not always suitable because it may waste hard-to-obtain seed.

If the seed has been properly dried, you can stratify—store and chill—the seed in a vegetable crisper or an unheated shed or garage until enough time has elapsed to permit germination, which varies with species. Place them in moist sand or sawdust and keep them for four months at 32 to 41 degrees Fahrenheit. Mix seed with an equal volume of damp fine sand and store in plastic bags in the refrigerator at 32 to 38 degrees Fahrenheit. Dry seed fully before storing in mouse-proof sealed plastic containers such as garbage cans or in plastic bags hung on thin string from the ceiling. Many early-spring bloomers, however, do best when sown fresh.

Sowing Seed

In spring, to get a head start on planting, you can sow seeds in a growing medium indoors. Try mixing the seed with wet sand for more even distribution.

Do not let the soil dry out completely until shoots appear. When transplanting outside, use no fertilizer and be sure, if at all possible, to water during times of drought until the new plants are well established. Plant fleshy fruits and nuts and acorns immediately in containers or in the field (if they are not dug up by squirrels). Do not expect quick results. Many woodland species take years to germinate or show any discernible aboveground growth.

Propagation by Division and Cuttings

You can often propagate woodland plants vegetatively, through division and cuttings as well as other methods. Fleshy rhizomes and rootstocks that form clumps of stems usually can be divided. The general rule is to divide when plants are not blooming; thus a plant that blooms in late summer or early fall would be divided in the spring, a spring ephemeral in the fall. If winters are cold in your region, divide spring bloomers in early fall so that the roots can develop before cold weather sets in. To make a division, break up the segments of the root mass or cut them apart with a sharp knife.

To divide bulbs, first let their foliage die back, then dig up the bulblets produced around the mother bulb. You can store them until the appropriate planting time and then replant them at the same depth as the parent bulb, with the growing tip pointed upwards.

Plants with a single taproot can be propagated by stem cuttings. Many woodland plants, such as bleeding heart, can be propagated by division and as cuttings. Because a cutting will be identical to the parent plant, use as many different source plants as you can to maximize the biodiversity of the species you are propagating.

To make a cutting of a softwood plant, like hollies and willow, cut a stem below a leaf during the growing season; then plant in shade with a rooting hormone. Remove the lower leaves or cut off half of the large leaves. You'll need to keep the cutting moist and surrounded by high humidity until it roots. One way is to cover it with a plastic-bag "greenhouse." Each day or so, remove the plastic bag for a few minutes.

To make a woody cutting, cut a stem from the previous season's growth below the leaf bud while the plant is dormant. Refrigerate the stem during the winter, and then plant it on the site in spring. A rooting hormone is usually helpful. To be sure you plant it right side up, make a slanted cut on the bottom end and a flat cut at the top when you first take the cutting. Set the cutting in the soil at a 45-degree angle to maximize rooting. Hardwood cuttings work well for elderberries, buttonbush, and willows, for example.

You can plant rooted or unrooted cuttings directly in the ground, protecting and keeping them moist with a plastic-bag greenhouse. Partial shade is usually necessary to maintain adequate soil moisture.

Layering is also a very adaptable and simple method of outdoor propagation that requires no plastic. One layering method simply requires you to bend a stem until it touches the ground and bury about a 4-inch section. Once new growth has emerged, you can separate the new plant from the parent and transplant it.

According to the New England Wildflower Society, the following species of wild herbs are relatively easy to cultivate from seed or divisions:

SCIENTIFIC NAMES	COMMON NAMES
Adiantum pedatum	maidenhair fern
Allium tricoccum	wild leek
Aralia spp.	sarsaparilla
Asarum canadense*	wild ginger
Cimicifuga racemosa	black cohosh
Collinsonia canadensis	stoneroot
Gaultheria procumbens	wintergreen
Hydrastis canadensis*	goldenseal
Matteuccia struthiopteris	ostrich fern
Mitchella repens	partridgeberry
Panax quinquefolius	ginseng
Podophyllum peltatum	mayapple
Polygonatum spp.*	Solomon's seal
Sanguinaria canadensis	bloodroot

* Sow seeds of these species immediately upon ripening. If the seeds dry out, they will take many years to germinate, if ever.

Some species are notoriously difficult to propagate because of symbiotic relationships with mycorrhizal fungi. Do not jeopardize their natural success by collecting seed or try to propagate them without adequate expertise. These include many orchids such as lady's slipper, spotted wintergreen, and ground pines. Some others are very slow but can be cultivated with care, including trillium and blue cohosh.

The following is a list of some plant hosts to butterflies:

TREES, SHRUBS, AND VINES	BUTTERFLIES AND MOTHS
beach plum	cecropia moth
black cherry	tiger swallowtail, red-spotted purple butterfly, Prometheus moth, cecropia moth, hummingbird moth, viceroy
flowering dogwood	spring azure butterfly
hackberry	hackberry butterfly, snout butterfly, question mark butterfly
persimmon	luna moth
sassafras	spicebush swallowtail, imperial moth
sumac	spring azure butterfly
sweetgum	luna moth, royal walnut moth or regal moth
trumpet honeysuckle	hummingbird moth
tulip poplar	tiger swallowtail, spicebush swallowtail
Virginia creeper	pandorus sphinx, lettered sphinx, hog sphinx
white pine	imperial moth
willow	mourning cloak, red-spotted purple

NATIVE FORBS	BUTTERFLIES AND MOTHS
aster	pearl crescent
bidens	variegated fritillary
butterfly weed	monarch
goldenrod	monarch
milkweed	monarch
violets	fritillaries

Unfortunately, a number of invasive exotic species continue to be recommended for use in butterfly gardens even though there are native alternatives that do not pose threats to other species. Commonly recommended exotics that should be avoided include autumn and Russian olive and Japanese and Tartarian honeysuckles.

Do not be tempted to plant only one species, such as all beebalm, because it so effectively draws butterflies. Rather, plant a variety of species that will flower throughout the season.

There are many books on propagating native species. Here are some of the best:

The Wildflower Gardener's Guide: Northeast, Mid-Atlantic, Great Lakes, Eastern Canada, by Henry Warren Art (1987) Storey Communications, 105 Schoolhouse Road, Pownal, VT 05261, 802-823-5810.

Collecting, Processing and Germinating Seed of Wildland Plants, by James A.Young and Cheryl G. Young (1986). Timber Press, Portland, OR.

Growing and Propagating Wildflowers, by Harry R. Phillips (1985). University of North Carolina Press, Chapel Hill, NC.

The Reference Manual of Woody Plant Propagation: From Seed to Tissue Culture, by Michael Dirr and Charles Heuser (1987). Varsity Press, Athens, GA.

Several publications are available directly from the New England Wildflower Society, 508-877-9348, including "Propagation of Native Plants," "Garden in the Woods Cultivation Guide," and "Sources of Propagated Plants."

Contract Propagation

Consider contracting with local propagators to grow native species that are otherwise unavailable. Plugs of native grasses that persist in light woodland, such as switchgrass and little bluestem, are easily produced at fairly low cost, as are "sods" or "mats" of shrub and/or herbaceous communities. Be sure to review appropriate guidelines with all collectors of any plants or propagules, professional and amateur alike.

Transplanting and Salvage

Do not overlook opportunities to rescue and transplant plants from areas where they are about to be disturbed. Even if you have been unsuccessful with transplants, try again: it's important not to let any plant resource go to waste.

Obviously, you should not deplete a native habitat, nor should you collect rare or endangered species without special expertise and permits. Consider collecting plants from an area being developed and relocating them to analogous protected habitats. Where disturbance levels are fairly low, consider collecting from elsewhere on your site to revegetate small patches.

A tree-mover, or tree spade (Figure 24.13), is often useful for moving larger trees and shrubs that are harvested on a site or, at least, very locally. Small mats of shrubs can often be collected quite easily to expand on-site populations. John Monro, a Pennsylvania ecologist, developed a very effective tool for plant salvage that is now available for sale in several sizes. His soil moving machine lifts and removes large blocks of soil as well as the plants, propagules, and soil organisms inhabiting them in uniform-size units that can be reassembled at a restoration site. Mycorrhizal- and fungal-dependent species typically survive because the soil block remains intact. The tool, called a soil mat lifter, was developed by Monro Ecological Services, 990 Old Sumneytown Pike, Harleysville, PA 19438-1215, 610-287-0671, and is owned by Resource Conservation Corporation. It is available through Bentley Development Co., P.O. Box 338, Old Route 22, Blairsville, PA 15717, 412-459-5775. Monro also teaches courses in plant salvage for the Society for Ecological Restoration.

Figure 24.13. A tree spade allows very large plants to be moved with limited losses.

Exclosures

Often when you attempt to reestablish vegetation in urban and suburban areas, you probably will need to use exclosures to protect the site from trampling, browsing, and/or vandalism. The particular exclosure you use should be tailored to the needs of the individual site as well as the surrounding environment. The following describes some possible exclosure strategies that might be integrated into a larger, coordinated program:

- **Temporary large-gate or large-opening exclosures** allow for the regeneration of the forest understory and eventual canopy replacement. Keep the exclosure in place until the woody species are large enough to withstand some deer browse. Remove invasive species as they appear to ensure that your effort is directed to the most desirable species, although exotics often benefit less than native species from the use of exclosures (Pettit, Froend, and Ladd 1995). Where the litter layer has been disturbed, a thin cover of woodchips covering from 50 to no more than 80 percent of the surface is recommended to foster native woody regeneration. Where important herbaceous species appear, consider making that exclosure permanent.
- **Permanent large-gate exclosures** provide optimal long-term protection for less-common woody plants as well as many declining herbaceous species. In some cases you may wish to remove native woody seedlings as well as exotic invasives to provide maximum benefit to selected desirable herbaceous species.
- **Temporary small-gate exclosures** foster the establishment of many small woody seedlings that are browsed by rabbits and other small mammals. Not all of these species will be highly palatable to deer and will persist after the fencing is removed, contributing to site diversity.
- **Permanent small-gate exclosures** will protect those species that are already established but will also limit the ability of new seeds to be brought to the site. In many urban parks where there are no deer, you can use such exclosures to evaluate the long-term impacts of such smaller species as gray squirrels or even rats.

At the smallest scale, repellents can be used to protect selected plants from deer browse and areas as long as population densities are not too high. There are both commercial repellents and numerous homemade versions. One repellent brewed by painter Ed Baynard is made in a blender with 1 quart of water, an egg, and 1 teaspoon of Szechuan hot oil. It must be sprayed throughout the season at least once a week and after every rain. His hypothesis is that the egg repels the deer and the hot oil repels raccoons.

According to the Eastern Native Plant Alliance the following species are especially desirable to deer and may merit special protection:

SCIENTIFIC NAMES	COMMON NAMES
Anemone virginiana	tall anemone
Arisaema dracontium	green dragon
Aster novae-angliae	New England aster
Athryium felix-femina	lady fern
Boltonia asteroides	boltonia
Chelone lyoni	pink turtlehead
Chrysopsis mariana	Maryland golden aster
Diphylleia cymosa	umbrella leaf
Disporum lanuginosum	yellow mandarin
Gentiana clausa	closed gentian
G. crinita	fringed gentian
G. linearis	narrow-leaved gentian
Habenaria ciliaris	yellow fringed orchid
Lilium canadense	Canada lily
L. philladelphicum	wood lily
L. superbum	Turk's cap lily
Panax quinquefolius	American ginseng
Penstemon smallii	pink beardtongue
Polygonatum biflorum commutatum	great Solomon's seal
Polysticum acrostichoides	Christmas fern
Rudbeckia triloba	black-eyed Susan
Ruellia humilis	wild petunia
Smilacina racemosa	false Solomon's seal
S. stellata	Solomon's seal
Streptopus roseus	twisted stalk
Trillium cernuum	nodding trillium
T. erectum	purple trillium
T. grandiflorum	great white trillium
T. luteum	yellow trillium
T. recurvatum	prairie trillium
T. sessile	toad trillium
T. stylosum [catesbeai]	rose trillium
T. viride	green trillium
Uvularia grandiflora	great merrybells
Zizia aurea	golden Alexanders

CHAPTER 25

Meadow Management

Early archival photographs of parks designed by Frederick Law Olmsted, including Central Park in New York and Cherokee Park in Louisville, Kentucky, reveal a richness of meadow landscapes overflowing with native grasses and wildflowers that today is entirely reduced (Figures 25.1 and 25.2). And until this century, infrequently mown remnant places and areas managed with fire or let go all together were more common in rural areas before gang mowers became ubiquitous. Turf and asphalt have gradually replaced these images of the past. Many meadow and grassland landscapes today serve as sites for dumping, parking, and uncontrolled off-trail-vehicle use. Moreover, with no native seed nurseries east of Wisconsin producing regional ecotypes, we are on the verge of losing the very landscapes we now hope to reestablish.

Meadows and grasslands are integral, even though usually temporary, aspects of the eastern forest. Many species, especially many rarer species, are characteristic of such landscapes. Many were never common in the forest and are even less abundant now. Birds such as savannah sparrows and meadowlarks depend on large-scale grasslands and are particularly affected by the rapid suburbanization of abandoned agricultural lands. There are also social and cultural reasons that meadows continue to be important and many sites where forest cannot be appropriately reestablished. Small meadows and grassland strips can be very important in managing runoff even though habitat opportunities are limited.

Because many early efforts in restoration were directed toward grasslands and prairie landscapes, there is a sizable body of documentation on restoring them. Many of the plant species are shorter-lived, making grasslands somewhat more convenient research subjects than forests. The fact that maintenance is

Figure 25.1. This archival painted photograph of this grotto in Louisville's Cherokee Park, circa 1910, documents the rich herbaceous communities of the historic landscape. (Photo used by permission of the Lake County Museum)

Figure 25.2. The death of some trees and modern turf management have impoverished this once-diverse landscape. Note amount of sediment from erosion that has filled this low spot.

required to sustain meadows itself serves to focus attention on the successional trends in these landscapes.

The central task of meadow management in naturally forested regions is to prevent the gradual takeover of the landscape by woody vegetation. All the current methods for controlling woody growth in use present both problems and opportunities for restoring meadows. Mowing and other mechanical methods for removing vegetation, weeding, selective herbicides, and prescribed burning—each has different impacts on natural succession in the landscape as well as a variety of technical limitations that can be difficult for managers to evaluate. The method of management you use to control succession will also affect exotics control and the success of different native species.

Meadow management is complicated simply because herbaceous landscapes do not persist in the same place under natural conditions in the eastern forest. Upland grasslands and meadows were always somewhat ephemeral and rarely occupied extensive areas. Rather, a shifting mosaic of wildfire, blowout, insect infestations, and other recurring events created a forested landscape with patches at every stage of succession. Consider managing successional landscapes in patches that shift over time in location to better mimic historic conditions and to minimize the impact to local populations by always sustaining some portion of the larger landscape and populations.

Reestablishing the more durable and lasting native species that were so vigorously eradicated in the past is severely hampered by competition from annual Eurasian grasses, whose early dominance consumes most of the available moisture before the native species even get started. In the more mesic Northeast, however, the moisture available eventually favors woody species. The annuals are usually overtaken anyway, except on the most degraded sites. More serious difficulties are presented by weedy perennials, woody vines, and especially long distance clonal species, such as knotweed, that can persist for very long periods.

This chapter discusses the following methods and strategies for managing lawn, meadows, grasslands, and prairies:

- Herbaceous landscape types
- Mowing
- Selective removals
- Herbicide management
- Prescribed burning
- Managed grazing
- Grassland establishment and management
- Savanna restoration

Mowing is the most common method of meadow management in the East, but it suppresses diversity compared to fire and other management tools, such as hand-weeding and selective herbicide use. Mowing mimics no natural process very closely; therefore few native species are adapted to the conditions it creates. Prescribed burning, on the other hand, closely re-creates the natural conditions of many historic meadows and native communities. Browsing by animals is usually more selective than mowing and leaves behind the plant parts as manure rather than mulch. Properly used, herbicides can be used to remove woody species either selectively or broadly to arrest succession as well as to control invasives. Mowing from two times a year to once every two years, like grazing and fire, can arrest succession of woody species and assist selected herbaceous species. It can be a useful management tool, especially where fire is not an option.

Herbaceous Landscape Types

Most of the maintained landscapes around us—flower beds, parks with lawns and occasional specimen trees and shrub beds, and agricultural lands—are largely herbaceous. In the East, unmaintained herbaceous landscapes often are areas that were once lawn and are now predominated by exotics, or tall grasslands and meadows in seminatural areas, once agricultural fields, that are a mix of native and nonnative species. Most of these unmaintained landscapes will eventually succeed to woodland if there is no control, deliberate or natural, of the woody vegetation. Shifting intensely maintained conventional landscapes, like lawn, to meadow, reduces the maintenance required to support them and lessens their negative environmental impacts while retaining an open character and views. The range of herbaceous landscapes that occur along the successional continuum have a sliding scale of costs and maintenance intensity.

High-Intensity, High-Impact Turf

Conventional lawn is suitable where use levels are very high, such as paths and playfields, because it takes relatively heavy foot traffic. By using other herbaceous landscapes, such as meadows, where durability is not required and limiting the extent of lawn to what can be maintained adequately, you can lower maintenance requirements and reduce the environmental impacts of turf by eliminating pesticides and weaning your lawn from chemicals. Contact your County Cooperative Extension Service for a soil test kit. Maintaining the proper pH reduces the need for fertilizer. Consider top dressing your lawn with compost or other fully decomposed organic matter as an alternative to fertilizer. Never fertilize in the summer or spring as this feeds weeds only. Fertilize only in the fall, well before winter when nutrients would simply wash into streams.

Low-Intensity, Low-Impact Turf, or Greensward

Greensward—a longer lawn that is cut to about 5 to 7 inches rather than 2 to 3 inches and maintained by organic methods—is a useful alternative to turf where you cannot use a taller grassland type (Figure 25.3), use is only occa-

Figure 25.3. Low-impact turf, or greensward, requires less maintenance and minimizes off-site impacts and energy use.

sional, and a slightly less wear resistant ground cover is acceptable. This type of turf is probably far closer to the historic models of "greensward" that typified the lawns of Victorian America as well as the pastoral English landscape that thrived on manure and was mown by sheep. A variety of small flowering forbs that are now seen as weeds, such as veronicas, were once important components of the lawn-seed mixes that preceded the monocultures of hybridized grasses that characterize many of today's lawns. Before World War II, a surprising amount of what is turf today was simply kept open by a few annual mowings, organic amendments, and lime. Inorganic fertilizers and pesticides and irrigation were not available.

Tall-Grass and Wildflower Meadows

The very difference between a tall-grass meadow and lawn has mostly to do with the frequency of mowing, which in turn has impacts on species composition as well. What makes the surface of turf so suitable for walking, sitting, and playing upon is its springy, dense carpet of new shoots, and their growth is promoted by frequent mowing. Reducing mowing to twice a year or less allows grass to grow long enough to be called a meadow, a pasture (although this usually implies grazing), or a tall grassland. The meadow may largely consist of nonnative species, especially if the area was recently a lawn, or mixed native and exotic species, as in the case of an abandoned farmfield, or, more rarely, mostly

329

native species, if it is a burn site in a natural area or relatively isolated from disturbance.

Prairie

Our word "prairie" derives from the French word for meadow and is inherited from early French explorers and colonizers. The word also tends to connote the indigenous grasslands that once spanned much of the continent, instead of the mixed native and exotic meadows and pastures that characterize much of the landscape today. Today many managers use the term to denote those areas where the management objective is to establish and sustain an all-native herbaceous plant community.

Both meadow and prairie are composed of both forbs and mixed grasses; however, warm-season (versus cool-season) grasses predominate in meadows that are largely native species. Mowing is generally inadequate for restoring or sustaining indigenous prairie communities. You can attempt the establishment of an all-native community incrementally, beginning with the removal of exotics and replanting native species gradually or all at once, with native seed or plugs. Fire is often the most effective management technique to favor both diversity and native species for prairie establishment, although alternatives, such as mowing, may be necessary in some highly developed areas.

Mowing

The term "mowing," as it is used here, refers to any mechanical management that cuts all aboveground stems. For management purposes, mowing is most frequently used to maintain open, that is, nonwoody, landscapes. It is likely to remain a common technique, especially in private landscapes and intensively used parklands.

Mowing can be a useful technique in meadow maintenance. In addition to inhibiting the establishment of new woody plants, mowing can be used to control invasive exotics or other undesirable species by exhausting their root systems' ability to recover after repeated cuttings. In general, the longer lived the plant, the more adversely it is affected by mowing or cutting. Another effect of mowing is to delay or reduce flowering. A plant that is cut just before flowering may be unable to develop new flower buds and set seed in the time it has remaining in that season, which may be useful in limiting dissemination of exotic seeds.

The frequency and timing of mowing largely determine the nature of vegetation on a site. Shrubs and trees may sprout repeatedly, but not indefinitely. Mowing only once every five years eliminates most woody vegetation, except for aggressive vines. This practice explains, in part, why so many roadsides are being reduced to vines and knotweed by current management practices. Lawn weeds, on the other hand, such as dandelion, as well as turf grasses, are sustained by more frequent mowing. Once mowing is reduced, taller, more vigorous herbaceous species quickly overwhelm them. In the temperate landscapes, mowing every few years up to nearly annually will tend to favor long-lived perennial warm-season grasses and wildflowers. Mowing annually to several times a year typically favors annuals and biennials as well as stoloniferous perennials, both grasses and forbs. The plants that do best in a given mowing regimen fare well because they are protected from competition with other species, such as trees, even though they lose some of their shoot growth.

As a rule, leave mown clippings on the ground. The old stalks contribute to erosion control and soil building, and would naturally, for the most part, have fallen to the ground with winterkill. They also act as mulch and play a role in succession by suppressing some competing vegetation. Remove the clippings, however, where you wish more rapid development of woody species and use them for seed hay or straw mulch elsewhere.

Mowing's greatest virtue—and also its biggest problem—is its familiarity. We take mowing for granted and often resist alternatives when reduced mowing is proposed. It is, however, always important to question the use of mowing and to make a serious effort to find alternatives, especially those that do not entail the use of motorized equipment, which compacts soils and does not allow for selective culling of undesirable plants. Simply reducing the size of equipment is also useful, as is minimizing the time equipment is left idling.

The lack of appropriate mowing equipment is the biggest limitation to using it in wildflower meadows and grasslands, especially in small areas. Mowers that cut to variable heights are often too large for many situations or budgets, and the smaller mowers that are available are not very adjustable and usually cut too close to the ground, exposing bare soil and desiccating vegetation. The modern convention of cutting to about 2 inches is as inappropriate for lawn care as it is for grasslands, except in the most specialized situations such as golf greens.

New push mowers that ride on a ball instead of a front wheel are intended for managing small-scale grasslands and other tall herbaceous landscapes. They come in several sizes and are designed to handle coarse stems but at present also cut far too close to the ground. These mowers cannot be modified with a larger wheel because of the ball, although the skids that adjust the height setting can be modified somewhat to raise the elevation. Manufacturers are apparently considering changes that would permit a higher cut. Let's hope such a product appears on the market soon. You can use a weed trimmer effectively to mow smaller-scale landscapes in the meantime.

The scythe is a traditional tool of meadow management that is greatly underutilized today. The scythe is ideal for managing smaller landscapes where there is no budget for heavy equipment. A swing blade is also useful but does not cut as cleanly. Consider using a scythe at larger scales as well because it is portable, pollution-free, and probably cheap compared to vehicular mowing when all costs are included. Perhaps fitness advocates will embrace restoration, saving themselves gym fees as they build muscle in the field the old-fashioned way. You should not be mowing areas that are excessively large, anyway. If you must use mechanized equipment, look for the new four-cycle models that are being developed in response to California's air quality requirements. Remember that operating a walk-behind, gas-powered, two-cycle lawn mower for twenty hours can pollute as much as one year's use of an automobile.

Experiment, too, by mixing mowing patterns in areas that were once all lawn. Use conventional short turf only along road and sidewalk margins, greensward for paths, and tall grass for open meadow where lawn is not required for walking and sitting upon (Figure 25.4). A broad band of tall herbaceous cover might be used as a filter strip to buffer natural areas from adjacent lawn or greensward areas (Figure 25.5). Broad greensward trails work well on dry ground for access through tall-grass and meadow landscapes. Adjust the width of the path to handle the level of traffic (Figure 25.6). Keep blades sharp and equipment well-maintained. Do not bring lawn up to the very edges of streams or you will encourage erosion. The closer turf is cropped, the shallower the roots and the poorer the soil protection.

Figure 25.4. You can use different mowing regimens depending on uses: lawn beside the main path, greensward for a meadow trail, and tall grass in the meadow.

Figure 25.5. This narrow strip of wildflower meadow at the edge of the forest assists the forest on the adjacent steep slope by reducing rates of runoff and nutrients.

Figure 25.6. This meadow at the Willowwood Arboretum in Gladstone, New Jersey, is selectively weeded, mown annually, and has mown turf trails for visitors.

Different mowing heights are recommended for the following meadow types:

Lawn or High-Use Turf

Try to cut no closer than 3 inches, especially in the warmer months when 3 1/2 inches is ideal. The taller height will reduce the need for water and help delay dormancy in cool-season grasses and the appearance of warm-season annual weeds such as crabgrass. Do not let the grass get taller than 5 inches before mowing it because it will be weakened if half or more of the blade is removed at once. Unfortunately, no modern equipment permits easy mowing between 3 and 5 inches without modification. Many people have put larger wheels on a mower to elevate the height of the blade. Such homemade modifications, however, can increase the hazards of operation, so use caution. A better solution is to convince manufacturers that there is a market out there for variable-height small mowers. Call often and inquire until they get the message and supply more suitable equipment.

Low-Impact Turf or Greensward

Try to cut greensward to a height of 5 inches, which will allow a variety of flowering forbs to persist in the matrix of grass, including small native species such as bluets. Do not be too concerned about lawn weeds except those that may imperil nearby native communities; they typically disappear gradually without mowing. Creeping fescues are also common, and many are hardly taller than 5 to 7 inches in some climates. Be careful, however, with tall and Chewings fescues, which are invasive and retard the return of native species in natural areas. In the upper Midwest and the cooler areas of the Northeast into Canada, you can use a mix of five fescues to create an almost no-mow lawn about 4 to 6 inches tall. It is not native, but it entails substantially less work and few impacts compared to conventional lawn. Be alert, however, for stilt grass, an annual invasive exotic that tolerates substantial shade, if it occurs in your area. It may be favored by reduced mowing as well as by fire.

Where you wish to sustain greensward, do not allow the grasses to get taller than 7 inches or so, or you will increase competition to the shortest species, the very plants you are trying to establish. This approach can reduce the number of total mowings to as few as five times a year versus over thirty times annually for turf. The 5-inch height does pose maintenance difficulties, however, especially in smaller areas. In only a few cases will it be possible to achieve it by elevating the mower's wheels. A weed trimmer can be used, but it is difficult to get an even texture without practice. Agricultural mowers may also be useful but can leave a fairly ragged cut. For large areas, large-scale mowers are usually sufficiently adjustable. In smaller areas, a weed trimmer is an interim solution.

Tall Grasslands and Wildflower Meadows

A single annual mowing typically will favor wildflowers; two mowings will favor shorter-lived grasses. More frequent mowing simply results in shorter grasslands such as greensward. (A reminder: erratic mowing is more likely to create bare soil and stands of aggressive weeds—it favors neither grasses nor wildflowers. Decide what you want—lawn, greensward, or meadows—and stick to a regimen.) Ideally, the height of the blade should be set at 12 inches and no lower than 8 inches. The meadow should still be tall enough after cutting to be easily distinguishable from turf and greensward. It is generally not necessary to remove cuttings, although there may be a particular management reason to do so, such as to foster more rapid woody growth.

Any time you reduce the frequency of mowing on a site, those species that require more frequent mowing will be eliminated, creating an opportunity for new species to become established. This temporary window is an excellent time to plant seeds of desirable species that can take advantage of the changing site conditions.

Meadow Mowing Guidelines

The most important thing about mowing is knowing when not to mow. Over-mowing is a greater problem than failing to mow.

- *Don't mow when ground-nesting birds and other wildlife are active.*

Be sure to review your proposed schedule with direct observations from your own area. Contact your local Audubon Society and invertebrate and wildlife specialists for further information on likely species to be concerned about and how to refine your schedule accordingly. In general, do not mow between the spring thaw and the Fourth of July. A good general rule of thumb for a single annual cut is to mow after at least two hard frosts in the fall to ensure that all species have completed the process of setting seed. You can also let the meadow stand all winter, which is very picturesque and increases wildlife opportunities. Do not wait too long in the late winter and early spring before mowing. The spring thaw comes on suddenly and can turn the surface to mud, restricting mowing or any other equipment use.

- *Don't mow when it's wet. Period.*

Mowing wet areas damages soils and spreads diseases among plants. If ruts are visible anywhere it means you need to change your schedule and/or technique and/or management goals. Ruts and compaction are not an acceptable result of management. Mow wet areas during a hard freeze.

- *Don't mow just because it's on a schedule.*

Understand why a schedule is given as a guideline, but make a judgment based on the condition of the field, the weather at the time, the species reproducing, and other specific factors on your site.

- *Modify the mowing schedule where desired to benefit or manage selected species.*

Especially in the early phases, meadows often have infestations of very undesirable species such as Canada thistle, stilt grass, or pigweed. To help control invasive species and ensure that invasive species do not colonize beyond the site, mow just before they set seed. Do not cut too soon or reflowering will occur. Some species need to be cut several times anyway to minimize the chance that they can complete the cycle to set seed again in the same season. You can favor many early-blooming species by mowing just after they have completed seed maturation. Similarly, you can favor late bloomers by mowing as early as local nesting schedules permit, and by setting the mower blade very high to cut the growing tips of the early taller plants to reduce competition with the shorter later species. For some species, such as Japanese honeysuckle, mowing is ineffective so be sure to monitor and evaluate your approach. An herbicide or other measure may be necessary in combination with mowing to control strongly invasive vines and clonal species and other unwelcome plants.

Consider changing the frequency of mowing where you find special site conditions. For example, it may be worth mowing three times in a single season in a meadow that is heavily infested with multiflora rose to reduce its extent. Wherever woody growth is extensive, especially exotic vines and shrubs such as multiflora rose, shrub honeysuckles, Japanese knotweed, and privet, try mowing two to three times a year rather than only once or twice for two or more years, or until woody growth is adequately controlled. If shrub honeysuckles or other exotics are present as patches, repeatedly mow the infested areas to control and reduce their extent.

A second mowing in summer to reduce the height of the meadow may also be desirable, even in meadows that are managed primarily for forbs if the season has been especially wet or the vegetation is especially rank. On the other hand, if woody growth is negligible, you can skip the annual mowing in some years. Even with these modifications, mowing is nonetheless a relatively crude and nonselective technique.

Selective Removals

To maintain a meadow in a way that is more selective than mowing, especially where woody recruitment is not significant, you can selectively weed manually or mechanically to remove exotics and to control woody vegetation. This cost-effective alternative to mowing does not alter the competitive relationships between herbaceous plants as much as mowing because only the woody species and invasives are removed in patches. A variety of methods are available, including all those used for exotics control such as weed pulling and brush hooking, cutting and stump or snag treatment, and root or shoot girdling. You can remove all woody saplings that become established or retain selected trees to create a savannalike character.

Selective weeding, like prescribed burning, is an underutilized technique for meadow management, possibly because it involves manual labor. But it also is very useful simply because of the work involved: there is greater incentive to minimize unnecessary removals and to monitor and assess the circumstances carefully before and after taking action.

Herbicide Management

Long-running field trials and experiments by William Niering at the Connecticut College Arboretum demonstrate that selective herbicide use is a highly effective method to remove individual plants while minimizing impacts to adjacent vegetation. Niering's work with herbicides minimizes the use of chemicals without eschewing their use altogether. The arboretum site includes parallel field trials and research on prescribed burning in woodlands, grasslands, and wildflower meadows.

The arboretum's woody oldfield and savanna landscapes have been continuously managed with selective use of herbicides to arrest succession for half a century. The landscapes have been designed to demonstrate the aesthetic potential of native, successional landscapes in residential settings. The object was to minimize the use of herbicides while relying primarily on natural successional patterns. The managers selectively remove woody plants, retaining only the most desirable ones, and often small-scale specimens of the native vegetation to keep portions of the meadow open and herbaceous.

At a much larger scale, Niering has demonstrated the cost-effectiveness of limited herbicide use to manage succession along transmission lines in the state of Connecticut. Selectively removing tree species to favor the development of shrublands reduced the cycle of maintenance from once every one to three years to once every twelve to fifteen years, with major cost savings and reduced environmental impact. Even these sites, however, are showing the impacts of severe exotics invasion as many former, largely native shrublands are succeeding to exotic vinelands throughout their rights of way (Niering and Goodwin 1974). Management for improved habitat for native communities could include managing invasive exotics as well as trees along these rights of way.

Prescribed Burning

Restorationists from ecologists and biologists to wildlife managers have observed that intentional fire as a landscape management tool often is the most direct route to greater species diversity in the successional landscape. Species dependent on natural fire regimes are not likely to reappear until historic conditions recur.

A particularly important aspect of fire, one less characteristic of other techniques such as mowing or grazing, is its beneficial effect on seed germination. The rapid consumption of the ground layer by fire provides a rich bolt of nutrients at the same time that the resulting ash on the newly exposed bare soil temporarily elevates pH to create an ideal seedbed at a time when the competition from existing vegetation is also reduced.

Historically, many of the most predominant native grasses were warm-season grasses rather than the now more-common European cool-season grasses. Described in the literature as "golden," the warm-season grasses are beautifully adapted to the occasional summer fire.

The growth pattern of one common native grass, broomsedge, illustrates its history of association with fire. Like other warm-season grasses, broomsedge does not begin serious shoot growth until late spring. The stalk that will eventually bear seed appears at the end of summer, in time to set and drop its seed over a fresh burn in the event of summer drought and wildfire. Many European roadside weeds in the Northeast, on the other hand, come from agricultural landscapes where natural fire has been largely controlled for centuries and are poorly adapted to the wildfire regime of our native grasslands.

Prescribed burning experiments in grasslands at the Connecticut College Arboretum (Niering and Dreyer 1989) have yielded several additional observations about the use of fire to manage landscape succession in the temperate landscape:

• *Fire is effective at arresting the succession of woody species.*
Decades later the overall proportion of woody to herbaceous species has remained essentially static in the fire-managed meadows. In some cases, the individual woody plants have remained the same; in others, only the site coverage of woody plants has remained static. Unless very hot, however, fire does not usually reduce woody coverage.

• *Some species are fire increasers—that is, their predominance increases on sites with fire management; others are fire decreasers.*
Fire has variable effects on species composition, so it is worth knowing the responses of different species in the landscape before initiating a prescribed burning program. Burning a meadow with sassafras and black locust, for example, will likely stimulate aggressive sprouting from both of those species, which are fire increasers. Thus, although woody vegetation is typically arrested by burning, in this instance the opposite is true.

The placement of firebreaks is an important consideration in prescribed or controlled burning. The most common firebreak is simply a plowed furrow of a specified width depending on the landscape's character; however, you can avoid repeatedly disturbing the soil in this way if other alternatives are fully utilized. Evaluate every feature of the landscape that can be used as a firebreak: wetlands, streams, turf areas, and built elements such as roads and paths. Or, where appropriate, simply plant a 12-foot margin of cool-season grass where needed along the perimeter of an area proposed for regular burning, which, when mown, can also serve as a path.

Schedule the timing of fire, like mowing, to minimize negative effects on wildlife. Review your proposed schedule with invertebrate and wildlife specialists to ensure appropriate timing. Burn in patches where the entire local habitat would otherwise be affected to ensure some remaining cover at all times. And, as discussed in Chapter 15, "Managing with Succession," work with trained crews and obtain permits to implement a burning program.

Managed Grazing

Historically, grazing was the most common method of meadow management, even for largely aesthetic grasslands such as ornamental lawns. During World War I, sheep provided effective turf management on the White House lawn. Participants at a recent conference on energy conservation at the White House suggested reviving this practice.

In general, we tend to associate grazing with grassland degradation because of the unfortunate tendency to overgraze on both public and private land. Indigenous wildlife, however, also graze and clearly play a vital role in landscape succession.

Allan Savory (1988) studies the natural ecological roles of grazers in an effort to develop grazing-based strategies for grassland restoration. In grasslands from Africa to North America he observed the difference between the impacts of native, and often migratory, wildlife and the more static patterns of modern livestock. American buffalo, for example, would regularly overgraze and over-trample the patches of land they used. The difference was that these areas were subsequently given an extended period of rest and recovery. The actions of the buffalo also set the stage for renewal as their hooves tilled and exposed the soil, temporarily minimizing plant competition. They also brought with them new seeds packaged in manure. When Savory adapted his observations to the management of overgrazed landscapes he was able to achieve a remarkable increase in the diversity of the grasslands despite continued use for grazing. Using smaller-scale and movable fencing he more tightly confined the animals and then moved them to another rotational paddock to allow the land just grazed to recover. The meadows he managed began to renew themselves rather than degrade.

Grazing is going to continue, but grazing practices can and should be modified to foster native grassland restoration as well as better site stability. Native grasses, rather than introduced pasture grasses, provide better habitat for native wildlife. This approach could create very high quality lands for agricultural purposes as well. After years of supporting patented varieties of exotic Eurasian grasses, researchers are now recognizing the high grazing quality of native grasses. Moreover, tall fescue including K-31, a heavily subsidized, patented grass hosts a fungus that is toxic to livestock. Even more important for farmers is that livestock gain weight faster on the native switchgrass, Indian grass, and little bluestem than they do on patented fescues and other agricultural varieties, and no fertilizer or lime must be used to maintain them. Because warm-season grasses should be grazed no lower than 12 inches, runoff is reduced in warm-season pastures and wildlife shelter is greater than in more closely cropped pastures (Capel 1992).

Grassland Establishment and Management

There are a variety of ways to establish a meadow. You can manage existing vegetation to effect a transition to tall grass, or, where you are dealing with bare ground or a completely degraded habitat, you can plant a meadow from scratch.

Conversion of Turf to Meadow

When your goal is to convert existing turf to meadow, you must decide whether to achieve a meadow incrementally by changing management or by killing and removing the existing cover and replanting. Each route has different opportunities and constraints. By gradually reducing the frequency of mowing you can slowly convert a lawn or greensward to a tall-grass and wildflower meadow. Selective removals of exotics and additions of natives are also usually necessary. A general rule of thumb is to start with five annual mowings the first year, four the second year, three the third, two the fourth, and one by the fifth year. That approach allows for the gradual replacement of shorter-lived and mower-dependent vegetation by longer-lived species. Without resorting to denuding and reseeding the site, however, exotics are not fully controlled and diversity may be limited.

Interseeding, sowing seed in the matrix of existing vegetation rather than removing it, can bring greater diversity to the meadow more rapidly than would occur naturally and helps compensate for reduced seed sources. Use locally propagated and/or collected seed, from analogous habitats with similar environmental conditions if possible. Where somewhat larger scale seed collection is desired, you might try a seed stripper. One handy model that attaches to a weed trimmer is called the Grin Reaper. With this equipment the operator can harvest seed wherever he or she can walk. It also limits soil compaction by eliminating the use of vehicles. The boxlike attachment hooks to a rotary line trimmer and collects all seed ripe at one time in the meadow. Repeat the process several times in the summer and fall to ensure a broad representation of plants. The Grin Reaper was developed by and is available from Environmental Survey Consulting, Inc., 4602 Placid Place, Austin, TX 78731, 512-458-8531.

You can interseed at almost any time of year because minimal soil preparation and aftercare are required. But timing, as they say, is everything. Replicating natural timing is likely to produce the best results, especially for field-sown plants. Typically, in the temperate East, seed will be available in the fall after it has ripened and can be field-sown immediately, which is similar to what happens under natural conditions. You can broadcast the seed manually or mechanically. Interseed on a continuous basis as long as appropriate seed material is available. For many species the right conditions for germination may occur only infrequently, so do not expect immediate or consistent results from year to year.

In areas of limited seed supply, you can take additional measures to increase the likelihood of germination, such as timing interseeding to follow immediately after a fire or controlled burn. Where burning is not possible or desirable, you can scarify the soil minimally to increase soil and seed contact. You can also use a mechanical seeder, such as a slit-seeder or drill, but use of the equipment presently available is sometimes restricted by the roughness of meadows and grasslands. Equipment modifications to allow interseeding are also needed for more effective native grassland establishment.

In many instances, weedy species grow more rapidly than the newly seeded native species. After interseeding you sometimes can use mowing to reduce competition from exotics and generalists. Mow typically when the weedy species reach over 1 to 1 1/2 feet in height and cut to 6 to 8 inches in height. Do not cut so much top growth that it smothers the new seedlings below. Do not mow where weedy species are not present.

Focus your efforts on establishing less-common species that do not recolonize disturbed areas and are more remnant-reliant, rather than on those species that are likely to establish themselves easily. This will also ensure that you do not plant aggressive native species that, like invasives, can overwhelm your landscape. As you deal with increasingly conservative species, you may not have enough seed for field sowing and may wish to work with seedlings and one- or two-year-old plants as plugs.

Plugs

When very rapid cover is needed, you can use vegetative plugs to hasten the establishment of selected species. Although costly at the outset because it is so capital and labor intensive, this method requires the least maintenance following installation because nearly complete cover is achieved in the first season.

Vegetative plugs are a good way to add forbs to an established grassland. Plugs are also useful where seed is especially limited but vegetative stock is available to establish a future seed source and for species that do not germinate well. Younger plugs (one year or less) are generally easier to transplant and do not experience as much transplant shock as those with more established root systems. Once installed, plugs spend the first few weeks putting on root growth but develop rapidly with the heat of summer. This approach is at present still fairly new: plugs are available commercially only on a limited basis. You may wish to negotiate with a local propagator to grow grass and wildflower plugs at least one year prior to a spring planting (Figures 25.7 and 25.8). Time the planting of plugs to coincide with reduced mowing or to follow a burn to take advantage of gaps created by a change in regimen.

Figure 25.7. Native grass plugs from North Creek Nurseries in Landenberg, Pennsylvania. The smaller grass plugs were grown from seed; the larger plugs are divisions of clones.

Figure 25.8. The National Park Service used plugs of native broomsedge to provide low-maintenance stabilization at Civil War–era earthen fortifications. By the end of the first summer the bluestem was tall and lush with minimal maintenance. (Photo used by permission of the Richmond National Battlefield Park)

According to the New England Wildflower Society, the following meadow species are fairly easy to propagate from seed or by division:

LATIN NAMES	COMMON NAMES
Agastache foeniculum	anise hyssop
Asclepias tuberosa	butterfly weed
Baptisia spp.	false wild indigo
Echinacea spp.	coneflower
Eupatorium spp.	Joe Pye-weed
Fragaria virginiana	wild strawberry
Geranium maculatum	wild geranium
Hypericum spp.	Saint John's wort
Ligusticum porteri	osha root
Lobelia spp.	lobelia
Monarda fistulosa	bergamot
M. didyma	beebalm
Pycnanthemum spp.	mountain mint
Scutellaria spp.	skullcap
Spigelia marilandica	pink root
Veronicastrum virginicum	Culver's root

You can also grow sods of wildflowers (Figure 25.9) from seed under field conditions just like turf where an instant meadow patch is desired, but this method is also costly. It is not applicable in woodlands where neither the soil nor the available light will support such dense herbaceous growth; however, sods are very effective at stabilizing swales and other open wetlands subject to higher velocity of runoff.

Figure 25.9. Almost any herbaceous vegetation, such as Joe Pye weed, meadowsweet, and sensitive fern, shown here, can be grown like turf.

Seeding

Seed is a very cost effective method for establishing warm-season meadows. You can use seed on bare soil as well as interseed areas with existing vegetation. Which approach you use will depend on the conditions existing on the site. Former corn or soybean fields may be largely weed-free, for example, and seed will be easy to establish. Complete reseeding may be necessary in areas where vegetation is nonexistent or where entrenched exotic cover has been completely removed. The success of a seeding project may hinge on competition with weed species. Do not underestimate the problem of weeds. Seed works best in bare soil conditions or on relatively impoverished sites where the developing native species have little competition. Richer soils often contain abundant weed seeds that may overwhelm newly seeded material. That is why a lot of meadow projects begin with a serious dose of herbicide. There are, however, alternatives to this approach. Where exotic seed is abundant, consider shallow disking several times to exhaust the seed stock, but do not plow, unless you are planning to do so repeatedly. Consider cropping the site to exhaust perennial weeds. A season planted in buckwheat also depletes the soil of nitrogen to further reduce weedy competition.

Complete removal of the existing vegetation may be desirable when it is entirely exotic, but you can often accomplish adequate weed control without it, except for a few invasive species that may require herbicide to be adequately controlled. You can then use prescribed burning to prepare a seedbed, especially if there is an historic precedent. In almost all other cases where natives are mixed with exotic weeds, a more selective approach, such as hand-weeding, is preferred.

A common mistake people make in meadow establishment is to overprepare the seedbed by making it too smooth, too neutral, and/or too fertile. Rather, leave the ground surface relatively rough to hold moisture better and to improve soil-seed contact. Also, do not try to seed everything all at once. It is a good idea to use a rich seed palette, but real success hinges on introducing seed and/or other propagules (from the same genetic neighborhood) to the site on a regular basis. The opportunities for colonization will vary each year with climatic variations as well as the level and kind of competition and changing environmental conditions in the meadow. In that sense, interseeding is as important on a continuing basis in newly seeded meadows as it is in existing meadows where you are trying to shift to native species.

Seed and plant material should consist of 50 to 80 percent native warm-season grasses like little bluestem, switchgrass, and Indian grass. Because they are bunch, not sod-forming, grasses, there is room between the clumps to interplant with forbs from seeds or seedlings. The grasses help physically support the wildflowers. Be very cautious about using exotic grasses as interim stabilizers. Even as little as 10 percent exotic grasses in the restabilization mix can

slow the recruitment and spread of native grasses. Legumes require special microbes called "rhizobia" and must be inoculated upon planting unless sold with
these root nodule associates.

Late-spring seeding and planting have been most successful in the Northeast; however, consider fall seeding for some plants. Be sure to buy warm-
season grass seed in amounts measured as pure live seed (PLS) rather than in
bulk, which contains a lot of chaff. Seeding rates vary considerably, from very
low rates of 3 to 5 pounds per acre for pasture to up to 10 or more pounds for
more rapid, dense cover. Seed is costly, so using lower volumes is an attractive
option, but it is a false economy where more maintenance in weeding may be
required. Some seed has long dormancy requirements. One-year-old switchgrass seed, for example, germinates far better than fresh seed. Because distributing small volumes of seed evenly over large areas of soil can be difficult, try
expanding the volume of seed with sand or vermiculite and practicing first.
Two experienced volunteers can hand-seed five acres in a morning.

For larger areas where you intend to use mechanical equipment, consider a
no-till drill that embeds the seed directly in the soil, especially for small,
smooth seed such as switchgrass. You can use a native-grass drill such as ones
made by Truax, 3609 Vera Cruz Avenue North, Minneapolis, MN 55422,
612-537-6639. Seed drills can often be rented from your local Natural Resources Conservation Service (formerly the Soil Conservation Service) center.
You will require clean seed for drilling. A drill can also be used for interseeding
after mowing or burning a meadow to reduce the volume of surface vegetation
for easier drill operation.

Try broadcasting fluffy seed like bluestem and Indian grass and then drag
lightly over the surface to cover the seed. It is often very helpful to lightly tamp
down the seed to compact the surface soil, reduce moisture loss, and improve
soil-seed contact. Consider using a Land Imprinter for seeding as well as erosion reduction. Developed by Robert Dixon to help solve problems of global
desertification, the Imprinter roughens and loosens the soil and notches the
surface, leaving behind a funnel-shaped waffle pattern on the ground that effects infiltration of substantial amounts of water without runoff at the same
time that it delivers seed. Although designed for arid conditions, it also performs well in temperate regions. The Land Imprinter is available from the Im-
printing Foundation, 1231 East Big Rock, Tucson, AZ 85718, 602-297-6165.

In more rural areas, where you can sometimes still find largely native
meadows on abandoned farmland, you can consider arranging to bale meadow
growth, just as you would a hayfield, at the end of the year as a source of seed.
Apply hay, if possible, just after a burn or after mowing. Broomsedge or switchgrass hay, for example, can be directly seeded on bare ground several inches
deep and secured with a straw crimper. Mulching and seeding are accomplished in a single step. It is a somewhat profligate use of seed and therefore

only appropriate where supply is not an issue, but it is also often cheaper than buying prepared seed. This method does not require seed cleaning or specialized seeding equipment. Rather, you can often contract this kind of work with an area farmer. You will need about 60–80 bales or 2–3 tons per acre to spread hay about 2 inches deep.

Weed control the first year after seeding is crucial. Weeds typically grow faster than native grasses and forbs, so you should mow the tops of the weeds to reduce competition with the warm-season grasses. Do not cut the growing tips of the new grass shoots, though, or you will set them back severely. Early-spring mowing is sometimes effective on both weeds and woody species. However, hand-weeding still will be essential for at least the first few years after establishment. If all goes well, you will be able to harvest bales and/or seed from the site for use elsewhere within two years, although it may take several more before full production (about 50 pounds of seed per acre) occurs. As meadow establishment in the East becomes more common, we are likely to see more specialized seed harvesting equipment, such as that found in the Midwest.

Controlled burning is a necessity for maintenance and weed control in prairies. Begin after the third year of establishment. Consider burning only one-third of the grassland annually to maintain cover for wildlife. A spring burn tends to set back woody species and benefit forbs; a later burn more typically benefits grasses.

Savanna Restoration

Savannas are open woodlands in which trees grow in parklike stands. They are plant communities midway between forests and prairies, where large trees and tall grasses are the dominant features. Savanna landscapes in the Northeast are ephemeral and, like meadows, require management to persist (Figure 25.10). As with prairies, removing and controlling dominant exotics, including grasses, such as fescue and nonnative forbs, is a challenging aspect of savanna restoration. Where woody recruitment is not occurring naturally, consider planting trees. Twenty to forty stems per acre is typical, but you may vary the density based on your design goals and historic conditions. For the best success, mimic historic and species-characteristic planting patterns rather than geometric patterns.

In *Miracle Under the Oaks: The Revival of Nature in America* (1995), William Stevens tells of the rediscovery of a savanna ecosystem that was all but lost to history. By the late 1970s, the Illinois Nature Conservancy's inventory of natural areas showed that only 11.2 acres of savanna remained in the entire state. Exotic buckthorn as well as native dogwoods, cottonwoods, and hawthorns were overwhelming landscapes where both natural and managed fire had been suppressed for decades. The project began in the 1970s as a prairie restoration under the direction of Steve Packard, Science Director of the Illinois Nature Conservancy, who coordinated the clearing of brush and trees and the sowing

Figure 25.10. A restored savanna and ephemeral pool on former lawn areas in Iroquois Park in Louisville, Kentucky, enrich the landscape and help address severe erosion on steep slopes. (Photo used by permission of Ted Wethen/Quadrat).

of hand-collected native prairie seeds. The effort has now grown to over 5,000 volunteers on 200 sites covering over 30,000 acres. Knowing that the prairie doesn't thrive without fire, by the 1980s they began fire management in earnest and developed many of the techniques used today.

When they started work on a grove of bur oaks choked with buckthorn and Tartarian honeysuckle (a shrub honeysuckle) about a half hour northwest of Chicago's Loop, there was no reference ecosystem to study. The restorationists found a useful reference in Floyd Swink's and Gerould Wilhelm's survey *The Plants of the Chicago Region* (1994), which provided exhaustive lists of plant associates. By cross-referencing the sites and species lists, they were able to identify 122 plant species that are associated with savanna as distinct from forest and prairie. The species most specialized to the rare savanna habitat became the focus of propagation for savanna restoration. They followed a successional sequence in their planting, initiating it with relatively competitive species that served as the basic matrix of the herbaceous layer. Now named the Vestal Grove, this landscape today gives us a glimpse of a forgotten ecosystem that Stevens describes as

> a far-flung confection of lush grasses, wildflowers, trees, butterflies, songbirds, turkeys, deer, elk, wolves, panthers and innumerable other creatures from the magnificent to the microscopic. Bison trooped through the oak groves and Indians hunted, camped, and foraged in them (1995, 18).

This restoration project has also given rise to a newly published restoration handbook. *The Tallgrass Restoration Handbook: For Prairies, Savannas, and Woodlands,* edited by Stephen Packard and Cornelia F. Mutel (1997) and published by Island Press, is invaluable to anyone interested in these landscapes. It reflects years of hands-on experience in these very specialized landscapes. With some more time and effort, restorers and managers will develop the same kind of detailed information for the eastern forest.

Most of us working in restoration have never even seen pristine examples of the habitats we are trying to restore. What we have seen are the remnants, and we have observed them dying. Anyone paying any attention to the environment is struck by its accelerating rate of deterioration.

With each local extinction we lose some of the intelligence of the universe. At the same time the supercompetitors thrive, overwhelming local populations just like the killer superstores that are eliminating their more specialized competition in the commercial marketplace.

This situation did not arise overnight: it is the result of millions of individual actions. And renewal will not happen overnight. Our only hope is to begin *right now* to make a habit of restoration. There is no simple set of answers that will work everywhere and at all times. No population or organism is fitted perfectly to its environment, and no environment is unchanging. There is no Eden, no fixed, harmonious garden.

What we can expect is to learn a process—a framework for learning that enables us to better observe the world around us and to interact with it in a manner that is more satisfying to us as well as less destructive than what we have done in the past. The intent is not so much to restrict what we can do but to use knowledge to better fulfill our aspirations to preserve, rather than destroy, the natural functions of the landscapes in which we dwell. We must open more doors than we close and listen to people who seem unsympathetic. We had better be open to criticism; we will get a lot of it and can learn from it.

To confront the biodiversity crisis we must live more sustainably. If we want to sustain and restore healthy ecosystems we must acknowledge that there are limits to growth. Ultimately we will have no choice but to manage our total population as well as the relative impacts of different lifestyles and rates of consumption and waste. Perhaps the most difficult challenge is our routine use of the automobile, which is wrapped up in perceptions of power and sexuality. Our landscapes are designed to accommodate the automobile at the expense of everything else. In order to do this we had to find ways to keep the price of energy artificially low, by overexploiting potentially renewable resources, by transferring costs to the public sector and to future generations, and by going to war, economically as well as with arms.

Another difficult challenge is the loss of physical isolation, a consequence of the globalization of human economies. Isolation is vital to conserving many subspecies and will force us to face everything from the impacts of tourism to trade policies. Sufficient resources in land, water, and energy must be preserved in their natural forms to sustain conservative species. In many landscapes historic management practices such as indigenous horticulture and prescribed burning will be necessary to sustain local ecotypes.

Restoration must be founded in good science and accurate information. This is a huge hurdle because there is often strong resistance to monitoring. Although we tend to think of ourselves as having access to more information than ever, we are shutting down stream gauging stations across the country and we do not even have baseline biological inventories for most of our national parks and national forests, much less most of our country. Information, including genetic data, is increasingly being privatized, restricting our access to it. And most of us would truly rather not know because the real state of things is difficult to confront.

There is great joy in the act of restoration that makes it easier to face the reality of the landscape, which gives global meaningfulness to our local actions. More and more people are getting involved in restoration, in part because restoration holds out some hope and makes people part of the solution. Sooner or later we will face the consequences of our actions. Restoration is the best preparation for the unknown that lies ahead. The real question is not whether we will restore the landscapes around us, but when will we start and how much will be left to work with when we do.

Nearly everyone makes decisions or takes actions that affect our environment—in the yards we maintain around our houses or in the way we run our businesses. The forester, the hunter, the gardener and propagator, the landowner, the walker—we all have new options that are vitally important to restoring and sustaining our rich forest heritage. All who become involved with ecological restoration will deepen their ties to the community. They will work side by side with many extraordinary people whose understanding of the connectedness of all things gives them tolerance and generosity. Their mission will fill them with curiosity, and they will experience the wonder and despair of more keenly observing the living world.

The value we place on the wildness around us and the uniqueness of each place are reflected in the gardens and all the landscapes we create. When we begin to see the natural world through the lens of ecology we find a new way of looking at and interacting with the world that forms the basis for a new kind of aesthetic outlook. The ecological aesthetic reexamines the traditional issues like form, texture, color, and line in the context of their relationships to the actual processes and patterns of living organisms. Form and the other qualities are seen as important in new and different ways because in organic structures form is inseparable from function. Thus the shape of a flower is directly related to its way of gathering light energy to reproduce. We become aware of a beauty that goes beyond a pretty color or graceful form when we perceive the unity of form and process, when we see the flowers on the trumpet vine being pollinated by the hummingbird or the dangling catkin of the pussy willow distributing its pollen on the wind. You will never be quite the same after you have grasped the ecological aesthetic. The landscape will be forever transformed in your eyes.

Driving down a country road, a city street, or an interstate highway will be different. Walking into a wood or into a backyard garden will be different. To see the world whole is to appreciate the integration of form between the bee and the flower, the shape of a blade of grass, and the character of the prairie or of the rhythm of growth in a forest. It is to see the need and the opportunities we have to reinhabit each place. It is to recognize that while many worry about the cost of "going green," the real issue is the price we have already paid and will continue to pay for exploiting and degrading our environment.

—Leslie Jones Sauer and James Amon

SOURCES CITED

Chapter 2. The Once and Future Forest

DiGiovanni, D. M., and C. T. Scott. 1987. Forest statistics for New Jersey. *Resource Bulletin NE-11*. Radnor, Pa.: U.S. Department of Agriculture, Forest Service, Northeastern Forest Experiment Station.

Drayton, B., and R. B. Primack. 1996. Plant species lost in an isolated conservation area in metropolitan Boston from 1894 to 1993. *Conservation Biology* 10: 30–39.

Duffy, D. C., and A. J. Meier. 1992. Do Appalachian understories ever recover from clearcutting? *Conservation Biology* 6: 196–201.

Little, C. E. 1995. *The Dying of the Trees: The Pandemic in America's Forests*. New York: Viking Press, 1995.

Noss, R. F., E. T. La Roe, and J. M. Scott. 1995. *Endangered Ecosystems of the United States: A Preliminary Assessment of Loss and Degradation*. Washington, D.C.: U.S. Fish and Wildlife Service.

Petranka, J. W., M. E. Eldrige, and K. E. Haley. 1993. Effects of timber harvesting on southern Appalachian salamanders. *Conservation Biology* 7: 363–70.

Whitney, G. G. 1994. *From Coastal Wilderness to Fruited Plain: A History of Environmental Change in Temperate North America from 1500 to the Present*. New York: Cambridge University Press.

Chapter 3. Fragmentation

Brittingham, M. C., and S. A. Temple. 1983. Have cowbirds caused forest songbirds to decline? *Bioscience* 33: 31–35.

Davis, M. B., and C. Zabinski. 1992. Changes in geographical range resulting from greenhouse warming: Effects on biodiversity in forests. In *Global Warming and Biodiversity*, ed. R. L. Peters and T. E. Lovejoy, 297–308. New Haven, Conn.: Yale University Press.

Fahrig, L., and G. Merriam. 1990. Conservation of fragmented populations. In *A Landscape Perspective: Readings in Conservation Biology*, ed. D. Ehrenfeld, 16–25. Cambridge, Mass.: Society for Conservation Biology and Blackwell Science.

Gates, J. E., and L. W. Gysel. 1978. Avian nest dispersion and fledgling success in field–forest ecotones. *Ecology* 59: 871–83.

Greenfeld, J., L. Herson, N. Karouna, and G. Bernstein. 1991. *Forest Conservation Manual: Guidance for the Conservation of Maryland's Forests During Land Use Changes, Under the 1991 Forest Conservation Act*. Annapolis: Maryland Department of Natural Resources. (This manual can be ordered from the Maryland Department of Natural Resources, Resource Conservation Service, Forestry Division, Tawes State Office Building, 580 Taylor Avenue, Annapolis, MD 21401.)

McCloskey, J. M., and H. Spalding. 1990. The world's remaining wilderness. *Geographical Magazine* 62 (August): 14–19.

Chapter 4. Succession and Recruitment

Guldin, J., J. R. Smith, and L. Thompson. 1990. Stand structure of an old-growth upland hardwood forest in Overton Park, Memphis, Tennessee. *New York State Museum Bulletin* 471: 61–66.

Handel, S. N., and A. J. Beattie. 1990. Seed dispersal by ants. *Scientific American* 263, no. 2: 76–83A.

Marquis, D. A. 1993. Silviculture as an alternative in deer management. In *Deer Management in an Urbanizing Region: Problems and Alternatives to Traditional Management*, proceedings of the Humane Society of the United States conference on April 13, 1988, at East Windsor, N.J., and updated in 1993 (HSUS stock no. GR 3184), 38–45.

Reidel, C. 1995. The great green East: Lands everyone wants. *American Forests* 101, no. 9: 12–64.

Yazvenko, S. B., and D. J. Rapport. 1996. A framework for assessing forest ecosystem health. *Ecosystem Health* 2: 40–51.

Chapter 5. Water Systems

Ashby, W. C. 1987. Forests. In *Restoration Ecology*, ed. W. R. Jordan III, M. E. Gilpin, and J. D. Aber, 89–108. New York: Cambridge University Press.

Dobson, J. E., R. M. Rush, and R. W. Peplies. 1990. Forest acidification and lake acidification. *Annals of the Association of American Geographers* 80: 343–61.

Flynn, J. 1994. The falling forest. *The Amicus Journal* 16, no. 4: 34–38.

Laurance, W. F., and E. Yensen. 1991. Predicting the impacts of edge effects in fragmented habitats. *Biological Conservation* 55: 77–92.

Meier, M. F. 1990. Reduced rise in sea level. *Nature* 343: 115–16.

Stephenson, N. L. 1990. Climatic control of vegetation distribution: The role of the water balance. *The American Naturalist* 135: 649–70.

Chapter 7. Invasive Exotics

Fairchild, D. 1938. *The World Was My Garden: Travels of a Plant Explorer*. New York: C. Scribner's Sons.

Holloway, M. 1994. Nurturing nature. *Scientific American* 270: 76–84.

Kloeppel, B. D., and M. C. Abrams. 1995. Ecophysiological attributes of the native *Acer saccharum* and the exotic *Acer platanoides* in urban oak forests in Pennsylvania, USA. *Tree Physiology* 15: 739–46.

Marinelli, J. 1996. Introduction: Redefining the weed. In *Invasive Plants: Weeds of the Global Garden*, ed. J. M. Randall and J. Marinelli, 4–6. New York: Brooklyn Botanic Garden.

Sasek, T. W., and Boyd R. Strain. 1990. Implications of atmospheric CO_2 enrichment and climatic change for the geographical distribution of two introduced vines in the U.S.A. *Climatic Change* 16: 31–51.

Shurtleff, W., and A. Aoyagi. 1985. *The Book of Kudzu*. Wayne N.J.: Avery Press.

Swink, F., and G. Wilhelm. 1994. *Plants of the Chicago Region: An Annotated Checklist of the Vascular Flora of the Chicago Region, with Keys, Notes on Local Distribution, Ecology, and Taxonomy, a System for the Qualitative Evaluation of Plant Communities, A Natural Divisions Map, and a Description of Natural Plant Communities*. Indianapolis: Morton Arboretum and Indiana Academy of Science.

Wyckoff, P. H., and S. L. Webb. 1996. Understory influence of the invasive norway maple (*Acer platanoides*). *Bulletin of the Torrey Botanical Club* 123: 197–205.

Chapter 8. Opportunistic Natives

Brothers, T. S. 1992. Postsettlement plant migrations in northeastern North America. *American Midland Naturalist* 128: 72–82.

Gehring, J. L., and Y. B. Linhart. 1992. Population structure and genetic differentiation in native and introduced populations of *Deschampsia caespitosa* (Poaceae) in the Colorado alpine tundra. *American Journal of Botany* 79: 1337–43.

Kraft, J. C., H. Yi, and M. Khalequzzaman. 1992. Geologic and human factors in the decline of the tidal marsh lithosome: The Delaware Estuary and Atlantic Coastal Zone. *Sedimentary Geology* 80: 233–46.

Chapter 9. Wildlife Impacts

Chadwick, D. H. 1996. Sanctuary: U. S. National Wildlife Refuges. *National Geographic* 190: 1–35.

Trust for Public Land in conjunction with New York City Audubon Society. 1990. *The Harbor Herons Report: A Strategy for Preserving a Unique Wildlife Habitat and Wetland Resource in Northeastern Staten Island.* New York: Audubon Society.

Wilson, E. O. 1992. *The Diversity of Life.* Cambridge, Mass.: Harvard University Press.

Chapter 11. Atmospheric Change

Aber, J. D. 1993. Modification of nitrogen cycling at the regional scale: The subtle effects of nitrogen deposition. In *Humans as Components of Ecosystems: The Ecology of Subtle Human Effects and Populated Areas,* ed. M. J. McDonnell and S. T. A. Pickett, 163–74. New York: Springer-Verlag.

Bennett, J. P., R. L. Anderson, M. L. Mielke, and J. J. Ebersole. 1994. Foliar injury air pollution surveys of eastern white pine *(Pinus strobus L.):* A review. *Environmental Monitoring and Assessment* 30: 247–75.

Davis, M. B., and C. Zabinski. 1992. Changes in geographical range resulting from greenhouse warming: Effects on biodiversity in forests. In *Global Warming and Biodiversity,* ed. R. L. Peters and T. E. Lovejoy, 297–308, New Haven, Conn.: Yale University Press.

Eichhorn, J., and A. Hutterman. 1994. Humus disintegration and nitrogen mineralization. In *Effect of Acid Rain on Forest Processes,* ed. D. L. Godbold and A. Hutterman, 129–62. New York: Wiley-Liss.

Ellenberg, H. 1987. Immissionen-Produktivität der Krautschicht-Populationsdynamick des Rehwilds: ein Versuch zum Verstandnis Ökologischer Zusammenhänge. *Natur und Landschaft* 61: 335-40.

Environmental Protection Agency. 1994. *Deposition of Air Pollutants to the Great Waters: First Report to Congress.* EPA-453/R-93-O55. Research Triangle Park, N.C.: EPA, Office of Air Quality, Planning, and Standards.

Flynn, J. 1994. The falling forest. *The Amicus Journal* 16, no. 4: 34–38.

Granger, F., S. Kasel, and M. A. Adams. 1994. Tree decline in southeastern Australia: Nitrate reductase activity and indications of unbalanced nutrition in *Eucalyptus ovata* (Labill) and *E. camphora* (Baker, R. T.) communities at Yellingbo, Victoria. *Oecologica* 98: 221–28.

Graveland, J., R. Vanderwal, J. H. van Balen, and A. J. van Noordwijk. 1994. Poor reproduction in forest passerines from decline of snail abundance on acidified soils. *Nature* 368: 446–48.

Harris, J. A., P. Birch, and K. C. Short. 1993. The impact of storage of soils during opencast mining on the microbial community: A strategist theory interpretation. *Restoration Ecology* 1: 88–100.

Likens, G. E. 1992. Some applications of the ecosystem approach to environmental problems and resource management. In *Responses of Forest Ecosystems to Environmental Changes,* ed. A. Teller, P. Mathy, and J. N. R. Jeffers, 16–30, London: Elsevier Applied Science.

Likens, G. E., C. Driscoll, and D. Buso. 1996. Long–term effects of acid rain: Response and recovery of a forest ecosystem. *Science* 272: 244–46.

McClure, M. S. 1991. Nitrogen fertilization of hemlock increases susceptibility to hemlock wooly adelgid. *Journal of Arboriculture* 17 (8): 227–29.

McDonnell, M. J., S. T. A. Pickett, and R. V. Pouyat. 1993. The application of the ecological gradient paradigm to the study of urban effects. In *Humans as Components of Ecosystems,* ed. M. J. McDonnell and S. T. A. Pickett, 175–89, New York: Springer-Verlag.

McKibben, W. 1996. Future old growth. In *Eastern Old-Growth Forests: Prospects for Rediscovery and Recovery,* ed. M. B. Davis, 359–63. Washington, D.C.: Island Press.

McNulty, S. G., J. D. Aber, and R. D. Boone. 1991. Spatial changes in forest floor and foliar chemistry of spruce-fir forests across New England. *Biogeochemistry* 14: 13-29.

National Acid Precipitation Assessment Program (NAPAP). 1991. *Integrated Assessment Report.* Washington, D. C.: The NAPAP Office of the Director.

Paerl, H. W. 1993. Emerging role of atmospheric nitrogen deposition in coastal eutrophication: Biogeochemical and trophic perspectives. *Canadian Journal of Fisheries and Aquatic Science* 50: 2254–70.

Reidel, C. 1995. The great green East: Lands everyone wants. *American Forests* 101: no. 9, 12–64.

Sander, S. P., R. R. Friedl, and Y. L. Yung. 1989. Rate of formation of the ClO dimer in the polar stratosphere: Implications for ozone loss. *Science* 245: 1095–98.

Schwartz, J. 1994. Air pollution and daily mortality: A review and meta-analysis. *Environmental Research* 64: 36–52.

Steuben, L. 1992. Air pollution effects on heathland. In *Ecological Indicators,* ed. D. H. McKenzie, D. E. Hyatt, and V. J. McDonald, 841–64. New York: Elsevier Applied Science.

Thimonier, A., and J.-L. Dupouey. 1992. Changes in the forest ground layer vegetation: An example of cultural eutrophication. In *Responses of Forest Ecosystems to Environmental Changes,* ed. A. Teller, P. Mathy and J. N. R. Jeffers, 930–31. London: Elsevier Applied Science.

Thimonier, A., J.-L. Dupouey, and M. Becker. 1994. Simultaneous eutrophication and acidification of a forest ecosystem in northeast France. *New Phytologist* 126: 533–39.

Vitousek, P. M. 1994. Beyond global warming: Ecology and global change. *Ecology* 75: 1861–76.

Wilson, A. D., and D. J. Shure. 1993. Plant competition and nutrient limitation during early succession in the southern Appalachian Mountains. *American Midland Naturalist* 129: 1–9.

Chapter 12. Restoration in Theory and Practice

Anderson, M. K. 1996. Tending the wilderness. *Restoration and Management Notes* 14: 154–66.

Ludwig, D., R. Hilborn, and C. Waters. 1993. Uncertainty, resource exploitation and conservation. *Science* 260: 17–36.

Martinez, D. 1995. Karuk tribal module of mainstem watershed analysis: Karuk ancestral lands and people as reference ecosystem for ecological restoration in collaborative ecosystem management. Unpublished manuscript of presentation at Winds of Change Conference, American Indian Science and Engineering Society, fall 1994, Boulder, Colo.

Chapter 13. Community-Based Education, Planning, and Monitoring

Cheskey, E. D. 1993. *Habitat Restoration: A Guide for Proactive Schools.* Ontario, Canada: The Waterloo County Board of Education, Curriculum and Program Development, Outdoor Education Department.

Kaplan, R., and S. Kaplan. 1989. *The Experience of Nature: A Psychological Perspective.* New York: Cambridge University Press.

Kempton, W., J. S. Boster, and J. A. Hartley. 1995. *Environmental Values in American Culture.* Cambridge: Massachusetts Institute of Technology Press.

Mills, S. 1993. Consensus: Wave of the past…and future. *Natural Food News* (January). Traverse City, Mich.: Oryana Food Cooperative.

Chapter 14. Restoration at the Macro Level

Anderson, M. K. 1996. Tending the wilderness. *Restoration and Management Notes* 14: 154–66.

Davis, M. B. 1993. *Old Growth in the East: A Survey.* Richmond, Vt.: The Cenozoic Society.

Duffy, D. C., and A. J. Meier. 1992. Do Appalachian understories ever recover after clearcutting? *Conservation Biology* 6: 196–201.

Karr, J. R. 1992. Defining and assessing ecological integrity beyond water quality. *Environmental Toxicology and Chemistry* 12: 1521–31.

Kimmins, J. P. 1996. The health and integrity of forest ecosystems: Are they threatened by forestry? *Ecosystem Health* 2: 5–18.

Leopold, A. 1949. *A Sand County Almanac and Sketches Here and There*. New York: Oxford University Press.

Litvaitis, J. A. 1994. Response of early successional vertebrates to historic changes in land use. In *The Landscape Perspective: Readings in Conservation Biology*, ed. D. Ehrenfeld, 189–96. Cambridge, Mass.: Society for Conservation Biology and Blackwell Science.

Margulis, L., and D. Sagan. 1997. *Microcosmos: Four Billion Years of Evolution from Our Microbial Ancestors*. Berkeley: University of California Press.

McCloskey, J. M., and H. Spalding. 1990. The world's remaining wilderness. *Geographical Magazine* 62 (August): 14–19.

Noss, R. F., A. Y. Cooperrider, and Defenders of Wildlife. 1994. *Saving Nature's Legacy: Protecting and Restoring Biodiversity*. Washington, D.C.: Island Press.

Seaton, K. 1996. The Nature Conservancy's preservation of old growth. In *Eastern Old Growth Forests: Prospects for Rediscovery and Recovery*, ed. M. B. Davis, 274–83. Washington, D.C.: Island Press.

Zahner, R. 1996. How much old growth is enough? In *Eastern Old Growth: Prospects for Rediscovery and Recovery*, ed. M. B. Davis, 344–58. Washington, D.C.: Island Press.

Chapter 15. Managing with Succession

Brown, R. T., and J. R. Roti. 1963. The "Solidago Factor" in jack pine seed germination (abstract). *Bulletin of the Ecological Society of America* 14: 113.

Clements, F. E. 1916. *Plant Succession: An Analysis of the Development of Vegetation*. Publication no. 242. Washington, D.C.: Carnegie Institute of Washington.

Curtis, J. T. 1959. *The Vegetation of Wisconsin*. Madison: University of Wisconsin Press.

Egler, F. E. 1954. Vegetation science concepts; I: Initial floristic composition, a factor in old-field vegetation development. *Vegetatio* 4: 412–17.

Heisey, R. M. 1996. Identification of an allelopathic compound from *Ailanthus altissima* (Simaroubaceae) and characterization of its herbicidal activity, *American Journal of Botany* 83: 192–200.

McCormick, Jack S. 1968. Succession. *Via 1, Ecology in Design* (University of Pennsylvania, Graduate School of Fine Arts): 22–35.

National Fire Policy Review Team. 1989. *Final Report on Fire Management Policy*. Washington, D.C.: U.S. Departments of Agriculture and the Interior.

Chapter 16. Restoring Natural Water Systems

Gebhart, D. L., H. B. Shannon, H. S. Mayeux, and H. W. Polley. 1994. The CRP increases soil organic carbon. *Journal of Soil and Water Conservation* 45: 488–92.

McElfish, J. M., and R. J. Adler. 1990. Swampbuster implementation: Missed opportunities for wetland protection. *Journal of Soil and Water Conservation* 45: 383–85.

National Research Council, Committee on Restoration of Aquatic Ecosystems. 1992. *Restoration of Aquatic Ecosystems: Science, Technology and Public Policy*. Washington, D.C.: National Academy Press.

Smith, W. D. 1980. Has anyone noticed that trees are not being planted any longer? In *Trees for Reclamation.*. USDA Forest Service General Technical Report NE 61. USDA: 53–55.

Chapter 17. Soil as a Living System

Cordell, C. E., D. H. Marx, and C. Caldwell. 1991. Operational application of specific ectomycorrhizal fungi in mineland reclamation. Unpublished paper presented at annual meeting of the American Society for Surface Mining and Reclamation, May 14–17, 1991. Durango, Colo.

Harris, J. A., P. Birch, and K. C. Short. 1993. The impact of storage of soils during opencast mining on the microbial community: A strategist theory interpretation. *Restoration Ecology* 1: 88–100.

Hartman, J. M., J. F. Thorne, and C. E. Bristow. 1992. Variations in old field succession. Proceedings *(Design + Values)* of annual meeting, Council for Educators in Landscape Architecture, Charlottesville. Va., 55–62.

Ingham, E. R. 1995. Restoration of soil community structure and function in agriculture,
grassland and forest ecosystems in the Pacific Northwest. Proceedings, Society for Ecological Restoration Conference, Seattle, 31.

McDonnell, M. J., S. T. A. Pickett, and R. V. Pouyat. 1993. Application of the ecological gra-
dient to the study of urban effects. In *Humans as Components of Ecosystems: The
Ecology of Subtle Effects and Populated Areas,* ed. G. E. Likens and W. J. Cronon, 175–89.
New York: Springer-Verlag.

Muono, J., and I. Rutanen. 1994. The short-term impact of fire on the beetle fauna in boreal
coniferous forest. *Annales Zoologici Fennici* 31: 109–21.

Nixon, W. 1995. As the worm turns. *American Forests* 101, no. 9: 34–36.

Patrick, Ruth. 1968. Natural and abnormal communities of aquatic life in streams. *Via 1,
Ecology in Design* (University of Pennsylvania, Graduate School of Fine Arts,): 36–41.

Chapter 18. Plants for Restoration

Burchick, M. 1993. The problems with tall fescue in ecological restoration. *Wetland Journal*
5, no. 2: 16.

Eastern Native Plant Alliance. 1994. *Network News.* April 1994: 2–3.

Chapter 19. Living with Wildlife

Associated Press. 1993. Wildlife group wants Pa [sic] to be the home where the buffalo
roam. *Philadelphia Inquirer,* June 5, 1993, B-11.

Bratton, S. P., and A. J. Meier. In Press. Restoring wildflowers and salamanders in eastern deciduous forests: Some concepts from studies of biodiversity loss. *Restoration and Management Notes.*

Bryan, H. D., and J. R. MacGregor. 1988. A guide to the determination of the suitability of
abandoned mine portals as habitats for rare and endangered wildlife in Alabama. Unpublished report to the state of Alabama. Department of Industrial Relations, Mont-
gomery, Ala.

Daniels, T. J., D. Fish, and I. Schwartz. 1993. Reduced abundance of *Ixodes scapularis* (Acari,
Ixodidae) and Lyme disease risk by deer exclusion. *Journal of Medical Entomology* 30:
1043–49.

Diefenbach, D. R., L. A. Baker, W. E. James, R. J. Warren, and M. J. Conroy. 1993. Reintroducing bobcats to Cumberland Island, Georgia. *Restoration Ecology* 1, no. 4: 241–47.

Miller, S. G., S. P. Bratton, and J. Hadigan. 1992. Impacts of white-tailed deer on endangered
and threatened vascular plants. *Natural Areas Journal* 12: 67–75.

Pullin, A. S. 1996. Restoration of butterfly populations in Britain. *Restoration Ecology* 4,
71–80.

Stafford, K. C. 1993. Reduced abundance of *Ixodes scapularis* (Acari Ixodidae) with exclusion of deer by electric fencing. *Journal of Medical Entomology* 30: 986–96.

Van Luven, D. 1995. Restoring a fire-dependent habitat in a developed landscape. *Northeast
Chapter Newsletter* (Society for Ecological Restoration) 3, no. 2: 2–3.

Woodward, R. 1993. In Wales red kite flies again. *Philadelphia Inquirer,* December 12, 1993,
A-17.

Yunt, J. D. 1995. Where the wild things are: Wildlife corridors in use. *On the Ground* 1, no. 2:
23–24.

Chapter 20. The North Woods of Central Park

Burton, D. 1997. *The Nature Walks of Central Park.* New York: Henry Holt and Company.

Central Park Conservancy. 1985. *Rebuilding Central Park: A Management and Restoration
Plan,* New York: Central Park Conservancy and New York Department of Parks and
Recreation.

Chapter 21. Monitoring and Management

Bentham, H., J. A. Harris, P. Birch, and K. C. Short. 1992. Habitat classification and soil restoration assessment using analysis of soil microbial and physico-chemical characteristics. *Journal of Applied Ecology* 29: 711–18.

Bratton, S. P., and A. J. Meier. In Press. Restoring wildflowers and salamanders in eastern deciduous forests: Some concepts from studies of biodiversity loss. *Restoration and Management Notes.*

Falk, D. A., C. I. Millar, and M. Olwell. 1996. *Restoring Diversity: Strategies for Reintroduction of Endangered Plants.* Washington, D.C.: Island Press.

Harker, D., S. Evans, M. Evans, and K. Harker. 1993. *Landscape Restoration Handbook.* Boca Raton, Fla.: Lewis Publishers and New York Audubon Society.

Harris, J. A., P. Birch, and J. P. Palmer. 1996. *Land Restoration and Reclamation: Principles and Practice.* Essex, England: Addison Wesley Longman, Ltd.

Harris, J. A., and T. C. J. Hill. 1995. Soil biotic communities and new woodland. In *The Ecology of Woodland Creation,* ed. R. Ferris-Kaan, 91–112. Chichester, England: John Wiley and Sons.

Herman, K. D. et al. 1997. Floristic quality assessment: Development and application in the state of Michigan (USA). *Natural Areas Journal* 17: 265–88.

Luttenberg, D., D. Lev, and M. Feller. 1993. *Native Species Planting Guide for New York City and Vicinity.* New York: City of New York, Parks and Recreation, Natural Resources Group.

Magnin, F., and T. Tatoni. 1995. Secondary succession on abandoned cultivation terraces in calcareous Provence: The gastropod communities. *Acta Oecologia—International Journal of Ecology* 16: 89–106.

Panzer, R. 1995. Prevalence of remnant dependence among prairie-savanna-inhabiting insects of the Chicago region. *Natural Areas Journal* 15: 4.

Reschke, C. 1990. *Ecological Communities of New York State.* Latham, N.Y.: New York State Department of Environmental Conservation, Natural Heritage Program.

Smith, K. D. 1990. Standards developed for white oak–hickory forest restoration (Ohio). *Restoration and Management Notes* 8: 108.

Swink, F., and G. Wilhelm. 1994. *Plants of the Chicago Region: An Annotated Checklist of the Vascular Flora of the Chicago Region, with Keys, Notes on Local Distribution, Ecology, and Taxonomy, a System for the Qualitative Evaluation of Plant Communities, A Natural Divisions Map, and a Description of Natural Plant Communities.* Indianapolis: Morton Arboretum and Indiana Academy of Science.

Williams, K. S. 1993. Use of terrestrial arthropods to evaluate restored riparian wetlands. *Restoration Ecology* 1, no. 2: 107–16.

Chapter 23. Controlling Invasives

Schwegman, J. Undated. *Illinois Garlic Mustard Alert.* Springfield, Ill.: Illinois Department of Natural Resources, Division of Natural Heritage.

Chapter 24. Planting

McIninch, S. 1993. Growing pains: The ethics of collecting from the wild. *Maryland Native Plant Society Native News* 1, no. 3 (December): 5.

Pettit, N. E., R. H. Froend, and P. G. Ladd. 1995. Grazing in remnant woodland vegetation: Changes in species composition and life form groups. *Journal of Vegetation Science* 6: 121–30.

Stevens, W. K. 1995. *Miracle Under the Oaks: The Revival of Nature in America.* New York: Pocket Books.

Chapter 25. Meadow Management

Capel, S. 1992. *Warm Season Grasses for Virginia and North Carolina: Benefits for Livestock and Wildlife.* Richmond, Va.: Virginia Department of Game and Inland Fisheries.

Niering, W. A,. and G. D. Dreyer. 1989. Effects of prescribed burning on *Andropogon sco-parius* in postagricultural grasslands in Connecticut. *American Midland Naturalist* 122: 88–102.

Niering, W. A., and R. H. Goodwin. 1974. Creation of relatively stable shrublands with herbicides: Arresting succession on rights-of-way and pastureland. *Ecology* 55: 784–95.

Packard, S., and C. F. Mutel. 1997. *The Tallgrass Restoration Handbook: For Prairies, Savannas, and Woodlands.* Washington, D.C.: Island Press.

Savory, A. 1988. *Holistic Resource Management.* Washington, D.C.: Island Press.

Stevens, W. K. 1995. *Miracle Under the Oaks: The Revival of Nature in America.* New York: Pocket Books.

Swink, F., and G. Wilhelm. 1994. *Plants of the Chicago Region: An Annotated Checklist of the Vascular Flora of the Chicago Region, with Keys, Notes on Local Distribution, Ecology, and Taxonomy, a System for the Qualitative Evaluation of Plant Communities, A Natural Divisions Map, and a Description of Natural Plant Communities.* Indianapolis: Morton Arboretum and Indiana Academy of Science.

Species List

* denotes exotics.

alder — *Alnus* spp.
African cattle egret — *Bubulcus iris**
American buffalo — *Bison bison*
American chestnut — *Castanea dentata*
American elm — *Ulmus americana*
American ginseng — *Panax quinquefolius*
Amur cork tree — *Phellodendron amurense**
anise hyssop — *Agastache foeniculum*
annual rye — *Secale cereale**
arrowhead — *Sagittaria latifolia*
arrowwood — *Viburnum dentatum*
ash — *Fraxinus* spp.
ash yellows — a disease of ash and lilacs caused by a
    pathogen called a phytoplasma
aster — *Aster* spp.
auroch — *Bos primigenius*
autumn olive — *Eleagnus umbellata**
balsam woolly adelgid — *Adelges piceae**
barberry — *Berberis* spp.*
basswood — *Tilia americana*
bats — the suborder Microchiroptera, a group of
    flying mammals
beach plum — *Prunus maritima*
bear, black bear — *Ursus americanus*
beaver — *Castor canadensis*
beebalm — *Monarda didyma*
beech — *Fagus grandifolia*
beech bark disease — *Nectria coccinea* var. *faginata**
    and *Nectria galligena*, fungal pathogens of beech
beech scale — *Cryptococcus fagisuga**, spreads beech
    bark disease
bellwort — *Uvularia* spp.
bergamot — *Monarda fistulosa*
bidens — *Bidens* spp.
big bluestem — *Andropogon gerardii*
bison — See American buffalo
black birch — *Betula lenta*
black cherry — *Prunus serotina*
black cohosh — *Cimicifuga racemosa*
black fly — *Simulium jenningsi*

black gum — *Nyssa sylvatica*
black locust — *Robinia pseudoacacia*
black oak — *Quercus velutina*
black walnut — *Juglans nigra*
black-crowned night heron — *Nycticorax nycticorax*
black-eyed Susan — *Rudbeckia triloba*
bleeding heart — *Dicentra eximia*
bloodroot — *Sanguinaria canadensis*
blue cohosh — *Caulophyllum thalictroides*
blueberry — *Vaccinium* spp.
blue iris — *Iris versicolor*
bluestem — See little bluestem
bluets — *Hedyotis* species
bobcat — *Felis rufus*
boltonia — *Boltonia asteroides*
broad-leaved plantain — *Plantago major**
brook trout — *Salvelinus fontinalis*
box elder — *Acer negundo*
broomsedge — *Andropogon virginicus*
brown trout — *Salmo trutta*
Bt — *Bacillus thuringiensis*
buckthorn — *Rhamnus* spp.*
buckwheat — *Fagopyrum sagittatum**
bullfrog — *Rana catesbiana*
bur oak — *Quercus macrocarpa*
burrowing mole salamander — *Ambystoma
    talpoideum*
bush honeysuckle — *Diervilla lonicera*
butterfly weed — *Asclepias tuberosa*
butternut walnut — *Juglans cinerea*
butternut canker — *Sincoccus clavigignenti-juglan-
    dacearum**
buttonbush — *Cephalanthus occidentalis*
Canada goose — *Branta canadensis*
Canada hemlock — *Tsuga canadensis*
Canada lily — *Lilium canadense*
Canada thistle — *Cirsium arvense**
Carolina parakeet — *Conuropsis carolinensis*
catalpa — *Catalpa bignonioides*
catbird — *Dumetella carolinensis*

cattail — *Typha* spp.

cattle — *Bos taurus**

cattle egret — See African cattle egret

cecropia moth — *Hyalophora cecropia*

cheat grass — *Bromus secalina*

Cheat mountain salamander — *Plethodon nettingi*

cherry — *Prunus* spp.

chestnut — See American chestnut

chestnut blight — *Cryphonectria parasitica**

Chewings fescue — *Festuca rubra commutata**

chicory — *Cichorium intybus**

chipmunk — *Tamias striatus*

Christmas fern — *Polystichum acrostichoides*

closed gentian — *Gentiana clausa*

common milkweed — *Asclepias syriaca*

common reed — *Phragmites australis*

coneflower — *Echinacea* spp.

coontie — *Zamia pumila*

cottontail — *Sylvilagus floridanus*

cottonwood — *Populus deltoides*

cowbird — *Molothrus ater*

coyote — *Canis latrans*

crabgrass — *Digitaria sanguinalis**

crane flies — family Tipulidae

creeping fescue — *Festuca rubra**

Culver's root — *Veronicastrum virginicum*

dandelion — *Taraxacum officinale**

devil's walkingstick — *Aralia spinosa*

dog — *Canis familiarus**

dog-strangling vine — *Polygonum perfoliatum**

dogwood — *Cornus* spp.

dogwood anthracnose — *Discula destructiva**

dogwood blight — See dogwood anthracnose

downy woodpecker — *Piscoides pubescens*

dragonflies — flying insects of the Order Odonata

duskywing skipper butterfly — *Erynnis persius persius*

Dutch elm disease — *Ceratocystis ulmi**

Dutchman's breeches — *Dicentra cucullaria*

eastern box turtle — *Terrapene carolina*

elderberry — *Sambucus canadensis* or *S. racemosa*

elk — *Cervus elephas*

elm — See American elm

elm spanworm — *Ennoma subsignarius**

English ivy — *Hedera helix**

Entomophaga mamaiga — a fungus that kills gypsy moth

fairy shrimp — *Streptocephalus* spp.

false Solomon's seal — *Smilacina racemosa*

false wild indigo — *Baptisia tinctoria*

fescue — *Festuca* spp., some native but most exotic*

fir — *Abies* spp.

Florida atala butterfly — *Eumaeus atala florida*

Florida panther — *Felis concolor coryi*

flowering dogwood — *Cornus florida*

flycatcher — insectivorous birds of the genera *Myiarchus* and *Tyrannus*

foxtail — *Setaria* spp.*

fringed gentian — *Gentiana crinita*

fritillaries — subfamily Argynninae of butterflies

frosted elfin butterfly — *Incisalia irus*

garlic mustard — *Alliaria petiolata**

giant tar spot — a fungal disease of Norway maple

ginseng — See American ginseng

glossy ibis — *Plegadis falcinellus*

gnat — *Simulium* spp. (see black fly) and other small swarming flies

golden Alexanders — *Zizia aurea*

goldenrod — *Solidago* spp.

goldenseal — *Hydrastis canadensis*

grape — *Vitis* spp.

gray wolf — *Canis lupus*

great egret — *Casmerodius albus*

great merrybells — *Uvularia grandiflora*

great Solomon's seal — *Polygonatum biflorum* var. *commutatum*

great white trillium — *Trillium grandiflorum*

greenbrier — *Smilax* spp.

green dragon — *Arisaema dracontium*

green trillium — *Trillium viride*

ground pine — *Lycopodium* spp.

gypsy moth — *Lymantria dispar**

hackberry — *Celtis occidentalis*

hackberry butterfly — *Asterocamper celtis*

hairgrass — *Deschampsia* spp.

hawthorn — *Crataegus* spp.

heath fritillary — *Melitaea athalia*

hemlock woolly adelgid — *Adelges tsugae**

hemlock — *Tsuga canadensis*

hepatica — *Hepatica nobilis*

hickory — *Carya* spp.

hobblebush — *Viburnum alnifolium*

hog sphinx moth — *Darapsa moron*

holly — *Ilex* spp.

honeysuckle — See Japanese honeysuckle

horse — *Equus caballus**

huckleberry — *Vaccinium* spp.

hummingbird, ruby-throated hummingbird — *Archilochus colubris*

hummingbird moth — *Macroglossum stellatarum*

hydrilla — *Hydrilla verticillata**

imperial moth — *Eacles imperialis*

Indian grass — *Sorghastrum nutans*

Indiana bat — *Myotis sodalis*

jack-in-the-pulpit — *Arisaema triphyllum*

jaguar — *Felis onca*

Japanese honeysuckle — *Lonicera japonica**

Japanese knotweed — *Reynoutria japonica**

Joe Pye weed — *Eupatorium purpureum, E. fistulosum* or *E. dubium*

K-31 — horticultural variety of tall fescue (see tall fescue)*

Karner's blue butterfly — *Lycaeides melissa amuelis*

knapweed — *Centaurea* spp., *Centaurea maculata*** is widespread
knotweed — See Japanese knotweed
Korean dogwood — *Cornus kousa**
kudzu — *Pueraria lobata**
lady fern — *Athyrium felix-femina*
lady's slipper — *Cypripedium* spp.
lamprey eel — *Pteromyzon marinus**
large blue butterfly — *Maculina arion*
least Bell's vireo — *Vireo bellii* var. *pusillus*
lesser celandine — *Ranunculus ficaria**
lettered sphinx moth — *Deidamia inscripta*
lilac — *Syringa* spp.*
little blue heron — *Egretta caerulea*
little bluestem — *Schizachyrium scoparius*
little brown bat — *Myotis lucifugus*
liverleaf — *Hepatica nobilis*
lizardtail — *Saururus cernuus*
lobelia — *Lobelia* spp.
locust — See black locust
longleaf pine — *Pinus palustris*
loosestrife — See purple loosestrife
lowbush blueberry — *Vaccinium pallidum*
luna moth — *Antheraea luna*
lupine — *Lupinus perennis*
Lyme disease — *Borrelia burgdorferi*, a bacterium infecting humans, deer, and other mammals, spread by Lyme ticks.
Lyme tick — *Ixodes scapularis*
maidenair fern — *Adiantum pedatum*
magnolia — *Magnolia* spp.
maple — *Acer* spp.
Maryland golden aster — *Chrysopsis mariana*
mayapple — *Podophyllum peltatum*
meadowlark — *Sturnella magna*
meadowsweet — *Spiraea alba*
mice — *Microtus pennsylvanica*, meadow vole, and *Peromyscus leucopus*, white footed mouse
mile-a-minute — See dog-strangling vine and see also kudzu which is called this in some areas
milkweed — *Asclepias* spp.
miscanthus — *Miscanthus sinensis**
mockernut hickory — *Carya tomentosa*
monarch butterfly — *Danaus plexippus*
mountain mint — *Pycnanthemum* spp.
mourning cloak butterfly — *Nymphalis antiopa*
mugwort — *Artemisia vulgaris**
mulberry — See red mulberry
mullein — *Verbascum thapsus**
multiflora rose — *Rosa multiflora**
narrow-leaved gentian — *Gentiana linearis*
New England aster — *Aster novae-angliae*
New Jersey tea — *Ceanothus americanus*
nodding trillium — *Trillium cernuum*
Norway maple — *Acer platanoides**
Norway rat — *Rattus norvegicus**

oak — *Quercus* spp.
oriental bittersweet — *Celastrus orbiculata**
osage orange — *Maclura pomifera*
osha root — *Ligusticum porteri*
ostrich fern — *Matteuccia struthiopteris*
pandorus sphinx — *Eumorpha pandorus*
panther — *Felis concolor*
partridgeberry — *Mitchella repens*
passenger pigeon — *Ectopistes migratorius*
pear thrips — *Taeniothrips inconsequens**
pearl crescent butterfly — *Phyciodes tharas*
peregrine falcon — *Falco peregrinus*
persimmon — *Diospyros virginia*
pickerelweed — *Pontederia cordata*
pigweed — *Amaranthus hybridus**
pine bark beetle — *Dendroctonus frontalis* and *Ips typographus*
pink beardtongue — *Penstemon smalli*
pink root — *Spigelia marilandica*
pink turtlehead — *Chelone lyoni*
pitcher plant, yellow pitcher plant — *Sarracenia alata*
poison ivy — *Rhus toxicodendron*
pokeweed — *Phytolacca americana*
porcelain berry — *Ampelopsis brevipedunculata**
prairie trillium — *Trillium recurvatum*
princess tree — *Paulownia tomentosa**
privet — *Ligustrum* spp.*
Prometheus moth — *Callosamia promethea*
purple loosestrife — *Lythrum salicaria**
purple trillium — *Trillium erectum*
Queen Anne's lace — *Daucus carota**
question mark butterfly — *Polygonia interrogationis*
rabbit — See cottontail
raccoon — *Procyon lotor*
ragweed — *Ambrosia artemisiifolia*
rattlesnake master — *Eryngium aquaticum*
red cedar — *Juniperus virginiana*
red clover — *Trifolium pratense*
red kite (Britain) — *Milvus milvus*
red maple — *Acer rubrum*
red mulberry — *Morus rubra*
red oak — *Quercus rubra*
red spruce — *Picea rubens*
red wolf — *Canis rufus*
red-spotted purple butterfly — *Basilarchia astyanax*
redstart — *Setophaga ruticilla*
redtop — *Agrostis gigantea**
regal fritillary — *Speyeria idalia*
regal moth — *Citheronia regalis*
rice cutgrass — *Leersia oryzoides*
ring-necked pheasant — *Phasianus colchicus**
robin — *Turdus migratorius*
roe deer (Europe) — *Capreolus capreolus*
rose trillium — *Trillium stylosum catesbei*
royal walnut moth — *Citheronia regalis*
rubus — *Rubus* spp.

Russian olive — *Eleagnus angustifolia**
Saint John's wort — *Hypericum perforatum**
salamanders — amphibians of the order Caudata
saltwater cordgrass — *Spartina alterniflora*
sandhill crane — *Grus canadensis*
sarsaparilla — *Aralia* spp.
sassafras — *Sassafras albidum*
savannah sparrow — *Passerculus sandwichensis*
sawtooth oak — *Quercus acutissima**
sedges — Cyperaceae
sensitive fern — *Onoclea sensibilis*
shagbark hickory — *Carya ovata*
shrub dogwood — *Cornus* spp.
shrub honeysuckle — *Lonicera maacki,** *L. morrowi,**
    L.morrowi x tatarica,* and *L. tatarica**
skullcap — *Scutellaria* spp.
skunk — *Mephitis mephitis*
skunk cabbage — *Symplocarpus foetidus*
snout butterfly — *Libytheana bachmannii*
snowy egret — *Egretta thula*
Solomon's seal — *Polygonatum* spp., *P. biflorum* and
    Smilacina stellata
spicebush — *Lindera benzoin*
spicebush swallowtail butterfly — *Pterourus troilus*
spotted wintergreen — *Chimaphylla maculata*
spreading dogbane — *Apocynum androsaemifolium*
spring azure — *Celastrina ladon*
spring beauty — *Claytonia* spp.
spring peeper — *Pseudacris* spp.
spruce — *Picea* spp.
squirrel corn — *Dicentra canadensis*
starling — *Sturnus vulgaris**
stilt grass — *Microstegium vimineum**
stoneroot — *Collinsonia canadensis*
sugar maple — *Acer saccharum*
sumac — *Rhus* spp.
swallowtail butterfly — *Papilio* spp.
swallow — various small birds of family
    Hirundinidae
swamp pink — *Helonias bullata*
swamp white oak — *Quercus bicolor*
sweet flag — *Acorus calamus*
sweet gum — *Liquidambar styraciflua*
swift — *Chaetura pelagica*
switchgrass — *Panicum virgatum*
sycamore maple — *Acer pseudoplatanus**
sycamore — *Platanus occidentalis*
tall anemone — *Anemone virginiana*
tall fescue — *Festuca arundinacea**
tanager — *Piranga olivacea*
Tartarian honeysuckle — *Lonicera tatarica**
thrush — *Hylocichla mustelina*
tiger swallowtail butterfly — *Papilio glaucus*
ti-tree — *Melaleuca quinquenervia**
toad trillium — *Trillium sessile*

tree-of-heaven — *Ailanthus altissima**
trillium — *Trillium* spp.
trout lily — *Erythronium americanum*
trumpet honeysuckle — *Lonicera sempervirens*
tulip poplar — *Liriodendron tulipifera*
turkey — *Meleagris gallopavo*
Turk's cap lily — *Lilium superbum*
twisted stalk — *Streptopus roseus*
umbrella leaf — *Diphylleia cymosa*
variegated fritillary — *Euptoetia claudia*
veronica — *Veronica* spp.*
viburnum — *Viburnum* spp.
viceroy butterfly — *Basilarchia archippus*
violets — *Viola* spp.
vireo — *Vireo* spp.
Virginia creeper — *Parthenocissus quinquefolia*
walnut — See black walnut
warbler — small birds of subfamily Parulinae
water hyacinth — *Eichhornia crassipes**
water lily — *Nymphaea odorata*
white ash — *Fraxinus americana*
white cedar — *Chamaecyparis thyoides*
white man's footprint — See broad-leaved
    plantain
white oak — *Quercus alba*
white pine blister rust — *Cronartium ribicola**
white pine — *Pinus strobus*
white snakeroot — *Eupatorium rugosum*
white-tailed deer — *Odocoileus virginianus*
whooping crane — *Grus americana*
wild geranium — *Geranium maculatum*
wild ginger — *Asarum canadense*
wild leek — *Allium tricoccum*
wild petunia — *Ruellia humilis*
wild strawberry — *Fragaria virginiana*
willow — *Salix* spp.
winged euonymus — *Euonymus alata**
wintergreen — *Gaultheria procumbens*
wisteria — *Wisteria sinensis**
witch hazel — *Hamamelis virginiana*
wolf — See red wolf, see also gray wolf
woodchuck — *Marmota monax*
wood frog — *Rana sylvatica*
woodland aster — *Aster divaricatus*
wood lily — *Lilium philadelphicum*
woodrush — *Luzula* spp.
wood sorrel — *Oxalis stricta*
yellow fringed orchid — *Habenaria ciliaris*
yellow locust — *Gleditsia triacanthos*
yellow mandarin — *Disporum laguninosum*
yellow pitcher plant — *Sarracenia alata*
yellow poplar — See tulip poplar
yellow trillium — *Trillium luteum*
yellow-crowned night heron — *Nyctanassa violacea*
zebra mussel — *Dreissenia polymorpha**

Invasive Exotics of the Eastern Forest

This list should not be considered comprehensive but it signals potentially serious problems that you should be on the alert for throughout the East. All these species have already naturalized in several states.

Trees

Acer ginnala — Amur maple
Acer japonicum — Japanese red maple
Acer platanoides — Sycamore maple
Acer pseudoplatanus — Norway maple
Ailanthus altissima — Tree-of-Heaven
Alnus glutinosa — black alder
Broussonetia papyrifera — paper mulberry
Kolreuteria paniculata — golden rain tree
Melia azedarach — chinaberry
Morus alba — white mulberry
Paulownia tomentosa — princess paulownia
Phellodendron amurense — Amur cork tree
Quercus acutissima — sawtooth oak
Populus alba — white poplar
Ulmus pumila — Siberian elm
Sapium sebiferum — Chinese tallow tree

Shrubs or smaller trees

Albizia julibrissin — mimosa
Berberis japonica — Japanese barberry
Berberis thunbergii — Japanese barberry
Berberis vulgaris — common barberry
Cytisus scoparius — Scotch broom
Eleagnus angustifolia — Russian olive
Eleagnus pungens — thorny eleagnus
Eleagnus umbellata — autumn olive
Euonymus alatus — winged wahoo
Hibiscus syriacus — shrub althea
Ligustrum obtusifolium — blunt leaved privet
Ligustrum sinense — Chinese privet
Lonicera maackii — Amur honeysuckle
Lonicera morrowi — Morrow honeysuckle
Lonicera morrowi x tatarica — Bell's honeysuckle
Lonicera tatarica — Tartarian honeysuckle

Rhamnus frangula — glossy buckthorn
Rhamnus cathartica — buckthorn
Rosa multiflora — multiflora rose
Rubus laciniata — cut leaved blackberry
Rubus phoenicolasius — wineberry
Spiraea japonica — Japanese spirea

Vines and ground covers

Akebia quinata — fiveleaf akebia
Ampelopsis brevipedunculata — porcelain berry
Celastrus orbiculatus — oriental bittersweet
Hedera helix — English ivy
Humulus japonica — hops
Euonymus fortunei — winter creeper
Lonicera japonica — Japanese honeysuckle
Polygonum aubertii — silver fleece vine
Pueraria lobata — kudzu
Solanum dulcamara — bittersweet nightshade
Vinca minor — periwinkle
Wisteria floribunda — wisteria
Wisteria sinensis — Chinese wisteria

Annuals

Amaranthus hybridus — pigweed
Arthraxon hispidus — jointed grass
Bidens polylepis — beggar tick
Cardiospermum halicababum — balloon vine
Carduus acanthoides — curled thistle
Chenopodium album — lamb's quarters
Commelina communis — common day flower
Digitaria sanguinalis — crab grass
Fagopyrum sagittatum — buckwheat
Ipomoea coccinea — red morning glory
Ipomoea hederacea — ivy leaved morning glory
Ipomoea purpurea — common morning glory

Lactuca serriola — prickly lettuce
Lapsana communis — nipplewort
Lepidium campestre — field cress
Lepidium virginicum — pepper grass
Microstegium vimineum — stilt grass
Perilla frutescens — beefsteak plant
Polygonum caespitosum — smartweed
Polygonum perfoliatum — mile-a-minute
Polygonum persicaria — lady's thumb
Raphanus raphanistrum — jointed charlock
Senna obtusifolia — sicklepod
Setaria faberi — giant nodding foxtail
Setaria pumila — yellow foxtail
Stellaria media — chickweed
Sonchus arvensis — sow thistle
Xanthium strumarium — cocklebur

Biennials
Alliaria petiolata — garlic mustard
Arctium minus — burdock
Arctium nemorosum — woodland burdock
Carduus nutans — nodding thistle
Centaurea maculosa — spotted knapweed
Cirsium vulgare — bull thistle
Conium maculatum — water hemlock
Daucus carota — Queen Anne's Lace
Dipsacus laciniatus — cut-leaf teasel
Dipsacus sylvestris — common teasel
Melilotus alba — white sweet clover
Melilotus officinalis — yellow sweet clover
Pastinaca sativa — wild Parsnip
Verbascum thapsus — flannel leaved mullein

Herbaceous Perennials
Achillea millefolium — yarrow
Aegopodium podagraria — goutweed
Agrostis capillaris — Rhode Island bent grass
Agrostis gigantea — redtop
Ajuga reptans — bugleweed
Allium vineale — wild onion
Arrhenatherum elatius — oatgrass
Artemisia vulgaris — mugwort
Arundo donax — giant reed
Bromus inermis — smooth brome
Carex kobomugi — asiatic sand sedge
Centaurea jacea — brown knapweed
Centaurea nigrescens — knapweed
Cichorium intybus — chicory
Cirsium arvense — Canada thistle
Convolvulus arvensis — field bindweed
Coreopsis lanceolata — tickseed
Coronilla varia — crown vetch
Cynodon dactylon — Bermuda grass

Dactylis glomerata — orchard grass
Dioscorea batatas — Chinese yam
Elytrigia repens — quackgrass
Epilobium hirsutum — hairy willow herb
Eragrostis curvula — weeping lovegrass
Euphorbia cyparissias — Cypress spurge
Euphorbia esula — leafy spurge
Festuca arundinacea — tall fescue
Festuca elatior — fescue
Festuca ovina — sheep fescue
Foeniculum vulgare — fennel
Galium mollugo — field madder
Glechoma hederacea — ground ivy
Holcus lanatus — velvet grass
Humulus japonica — hops
Hypericum perforatum — St. John's wort
Imperata cylindrica — cogon grass
Iris pseudacorus — yellow iris
Lespedeza cuneata — Chinese lespedeza
Linaria vulgaris — butter and eggs
Lotus corniculatus — birdsfoot trefoil
Lysimachia nummularia — moneywort
Lythrum salicaria — purple loosestrife
Lythrum virgatum — purple loosestrife
Miscanthus sinensis — miscanthus
Phalaris arundinacea — reed canary grass
Phleum pratense — timothy
Plantago lanceolata — narrow-leaved plantain
Plantago major — broad-leaved plantain
Poa compressa — Canada bluegrass
Poa trivialis — rough bluegrass
Ranunculus ficaria — lesser celandine
Reynoutria japonica — Japanese knotweed
Rumex acetosella — sheep sorrel
Rumex crispus — curly dock
Rumex obtusifolia — broad leaved dock
Sorghum halepense — Johnson grass
Urtica dioica — stinging nettle
Vinca minor — periwinkle

Aquatics
Alternanthera philoxeroides — alligator weed
Butomus umbellatus — flowering rush
Cabomba caroliniana — fanwort
Eichhornia crassipes — water hyacinth
Egeria densa — Brazilian water weed
Hydrilla verticillata — hydrilla
Hydrocharis morus-ranae — european frogbit
Myriophyllum aquaticum — parrot's feather
Myriophyllum spicatum — european water milfoil
Nasturtium officinale — watercress
Trapa natans — water chestnut

Excerpt from "Tending the Wilderness," by M. Kat Anderson, Ethnoecologist, Natural Resources Conservation Service, in *Restoration and Management Notes*, Vol. 14, No. 2, pp. 154–166.

Excerpt from "Information Relating to Deer Damage," by Leslie A. Duthie, Norcross Wildlife Sanctuary, Massachusetts.

Excerpts from "Propagation of Native Plants," "Garden in the Woods Cultivation Guide," and "Sources of Propagated Plants." Reprinted with permission from the New England Wild Flower Society, Inc., Garden in the Woods, Framingham, Massachusetts.

Excerpt from "News on Lythrum in the White Flower Farm Catalog," *Newsletter of the Eastern Native Plant Alliance*, April 1994, Silver Spring, Maryland.

Excerpt from a letter to Leslie Sauer from Dr. Henry Art, Williams College Department of Biology, Williamstown, Massachusetts.

Figure 3.1, map of forest loss in Sligo Watershed, from *The Forest Conservation Manual: Guidance for the Conservation of Maryland's Forests Under the 1991 Forest Conservation Act*, p. 5, Figure 1.1.1. Reprinted with permission of the Metropolitan Washington Council of Governments.

Figures 21.1 and 23.1, maps of Prospect Park, from *A Landscape Management Plan for the Natural Areas of Prospect Park*, supported by a grant from the Prospect Park Alliance and an in part grant from the National Endowment for the Arts, a Federal Agency, August 1994.

Figures 8.1 and 19.1, photographs, by Ann F. Rhoads, the Morris Arboretum of the University of Pennsylvania, Philadelphia.

Figure 25.8, photograph, by the National Park Service, Richmond National Battlefield Park.

Figures 18.1–18.5, photographs, by Edward L. Blake, Jr., Hattiesburg, Mississippi, and the Crosby Arboretum, Picayune, Mississippi.

Figures 20.1, 20.4, and 20.5, photographs, by Sara Cedar Miller/Central Park Conservancy, New York.

Figures 20.3 and 22.21, photographs, by Dennis Burton/Central Park Conservancy, New York.

Figure 20.2, photograph, by Marianne Cramer/Central Park Conservancy, New York.

Figures 10.1 and 23.3, photographs, by Clare Billett.

Figure 20.2, photograph, by James Yap.

Figures 1.1, 3.1, 21.1, 22.7, 22.9, 22.16, 22.17, 22.20, 22.27, 24.1, and 24.12, drawings, by Marita Roos and Devinder Soin, Andropogon Associates Ltd.

Portions of this manual were reprinted from the following:

Franklin, Carol, "Fostering Living Landscapes," in *Ecological Design and Planning*, edited by George Thompson and Frederick Steiner, John Wiley and Sons, New York, 1997.

Sauer, Leslie, "Bring Back the Forests: Making a Habit of Reforestation Saving the Eastern Deciduous Forest," *Wildflower, North America's Wild Flora Magazine,* Vol. 8, No. 3 (Summer 1992), pp. 26–34.

Sauer, Leslie, "Soil and Water Conservation in a Landscape Perspective," *Journal of Soil & Water Conservation,* Vol. 46, No. 3 (May–June 1991), pp. 194–196.

Sauer, Leslie, "The North Woods of Central Park," *Landscape* (March 1993), pp. 55–57.

Sauer, Leslie, "Plants for Restoration," in *Perennial Plant Society Symposium Proceedings 1994,* edited by Steven Still, pp. 11–16.

Sauer, Leslie, "Native Plant Restoration," Parts 1 & 2. *American Nurseryman,* Vol. 192, Issue 1, pp. 90–99 and Issue 2, pp. 47–51 (1995).

Sauer, Rolf, "Master Plan for Renewing Louisville, Kentucky's Olmsted Parks and Parkways: A Guide to Sustainable Landscape Management," *Forum: The Journal of the George Wright Society,* Vol. 13, No. 1, pp. 64–75 (1996).

Portions of this manual appeared in reports prepared by Andropogon Associates, Ltd.:

Andropogon Associates, Ltd., "Landscape Management & Restoration Program for the Woodlands of Central Park, Phase One Report: Consensus of the Interviews, Key Issues & Initial Program Recommendations," prepared for the Central Park Conservancy, The Arsenal Building, Central Park, New York, NY 10021 (August 1989).

Andropogon Associates, Ltd., "Earthworks Landscape Management Manual," prepared for the Mid-Atlantic Regional Office, Philadelphia, Park Historic Architecture Division, Cultural Resources, National Park Service, U.S. Department of the Interior, Washington, D.C. (1989).

Andropogon Associates, Ltd., Landscapes, PDR Engineers, Inc., Eco-Tech, Inc., The Frederick Law Olmsted Papers, "Louisville's Olmsted Parks & Parkways Master Plan: A Guide to Renewal & Management," prepared for the city of Louisville, Kentucky, by the Louisville Olmsted Parks Conservancy, Inc., in conjunction with the Louisville and Jefferson County Parks Department (1994).

Portions of this manual were prepared under:

Grant No. 87-4251-0067 of the Design Arts Program, National Endowment of the Arts, Room 625, 1100 Pennsylvania Avenue, NW, Washington, DC 20506.

Grant No. 9298 from The Graham Foundation for Advanced Studies in the Fine Arts, 4 West Burton Place, Chicago, IL 60610.

INDEX